CAMBRIDGE LIBRARY COLLECTION

Books of enduring scholarly value

Monographs of the Palaeontographical Society

The Palaeontographical Society was established in 1847, and is the oldest Society devoted to study of palaeontology worldwide. Its primary role is to promote the description and illustration of the British fossil flora and fauna, via publication of an authoritative monograph series. These monographs cover a wide range of taxonomic groups, from microfossils, trilobites and ammonites through to Coal Measure plants, mammals and reptiles, and from all ages from Cambrian to Pleistocene. They form a benchmark for understanding the past life of the British Isles and many include the original descriptions of numerous key species. The first monograph (on the Crag Mollusca) was published in March 1848 and the Society still continues this work today. Notable authors in the series include Charles Darwin (fossil barnacles) and Richard Owen (dinosaurs and other extinct reptiles). Beginning in 2014, the Cambridge Library Collection and the Society are collaborating to reissue the earlier publications, focusing on monographs completed between 1848 and 1918.

A Monograph of British Graptolites

Prepared by Gertrude Elles (1872–1960) and Ethel Wood (1871–1946), this monograph was originally issued in ten instalments between 1901 and 1914, with the title page and index appearing in 1918. It is reissued here in two volumes. A systematic description and illustration of over 370 species of graptolites known at that time from the British Isles, and including nearly 100 new species, it was prepared under the editorship of Charles Lapworth (1842–1920), who also wrote a short introduction and some general discussion on classification. Elles wrote much of the descriptive text and Wood prepared the hundreds of wash-drawings. The resultant monograph is valued for the accuracy of these illustrations; those on the plates were printed at natural size, enabling a user equipped with a hand-lens to compare a rock specimen directly with the printed figures. The monograph was much admired and led to imitations, but none could match the quality of the original.

A Monograph of British Graptolites

VOLUME 2

GERTRUDE L. ELLES
ETHEL M.R. WOOD
EDITED BY CHARLES LAPWORTH

CAMBRIDGE
UNIVERSITY PRESS

CAMBRIDGE
UNIVERSITY PRESS

University Printing House, Cambridge, CB2 8BS, United Kingdom

Cambridge University Press is part of the University of Cambridge.
It furthers the University's mission by disseminating knowledge in the pursuit of
education, learning and research at the highest international levels of excellence.

www.cambridge.org
Information on this title: www.cambridge.org/9781108084413

© in this compilation Cambridge University Press 2015

This edition first published 1901–18
This digitally printed version 2015

ISBN 978-1-108-08441-3 Paperback

A MONOGRAPH

OF

BRITISH GRAPTOLITES.

BY

GERTRUDE L. ELLES, Sc.D.,

LATE GEOFFREY FELLOW, NEWNHAM COLLEGE, CAMBRIDGE ;

AND

ETHEL M. R. WOOD, D.Sc.

[MRS. SHAKESPEAR],

OF NEWNHAM COLLEGE, CAMBRIDGE ; AND THE UNIVERSITY OF BIRMINGHAM.

EDITED BY

CHARLES LAPWORTH, LL.D., F.R.S.,

LATE PROFESSOR OF GEOLOGY IN THE UNIVERSITY OF BIRMINGHAM.

HISTORICAL INTRODUCTION.

LONDON:

PRINTED FOR THE PALÆONTOGRAPHICAL SOCIETY.

1901—1918.

BRITISH GRAPTOLITES.

HISTORY OF RESEARCH.

THE History of Research among Graptolites commences in the early years of the eighteenth century. Examples of these fossils were first noticed in the works of von Bromell about 1727, and a few years later the name *Graptolithus* was first suggested by Linnæus in his 'Systema Naturæ.'

The History itself falls conveniently into four periods. In the first of these (1727 to 1850) papers upon Graptolites were comparatively rare, and their authors did little more than figure and describe the forms which they collected. During the second period (1850 to 1865) the brilliant papers of Barrande in Europe and Hall in America called universal attention to the abundance and variety of Graptolites in the older Palæozoic rocks, and these authors laid the foundation of our present ideas respecting their structure and their alliances. During the third period (1866 to 1880) the workers among these fossils were largely British, who devoted themselves firstly to the investigation and description of the British species and the determination of their geological horizons; and secondly to the discussion of the problems of the classification, zonal distribution, and probable mode of life of the Graptolites in general. Finally, during the fourth period (reaching from 1881 down to the present time) the workers among Graptolites have been mainly non-British, especially Swedish, and the advances made have been great, particularly as respects the intimate structure of the fossils themselves, their distribution in space and in time, and their probable conditions of existence.

A separate chapter is here devoted to each of these four periods. The publications of the several investigators are taken up in chronological sequence, and each generic or specific title when proposed or employed for the first time is distinguished by being printed in clarendon type.

a

CHAPTER I.

First Period, 1727 to 1850.

1727.

Von Bromell,

" Lithographiæ

Suecanæ,"

' Acta literaria Sueciæ

Upsaliæ,' vols. i, ii,

1720–9.

There seems but little doubt that Magnus von Bromell, of Upsala, Sweden, must be credited with the first notice and description of the fossils which we now call Graptolites, although he did not use the word " Graptolithus," nor did he give figures of these fossils.   In the years 1720–9 he brought out his work entitled " Lithographiæ Suecanæ."   In this he gives, among other things, a description of a collection of fossils belonging to himself.   This work was published in the ' Acta literaria Sueciæ Upsaliæ ' (1720–9), and also separately in two parts, the *Specimen primum* in 1724, and the *Specimen secundum* in 1727.

In the *Specimen primum* there is no reference which can be considered as applying to Graptolites ; but in the *Specimen secundum* the *Articulus primus* is entitled " Concerning a moss incrusted and delineated in stone."   The example No. 1 given in this section is described as " a rock of ashy colour, fissile, fœtid, called ,' Swinestone' (lapis suillus), exhibiting on the surface a black tangle of branched moss imprinted as with a fine pencil. . . . . The moss which is seen delineated on the above-mentioned rock is of a stony nature, hair-like, not penetrating the actual substance of the rock—as one can see in some of the Florentine and German dendrites,--but spreading its delicate form on the outer-most surface."

Tullberg, to whose valuable memoir, " On the Graptolites described by Hisinger and the older Swedish Authors," ' K. Svenska Vet. Akad. Handl.,' 1882, we are especially indebted, remarks that von Bromell elsewhere uses the term " lapis suillus " for real anthraconite or swinestone (concretions occurring in the alum shales of Sweden), and it may therefore be inferred with fair certainty that he refers to that rock in this case.   If so, the fossil which von Bromell describes as a " branched moss," and compares with a dendrite, is probably the *Dictyonema flabelliforme* of Eichwald, a fossil which frequently occurs in the balls of anthra-conite in the alum shales of Westrogothia.

Von Bromell's *Articulus secundus* is headed " Concerning the imprints and remains of leaves in various rocks."   Example 3 in this section is described as " Leafy impressions and traces of different plants in a black fissile rock from Mt. Dalaberg in Westrogothia ; " and the author remarks, " The true names of these leaves I cannot at present state ; for while some, by their pinnules, seem to recall ferns, others by their narrowness and length a kind of grass, some by their pointedness and tenuity a willow, others the heather and small water-lily, yet I dare not decide precisely to what they belong until I have received more and better specimens."

These observations of von Bromell are mentioned by Wahlenberg and later writers, who are of the opinion that the rock referred to was a shale containing Graptolites. Wahlenberg remarks that " when the shale is of a bluish or bluish-grey colour, the pictures of *Graptolithus* appear black, and at the same time their outlines so blend among themselves that their full figures appear only linear, and these Bromell took for leaves of grasses."

1735.

Linnæus,

'Systema Naturæ,'

edit. 1.

The great systematist, Linnæus, was the second to notice and describe examples of Graptolites; and we owe to him the title *Graptolithus*, which was subsequently adopted as the name of the genus that eventually became accepted as the type of the entire group. In the first edition of his 'Systema Naturæ,' section " Regnum lapideum," Classis III, Fossilia, he defines his third order as follows :—" **Graptolithus**, picture resembling a fossil." It would appear, therefore, from this description, and from the list of the specimens which he refers to as belonging to this order, that the title *Graptolithus* was originally proposed by Linnæus for those well-known markings—dendritic incrustations and the like— which frequently occur in rocks and which simulate fossils, but which had, even previous to his time, been generally acknowledged not to be fossils in the true sense of the word. Indeed, in the twelfth edition of his 'Systema Naturæ,' published in 1767, he asserts definitely, "A Fossil, properly speaking, is not a Graptolite."

1751.

Linnæus,

'Skånska Resa.'

Between the appearance of the first and twelfth editions of his 'Systema Naturæ,' Linnæus published, however, his 'Skånska Resa' ('Travels in Scania'). In that work he figured and described certain markings on a slab which are undoubtedly true fossils, and clearly belong to the group of organic remains now known as Graptolites. His description is as follows :—" Fossil or Graptolite, of a strange kind, ... which, in the grey rock with black characters, resembles a line imprinted with markings like those on the edge of a coin, and often passes into a narrower spiral end."

According to Tullberg (*op. cit. supra*), the slab bearing these fossils was obtained from a gravel hill, named Bybjer, near Herrestad. At this locality Tullberg asserts that no Graptolite shales exist *in situ*, but that loose blocks occur in the mass of gravel.

The exact specific identity of the fossils thus referred to by Linnæus in his 'Skånska Resa' became a matter of considerable controversy among palæontologists on both sides of the Atlantic a century later, but the consensus of opinion at the present day is in favour of the view that the fragments figured by him are true organic remains, and represent two distinct Graptolite species; one being the species now known as *Climacograptus scalaris* (Linnæus), and the other *Monograptus triangulatus* (Harkness), both of which species are, according to Tullberg, abundant in the blocks of shale occurring at this Scanian locality.

1768.
Linnæus,
'Systema Naturæ,'
edit. 12.

Following the same plan as that originally adopted by him in the first edition of his 'Systema Naturæ,' Linnæus, in his twelfth edition, still employed the term *Graptolithus* for inorganic markings or bodies which *simulate* fossils; and he further divided his order Graptolithus into various species.

Among these species of supposed false fossils, one (given as No. 6) is described as " *Graptolithus* **sagittarius,** with toothed impressions,—'Anonymum. Volkam. Siles.,' 3, p. 332, vol. iv, fig. 6,—found in hard rock, with imbricate impressions, toothed, without a pedicel, regularly arranged, pointed towards the apex." An examination of the figure which is given by Volkmann in his work ' Silesia subterranea,' and which is referred to by Linnæus in the above description, makes it quite clear (as has been pointed out by Tullberg and others) that it represents a *Lepidodendron* or *Sigillaria*, and not a Graptolite at all. Thus the Linnæan name, *Gr. sagittarius*, was employed for a Carboniferous plant, and cannot therefore be retained for a species of Graptolite.

Linnæus' species No. 7 of this twelfth edition of his ' Systema Naturæ,' however, is the same form as that originally described by him as *Graptolithus* in his 'Skånska Resa' of 1751. It is referred to in the text in the following words :—" *Graptolithus* **scalaris,** looking like a line and transverse markings. Found in the common shale of Scania." Linnæus, however, gives no fresh figure. (Compare also 13th edition, Gmelin, 1793.)

1771.
Walch,
' Naturgeschichte der
Versteinerungen zur
Erläuterung der
Knorr'schen Sammlung
von Merkwerdigkeiten
der Natur,'
suppl. iii.

We find therefore that von Bromell described certain forms now known to be Graptolites as mosses or leaves, while Linnæus described some others in his genus *Graptolithus*, believing them to be "false fossils" simulating true ones. Walch, however, was the first naturalist to recognise the animal nature of the organic remains of the type of the Graptolites of the ' Skånska Resa.' In the supplement to a work of his own on the fossils collected by Knorr, he figures two fossil forms or species which he considers to be minute Cephalopods. One of these is certainly the same form as that subsequently named by Hisinger *Prionotus convolutus*. Walch described this as a " unique species " with a testaceous body like a *Lituites*, and denticulated like the " denticulated Orthoceratites." " Its denticulated border proves that it was chambered, and the teeth mark the extremities of the chamber walls."

The second form figured by Walch is also certainly a Graptolite, but it is difficult to say from the figures to what species the examples given by him should be referred ; but it is possible that they belong to the form known at present as

Monograptus priodon (Bronn). Walch refers to them as a new species of "small denticulated Orthoceratites," distinct from *Dentalitæ geniculati*. "The number of chambers appears to be equal in all those of the same length, and those higher up are narrower than the basal ones;" but the siphon, he writes, has not yet been found. Both forms figured by Walch probably come from the same locality, near Stargard, Mecklenburg, from the grey limestone of the Northern Drift. Walch does not name either of his species.

1821.
Wahlenberg,
"Petrificata Telluris
Suecanæ," 'Nova Acta
Reg. Soc. Scientarum
Upsal.'

We find no further mention of Graptolites for the next half-century. But in the year 1821 Wahlenberg recognised the fact that at least one of the forms classed by Linnæus as *Graptolithus* (namely, *G. scalaris*) was a true fossil. But he agrees with Linnæus that none of the other forms (with the exception of *G. sagittarius*) embraced by him in his group *Graptolithus* (or bodies which "simulate fossils") are fossils in the true sense of the word.

Nevertheless, while agreeing with Linnæus in this general opinion, Wahlenberg boldly employed the term *Graptolithus* for the true fossils originally figured by Linnæus as *Graptolithus* in the 'Skånska Resa,' and established these as the original types of the palæontological genus *Graptolithus* or Graptolites in general. In this he has been followed by all subsequent palæontologists, and the term *Graptolites* has been consistently used ever since for all those fossils which are presumably identical with or allied to Linnæus' *Graptolithus scalaris*.

Wahlenberg had the same idea of the nature of the Graptolites as Walch had previously held, and believed them to be true Orthoceratites. He noticed their frequent association with undoubted Orthoceratites in the upper shales of Vestro-gothia, and believed that it was possible to trace all the intermediate stages between the large calcareous forms of *Orthoceras*, and the very small membranous and "apparently translucent varieties" which Linnæus had described as *Graptolithus scalaris*. He gives a description of what appears to be Linnæus' *G. scalaris* under the name *Orthoceratites tenuis* in the following terms:—"It has a maximum breadth of barely a line, a length of one inch, and a linear shape. Its joints here and there have been separated from each other alternately, and have been turned over so that they have imprinted in the shale circles smaller than mustard seed. Longitudinal types, instead of a siphon, show a definite medullary nerve (thread), to the sides of which dissepiments are attached, often opposite each other, as in their natural position, but sometimes alternating. This (alternation), however, might have been brought about by disturbance or by obliquity." These small fossils (the Graptolites), he writes, occur frequently alone, only rarely mixed with the larger forms (of Orthoceratites). He concludes from this that they "lived on as such through the period of the upper shales, after the extinction of all the large Orthoceratites."

Wahlenberg describes the occurrence of similar Graptolites also in Dalecarlia (at Osmundsberg and Furudal), and notices carefully the various conditions of preservation in which they are found. " In the shale of Scania, which always has a black colour, the Graptolites occur as impressions with a certain peculiar lustre, or when pyrites is present are rusty, and at the same time they are occasionally found as solid bodies filled up with pyrites." These, he contends, " show clearly the nature of Orthoceratites."

In addition to Linnæus' *G. scalaris*, Wahlenberg also notices the occurrence of certain one-sided forms which he erroneously refers to Linnæus' *G. sagittarius*. According to the Swedish geologists, the form identified by Wahlenberg with *G. sagittarius* is probably identical with the species subsequently named *Monograptus leptotheca* by Lapworth. At any rate, it is quite clear that it is not the same as Linnæus' *G. sagittarius* (see *ante*). He holds that there is but little doubt that these one-sided forms have their origin from the same minute Orthoceratites by a peculiar kind of decomposition. " Some of these Orthoceratites, which had a siphon or a lateral nerve, seem to have curved themselves after the destruction of the opposite wall, and so to have produced lines on one side like the points of an arrow." He considers that these arrow-like structures, which are curved, have arisen from the " interlocular dissepiments."

The result of Wahlenberg's work was to call general attention to these fossils, and the observers who succeeded him found no difficulty in recognising the organic remains thus fixed as " Graptolites," and they employed the term strictly in Wahlenberg's sense.

1822.
Von Schlotheim,
' Nachträge zur Petre-
faktenkunde,' pt. ii.

The views of Walch and Wahlenberg, with respect to the Cephalopodous nature of the Graptolites (as thus restricted by Wahlenberg), were adopted by von Schlotheim, who, in 1822, described and figured a Graptolite under the name *Orthoceratites* **serratus**. It is impossible to identify this species with certainty from Schlotheim's figure, but the " haarförmige Nervenröhre," which he notices running down the back of the fossil, is almost certainly the structure now known as the virgula. This form was obtained from the shales of Andrarum. According to von Schlotheim, his species is the same as that described by Schröter in his ' Einleitung,' Th. iv, but we have not been able to gain access to Schröter's paper.

1828.
Brongniart,
' Histoire des
Végétaux Fossiles.'

Von Bromell's primitive idea of the vegetable origin of these curious fossils, however, was not relinquished by all palæontologists, and in 1828 Brongniart described and figured two new species of Graptolites from the limestone of Pt. Levis, under the names of *Fucoides* **serra** and *F.* **dentatus**. The former is the species of *Tetragraptus* which was subsequently described by Hall as *T. bryonoides* (' Graptolites of the Quebec Group,' 1865), and the latter is a *Diplograptus* (*D. dentatus*).

1829.
Holl, Fr.,
'Handbuch der Petre-
fakten.,' vol. ii.

In 1829 Holl gave a short description of Schlotheim's *Orthoceratites serratus* in his own text-book, but added nothing to our knowledge of the structure of Graptolites.

According to Dr. Beck and Prof. Eichwald, Prof. Nilsson of Lund published about this time (in the 'Transactions of the Physiographic Society' in Lund) some

1830-35.
Nilsson.

notes on the nature of Graptolites; but according to Tullberg (*loc. cit. supra*) there is nothing written by Nilsson to that effect in the publications of the society. It seems fairly certain, however, that Nilsson (either in a letter or in a manuscript) distinctly formulated the opinion that Graptolites are polyparies belonging to the "Polypi ceratoporæ," or horny Polyps.

Nilsson proposed to change Linnæus' name of *Graptolithus* to *Priodon.* But this name had already been employed by Cuvier for a genus of fish, and Nilsson therefore, in a letter to Hisinger, dated December 27th, 1835, writes, " I have named the genus of *Graptolites* **Prionotus** (like a saw), both in my annotations (where a sketch of a monograph on this genus is to be found) and in letters to several foreigners. I regret that I gave the name wrongly through a slip of memory once when visiting you in Stockholm." (He probably here refers to the generic name originally proposed by him, namely, *Priodon.*) " The name *Prionotus* I consider characteristic, and I therefore intend to retain it." This name, however, was also preoccupied, and it has therefore subsequently fallen into disuse among graptolitho-logists.

1835.
Bronn,
'Lethæa Geognostica,'
vol. i.

The confusion in the nomenclature of the Graptolites was not diminished by Bronn, who in 1835, in his 'Lethæa Geognostica,' described a new species, and suggested a fresh generic name, **Lomatoceras**, instead of Nilsson's already pre-occupied name of *Priodon.* This choice was as unfortunate as that of Nilsson, for the name was already in use for a special genus of insects. The species figured and named by Bronn is his well-known *Monograptus* **priodon.** Bronn figured this form very accurately. He grouped *Lomatoceras* as one of the genera of the Polyparia; but in his description he speaks of the siphon and inner chambers as if the fossil were a Cephalopod.

1837-40.
Hisinger,
'Lethæa Suecica, seu
Petrificata Sueciæ.'

The Swedish naturalist Hisinger gave special attention to the Graptolites of his native country, and described and figured many new forms. He was originally (' Anteckningar i Physik och Geognosi,' p. 168) of the same opinion as Wahlenberg with regard to the zoological place of the Graptolites; but in 1837 (in his 'Lethæa Suecica, Supplementum ') he relinquished his earlier view of their alliance with the Cephalopoda, and agreed with Nilsson in referring the Graptolites to the "Polypi ceratoporæ." In this work he described and figured Linnæus' *Graptolithus scalaris,* and the form identified by Wahlenberg with Linnæus'

Gr. sagittarius, together with three new species of his own, namely, *P.* **pristis**, *P.* **folium**, and *P.* **convolutus**. Three years later he described and figured in his ' Supplementum secundum ' three additional forms, namely, *Pr.* **geminus**, *Pr.* **teretiusculus**, and a net-like form. This last was a *Dictyonema*, although he regarded it as the impression of a monocotyledonous plant. All Hisinger's species were classed by him under Nilsson's generic name of *Prionotus*.

1839.

Murchison,

' The Silurian System.'

In 1839, Murchison, in his ' Silurian System,' quotes the views of Dr. Beck, the Danish naturalist, as to the probable nature of the Graptolites. He writes, " Very different opinions have been entertained as to the place which the Graptolites hold in the series of living beings, but that of Professor Nilsson may come nearest to the truth, who conceives the Graptolite to be a polyparium of the ceratophydian family. Yet I am more inclined to regard them as belonging to the group *Pennatulinæ*, the Linnæan *Virgularia* being the nearest form in the present state to which they may be compared." Dr. Beck refers to the new names of *Priodon* and *Lomatoceras*, suggested by Nilsson and Bronn respectively, but considers both to be unnecessary.

Three species of Graptolites are figured in Murchison's ' Silurian System.' The descriptions of two of these are by Beck, namely, those of *Graptolithus* **ludensis** and *G.* **Murchisoni**. The name *G. Murchisoni* was given by Beck, but the name *G. ludensis* was substituted by Murchison for a form previously recognised by himself, which had been named by Beck in his MSS. *Graptolithus* **virgulatus**. One of the forms figured as *G. ludensis* is identical with Bronn's *Monograptus priodon;* the other is rather of the type of *M. colonus* of Barrande. *G. Murchisoni* is an example of *Didymograptus*. The third species, *G.* **foliaceus**, is a *Diplograptus*, and is described by Murchison himself.

Murchison emphasises throughout his work the fact that the range of the Graptolites is exclusively Silurian, and he records them as high up in the series as the lower Ludlow shales.

1840.

Quenstedt,

" Ueber die vorzüg- lichsten Kennzeichen der Nautileen," ' Neues Jahrb. f. Min.,'

Quenstedt, in a short paper in the ' Neues Jahrbuch ' in 1840, notices the structure of certain Graptolites. He adopts the views of the older palæontologists that they belong to the family of the Cephalopoda, and he states that he " does not see why Nilsson should place them among the sea-pens." But he evidently does not consider that the question of their affinities is definitely settled, for he says elsewhere, " A close examination of well-preserved specimens might, however, perhaps strengthen the view that they all belong to the class of the Foraminifera, and not either to the Cephalopoda or to the corals."

As regards the structure of the Graptolites, he notes that he has observed distinct transverse partition walls, as in the Cephalopoda, but no last chamber. He recognises also a siphon running down the back of the shell. He considers with Wahlenberg that the toothed appearance of one of the margins is not original, but a result of the mode of preservation. He describes and figures two species of Graptolites, *G. scalaris*, Linn., and *G. tenuis*, Wahl., and redescribes (without figuring) the *G. serratus* of Schlotheim. Quenstedt's drawings are figures of fragmentary forms, and represent two unidentifiable forms of *Monograptus*.

During the same year Eichwald (" Ueber das silurische Schichtensystem in Estland ") described a new species of Graptolite from the Silurian formation under the name of *Lomatoceras* **distichum**. This he considers to be quite distinct from any previously noted species of this genus, on account of its being denticulated on both sides. He dismisses the matter of the affinities of the group in a single sentence : " The Lomatocerases or Graptolites are absolutely pro-blematical bodies, which can scarcely belong to the Cepha-lopoda ; one might rather class them in a family among the Zoophytes."

Two years later (1842) appeared Geinitz's first paper on the Graptolites, entitled " Ueber Graptoliten." In this paper he gives a diagnosis of the genus *Graptolithus*, and includes under it forms both with two rows and one row of cells. He considers that all Graptolites actually consisted of two rows of cells spread out in one plane ; the single-rowed appearance of some being due to the fact that the two halves have got applied the one on the other. He does not think, however, that the " animal could fold itself up (' clap itself together ') at will, owing to its firm though thin shell." He describes and figures five species of Graptolites, viz. *Graptolithus foliaceus* (which may be a *Retiolites*), *Gr. priodon*, *Gr. serratus*, *Gr. scalaris* (which is a *Monograptus* of the type of *M. Hisingeri*), and a new form, *Gr.* **spiralis**. Some of those which he grouped under a single name, especially his *Gr. spiralis*, included more than one species as at present understood. The general structure of the fossils, however, is well represented in his figures. Geinitz notices in this paper the fact that the cells become larger and more closely set as the poly-pary grows older.

Geinitz also discusses the affinities of the Graptolites. He suggests that they bear great resemblance to the Chætopods, but he does not definitely class them with that group, and agrees with Walch, Wahlenberg, and earlier observers in placing them with the Cephalopods.

In the same year d'Orbigny (in his ' Voyage dans l'Amérique méridionale ') figured fragments of a single branch only of a new species from Bolivia, which he named *Gr. dentatus*. According to him it is a two-branched form identical with *Gr. foliaceus* and *Gr. Murchisoni*. It is difficult to understand, therefore, why he suggested a new specific name, especially as he considers that the specific names of Graptolites have been too much multiplied, varieties resulting from the effects of " alteration and deformity " being described as different species.

1842.
Eichwald,
' Die Urwelt Russ-
lands.'

Eichwald, in 1842 (' Die Urwelt Russlands '), described and figured under the new name of *Gorgonia* **flabelliformis** the same form as had been previously described as a moss by von Bromell, and as a monocotyledonous plant by Hisinger. Eichwald believed, however, that it was the impression of a coral resembling *Fenestella*.

1842.
Vanuxem,
'Geological Report of
the 3rd District of
New York.'

In the same year Vanuxem noticed the occurrence of *Gr. dentatus* and *Gr. scalaris* in the Utica slates of America, and figured an example of the former species. He distinctly advocates the vegetable nature of the Graptolites: " The ramose nature of two of the species shows that their origin is vegetable, not animal as conjectured by some naturalists. Their chemical composition confirms their vegetable nature ; no animal ever existed whose material was almost entirely carbon, as is the case with these fossils."

1843.
Portlock,
' Report on the Geology
of Londonderry and
of Parts of Tyrone.'

The appearance of Portlock's ' Report on the Geology of Londonderry and of Parts of Tyrone,' which was published in 1843, marks a distinct epoch in the history of our knowledge of the Graptolites. His work, however, was rather prophetic than conclusive, for his own personal researches on the various species of Graptolites, even when reinforced by the work done by previous observers, were wanting in that knowledge of the details of structure which Barrande subsequently obtained. But Portlock's acumen was so great that he deserves a place second only to Barrande, and he may be regarded as the precursor of the new era which Barrande subsequently founded.

Portlock describes and figures nine species of Graptolites : six of these had been named by previous authors, namely, *Graptolithus convolutus, Gr. foliaceus, Gr. folium, Gr. pristis, Gr. sagittarius, Gr. scalaris ?* whilst three—*Gr.* **Sedgwickii,** *Gr.* **distans,** and *Gr.* **tenuis**—were new forms named by himself.

Portlock strongly advocated and endeavoured to demonstrate the polyp-like character of the Graptolites, and he controverted the view that they were allied to the Orthoceratites. The presence of septa may, he considers, merely indicate the connection of the polyp cells with an internal axis. The double and single Graptolites, according to him, seem more analogous to Sertularia and Plumularia, but differ in having neither branches nor pinnæ. The cell-like structure seen in a scalariform view is analogous to that seen in the Cellaria. Portlock considers that it is probable that there are several Graptolite genera belonging to even more than one order. He does not, however, suggest new names for these genera, and the species described are all grouped by him under the single old generic name *Graptolithus* of Linnæus.

1843.
Hall,
'Geological Report of
the 4th District of
New York.'

During the same year Hall, while describing a new species—*Gr.* **clintonensis**—notes the fact that the shales in which Graptolites occur are black, as if from the carbonaceous matter derived from the fossils. This, he points out, would seem to argue against placing these bodies among the calcareous Polyparia, but he does not suggest their alliance with any other group of the animal kingdom.

1844.
Emmons,
'The Taconic System.'

In his 'Taconic System' Emmons gives figures of certain fossils which he names *Fucoides* **simplex** from the roofing slate of Hoosic, New York. These fossils, he says, "have much the appearance of the Graptolites of the Utica slates, but which I am now satisfied are marine vegetables." There can be but little doubt, however, that Emmons' fossils are really Graptolites of the type of *Diplograptus foliaceus* of Murchison; but, owing to the cleavage that the containing rock has undergone, the fossils present very different appearances according to their position on the slab.

1843–4.
Mather,
'Geological Report of
the 1st and 2nd Dis-
tricts of New York,'
pt. i.

The general view held by Americans about this time as to the vegetable nature of the Graptolites was also endorsed by Mather in his 'Geological Report' in 1843. Mather mentions that there are at least five species of these plants (Graptolites) in the Utica slates, and that they also occur in the Hudson River group, but he does not describe them. He copies Vanuxem's figures of *Gr. dentatus.*

1844.
Owen,
"Review of the New
York Geological
Reports," 'Amer.
Journal,' vol. xlvii.

Owen, however, in a review of Vanuxem's work (*op. cit. supra,* 1842), points out the differences between the views of Vanuxem and those of the European geologists as to the zoological affinities of the Graptolites, and seems himself to incline to those of the latter. He suggests that the carbonaceous matter almost invariably found in connection with the Graptolites " may have resulted from the peculiar conditions and circumstances attending their deposition;" and he asks, " might not, by the action of some chemical affinity, the less stable elements of the Polyparia have been removed and the carbon alone left ? "

1845.
Boubée,
'Bull. de la Soc. Géol.
de France,' ser. 2, t. ii.

In 1845 Boubée recorded in his paper " Sur les Graptolites des Pyrénées " the occurrence of *Graptolithus sagittarius* in the Silurian beds of the Pyrenees.

1845.
*Murchison, de Verneuil,
and Keyserling.*

Murchison, de Verneuil, and Keyserling, in their 'Geology of Russia and the Ural Mountains' (vol. ii), note the occurrence of *Graptolithus sagittarius* and *Gr. distichus* in the Silurian beds of Russia.

In the year 1846 Geinitz, in a second paper on Graptolites, suggested the first subdivision of the forms then included under the genus *Graptolithus.* He divides them into (*a*) straight, and (*b*) spirally curved forms. In the first group (*a*) he places (1) *Gr. foliaceus,* (2) *Gr. pristis,* (3) *Gr. folium,* (4) *Gr. dentatus,* (5) *Gr. priodon,* (6) *Gr. ludensis* (with its variety *Gr. virgulatus*), (7) *Gr. teretiusculus* (this he regards as synonymous with *Gr. priodon* and *Gr. sagittarius*), (8) *Gr. sagittarius* (which he believes to be identical with *Gr. scalaris*), and (9) *Gr. serratus.* He refers to the two branched forms, *Gr. geminus* and *Gr. Murchisoni,* and suggests with considerable hesitation that they may be identical with *Gr. sagittarius* and *Gr. priodon* respectively. In the second group (*b*) he places only one species, *Gr. convolutus.* He figures a few of these species, but some of his figures are merely copies of those of previous authors. His figure of *Gr. foliaceus* clearly represents an example of *Retiolites Geinitzianus,* and the network is well shown. He retains unmodified his previous views as to the structure and affinities of the Graptolites (*op. cit. ante,* 1842).

In 1847 Hall described in his "Graptolites of the Inferior Strata of the New York System" a number of forms of Graptolites from the Utica slate and the beds of the Hudson River group. The species described and figured include five forms identified by him with forms described or noticed by previous authors, and eight additional species for which new names are proposed. It is very doubtful if any of the forms referred by Hall to species previously described are identical with those species. Thus his (1) *Gr. pristis* (His.) is not Hisinger's form of that name ; (2) *Gr. scalaris,* Linn., is mainly *Climacograptus bicornis,* Hall; (3) *Gr. secalinus,* Eaton, is an ally of *Diplograptus foliaceus,* Murch., deformed by cleavage ; (4) *Gr. sagittarius* (His.) is almost certainly an example of *Didymograptus,* as is also (5) *Gr. tenuis* (Portlock). The new species described and named by Hall include (6) *Gr.* **sextans,** (7) *Gr.* **furcatus,** (8) *Gr.* **ramosus,** (9) *Gr.* **serratulus,** (10) *Gr.* **bicornis,** (11) *Gr.* **mucronatus,** (12) *Gr.* **gracilis,** and (13) *Gr.* **lævis.** This last is not a Graptolite, but an alga or a worm track. All the figures are good, and nearly all the species are recognisable.

Hall at that time grouped all these Graptolites under the single title *Graptolithus,* which was the only genus then recognised, but several of them have subsequently been made the types of distinct genera. Thus the forms there described as *Gr. bicornis, Gr. ramosus, Gr. sextans,* and *Gr. gracilis* became the accepted types of the genera *Climacograptus, Dicranograptus, Dicellograptus,* and *Cœnograptus* respectively.

Hall says little or nothing of the structure of the Graptolites, merely remarking that they had a "semi-calcareous body with a corticiform covering." As regards their zoological affinities, he considers that they have a close analogy with *Virgularia.*

1847.
Nimmo,
'Calcutta Journal of
Natural History,'
vol vii.

Nimmo, in the year 1847, expressed his belief that *Gr. foliaceus* is nothing more or less than the "serrated tail spines of the *Raja pastinaca*," or an allied species. Such a theory, founded as it was upon the most imperfect knowledge of the structure of a Graptolite, hardly merits serious consideration.

1848.
Phillips and Salter,
'Memoir of the
Geological Survey,'
vol. ii.

In 1848 Phillips and Salter recorded *Graptolithus pristis* (?), *Gr. ludensis*, and *Gr. Murchisoni* (?) from the Llandeilo rocks of Western England and South Wales; and in accordance with the general opinion of the palæontologists of that day, the Graptolites are placed by these authors among the organisms then classed as Polyparia.

1848.
Sedgwick and M'Coy,
"On the Organic
Remains found in the
Skiddaw Slates,"
'Quart. Journ. Geol.
Soc.,' vol. iv.

Sedgwick, in a paper on the 'Organic Remains of the Skiddaw Slates,' published the same year, records *Graptolithus sagittarius* from the Skiddaw slates. In an appendix to this paper M'Coy describes and figures a new species, *Gr.* **latus**. It is clear from the figured example that *Gr. latus* was in all likelihood founded on a branch of a Dichograptid, possibly the *Didymograptus hirundo* of Salter.

1849.
Bronn,
'Geschichte der
Natur,' vol. iii, part 2.

In 1849, Bronn gave a list of the Graptolites known up to that time. These include *Graptolithus convolutus, Gr. sagittarius, Gr. priodon, Gr. ludensis, Gr. Sedgwickii, Gr. distans, Gr. tenuis, Gr. dentatus, Gr. scalaris, Gr. distichus, Gr. pristis, Gr. foliaceus, Gr. folium, Gr. teretiusculus, Gr. Murchisoni, Gr. geminus*. As regards the position that the Graptolites occupy in the animal kingdom, he classifies the Graptolithina as one of the sub-families or groups of the Anthozoa.

1849.
Salter,
'Quart. Journ. Geol.
Soc.,' vol. v.

Salter, in his "Note on the Fossils from the Limestone on the Stinchar River, and from the Slates of Loch Ryan," gives figures of seven species of Graptolite: *Gr. folium, Gr. pristis* (which appears to resemble *Gr. mucronatus*), *Gr. pristis*, var. *foliaceus, Gr. ramosus, Gr. tenuis, Gr. sextans*, and a new form which he names *Gr.* **tænia**. This last it is impossible to identify with certainty. One of the figures given by Salter represents eighteen specimens of *Gr. sextans* (*Dicellograptus*) apparently suspended on a branch of this so-called *Gr. tænia*. This figure is of special interest from the modern point of view, as bearing on the mode of life of the Graptolites; but Salter himself does not draw any conclusions from the curious association.

1849.
Sharpe,
'Quart. Journ. Geol.
Soc.,' vol. v.

In 1849, Sharpe, in his paper on the "Geology of the Neighbourhood of Oporto," notified the occurrence of *Graptolithus Murchisoni?* in association with Trilobites of Lower Silurian age.

1849.

Hall,

" On Graptolites : their
Duration in Geological
Periods and their Value
in the Identification of
Strata," ' Proc. Amer.
Ass. for the Advance-
ment of Science.'

In the same year Hall discussed the geological range of the Graptolites in America in some detail.   He observes that they are peculiarly typical of the Lower Silurian; " few are known in a higher position, being less widely distributed and very limited in geological range."   That is to say, there are fifteen species found in the Lower Silurian, and three in the Upper Silurian.   Of these three species, two—*Gr. clintonensis* and *Gr. venosus*—are peculiar to America, and occur towards the base of the upper division in the Clinton group, and the third—*Gr. ludensis*— at a slightly higher horizon.

He describes the new form *Gr.* **venosus** (which is now known to be a species of Barrande's genus *Retiolites*) as " broad, with a central capillary axis and serratures on both sides."   The whole substance is finely veined or reticulated, like the skeleton of a leaf.   On account of its peculiar structure he thinks it may " very well form the type of a new genus in the future."   Hall does not discuss the affinities of Graptolites fully, but as it is impossible to trace the connection of these fossils with any living forms through the Devonian and Carboniferous rocks, he is " disposed to question their analogy with the Sertularidæ or Pennatulidæ."

1850.

M‘Coy,

" On some New Genera
and Species of Silurian
Radiata in the Collec-
tion of the University
of Cambridge,"

' Ann. Mag. of Nat.
Hist.,' ser. 2, vol. vi.

In 1850 M‘Coy gave a diagnosis of the family of the Graptolitidæ in the following words :—" Stem simple or branched, thin, usually linear, horny, unrooted; polyp cells divided at bottom by a transverse diaphragm."   M‘Coy was thus the first palæontologist to draw attention to the " free " nature of the Graptolites, and he points out distinctly that in this particular they differ from the Sertularidæ, " with which, however, they agree in the form of the polypidom and the cells."   His view of the presence of a transverse diaphragm at the base of the cells was probably obtained from an examination of Lake District specimens, as markings suggestive of septa were described later by Hopkinson from similar material.

As regards the method of reproduction of the Graptolites, M‘Coy acknowledges that no ovarian vesicles have been found so far, but he suggests that the ova were developed in naked sacs attached to the base of the tentacles of the polyps, and hence were not preserved : this position of the reproductive organs would be analogous to that in *Corymorpha*.   His views, he is careful to point out, are much the same as those previously published by Nilsson and Portlock.

An important advance as respects the classification of the Graptolites was made by M‘Coy in this paper, for he recognises that the structural differences between the uniserial and biserial forms ought to be regarded as of generic value, and suggests that the name *Graptolites* should be restricted to the former (the single-sided forms), and he proposes the new name **Diplograpsis** for those Graptolites which have denticles on both sides. This suggested nomenclature was soon generally adopted, but there has been considerable discussion among subsequent writers as to whether the original Graptolite described by Linnæus as *G. scalaris* was not in reality a biserial form, and that consequently the generic name *Graptolites* ought rather to be used for biserial than for uniserial forms.

M'Coy gives figures of three new species of Graptolites, namely, *Gr.* **lobiferus,** *Gr.* **millepeda,** and *Diplograpsis* **rectangularis.** He does not, however, give any description of these forms.

1850.
Nicol,
" Observation on the Silurian Strata of the S.E. of Scotland," 'Quart. Journ. Geol. Soc.,' vol. vi.

Nicol, in a paper which is mainly stratigraphical, described and figured a new species of *Monograptus, Graptolites* **griestonensis.** Descriptions are also given of *Graptolites convolutus* (which he considers to be identical with *Gr. spiralis* and *Gr. ludensis*), and a " foliaceous species " which, " if a true Graptolite, seems undoubtedly a new species." To this he gives the name **Gr. laxus.** He points out " that it so closely resembles some plants of the moss tribe (*Hypnum*) as to render its real character doubtful."

Nicol gives a brief review of the opinions previously held by palæontologists as to the true character of Graptolites, and asserts that while the general opinion is in favour of their being Polyparia, he finds that some are carbonaceous, some horny; and he throws out the suggestion that these remains may have belonged to animals of more than one class, and that " some of them may have been internal organs, rather than the external axis of a variety of polypifer."

1850.
Naumann,
' Lehrbuch der Geognosie,' vol. i.

In 1850 Naumann gave four woodcuts of Graptolites, and also figured seven species in an illustrative lithographic plate. All these, however, are bad copies of figures taken either from Murchison's ' Silurian System' (*Gr. ludensis, Gr. Murchisoni*) or from Geinitz's illustrations in the ' Neues Jahrbuch ' for 1842 (*Gr. folium, Gr. convolutus, Gr. sagittarius, Gr. priodon,* and *Gr. scalaris*), and they give no new information.

1850.
Richter,
" Aus der Thüringischen Grauwacke," ' Zeitsch. d. Deut. geol. Gesell.,' Bd. ii.

Richter's first contribution to the study of the Graptolites was published in 1850, and dealt mainly with questions of classification and structure. His classification is very similar to that suggested by Geinitz in 1846. He divides Graptolites into two groups: (I) curved (or single-toothed), and (II) straight (or double-toothed); and suggests a possible third class, or " closed." Under the first group (curved forms) he includes two species : (1) *Graptolithus sagittarius,* of which he gives seven figures ; and (2) *Graptolithus,* species unnamed. Three of his figures of *Gr. sagittarius* are probably drawings of examples of *Monograptus cyphus,* while three others are clearly of forms belonging to the Diplograptids. The species which he refers to as *Graptolithus,* sp., seems to be Barrande's *Rastrites peregrinus.*

In his second group (*straight* or double-toothed Graptolites) Richter includes four species: (3) *Gr. folium;* (4) *Graptolithus,* sp., for which he suggests the new name *Gr. mucronatus* (a name already preoccupied); (5) *Gr. priodon,* of which he gives no figure, and which he thinks might belong to a third group—the *closed* Graptolites; and (6) *G. scalaris,* under which he apparently includes the two forms usually known in recent years as *Climacograptus normalis* and *C. rectangularis.*

The figures given by Richter are for the most part good, and show well the typical graptolitic structure. He notices such details of structure as (1) the alternating arrangement of the cells in double Graptolites, a character which he considers invariable; (2) the thickening of the wall at the aperture; and (3) the prolongation of the virgula (or the "siphon," as he calls it) in a proximal direction. This last phenomenon he holds may be explained either by the fact that the cells have dropped off, or that they were very small during life.

Richter notes that the species of Graptolites described were all obtained by him from the Silurian formation of the Thüringer Wald, in the neighbourhood of Saalfeld, but gives nothing as to their further range or geographical distribution.

CHAPTER II.

SECOND PERIOD, 1850 TO 1865.

1850.

Barrande,

'Graptolites de Bohème.'

IN the year 1850 appeared Barrande's epoch-making work on the 'Graptolites de Bohème.' This added so largely to our previous knowledge of the Graptolites, and was marked throughout by such an admirable method of treatment, simplicity and clearness in the presentation of the facts, and brilliance of inference and generalisation, that it gave an impetus to their study, the importance of which can hardly be over-estimated. The work is one of much detail, and it will be best to discuss it in the order of arrangement that Barrande himself adopts.

General Sections.

1. *Nature of the Graptolites.*—Barrande considered that the evidence he had himself accumulated as to the nature of the Graptolites was strongly opposed to the view held by the majority of previous observers, namely, that the Graptolites were

allied to the Cephalopods. He cites the following arguments in support of his opinion :

(1) In some species there is a single row of cells, in others a double series.

(2) No Graptolite has a large terminal chamber like that of the Cephalopoda.

(3) Each Graptolite cell has its independent opening.

(4) Certain species of Graptolites were attached to the sea floor, as appears to be evident from Hall's figures.

Barrande agrees with Nilsson that the Graptolites must be regarded as belonging to the class of the Polyparia, but that not enough is known to determine exactly to what family they belong. He is inclined, however, to agree with Dr. Beck that they should be grouped with the Pennatulinæ, near the genus *Virgularia*.

2. *General Form.*—As regards the general shape of the Graptolites, Barrande recognises only two structural forms, viz. (1) the single-celled species, (2) the double-celled species ; the former being straight or curved either spirally or conically, and the latter being always straight.

3. *Solid Axis or Virgula.*—Barrande, although not the first to observe the important body known as the *solid axis,* was the first to discern its invariable presence in Graptolites, to describe its constitution, and to give it its name. He writes, "Graptolites are always provided with a solid axis. This axis is cylindrical and fibrous, and may be prolonged beyond the cellular part in certain mono-prionidian forms (*Gr. colonus*). In the double-celled forms also the axis is similarly prolonged, and is probably double, the two portions possibly forming a much flattened tube between the two series of cells (*Gr. palmeus*)." He considers it possible, judging from the upper part of the axis in *Gr. palmeus*, that each of these portions may consist in its turn of two layers, which have become separated by decomposition. In *Gladiolites*, however, Barrande notes that he could discern no true solid axis.

4. *Common Canal.*—Barrande first applied the name of "common canal" to that tube-like portion of a Graptolite which lies between the axis and the individual cells. In the double Graptolites there appear to be two common canals, quite independent of one another, and Barrande gives drawings of cross-sections of *Gr. priodon* and *Gr. palmeus* to illustrate the simple and double nature of this body. In *Gladiolites* there is only one median canal. As regards the function of this canal, Barrande is quite definite. He asserts that "this space enclosed the body of the polyp, and from it arose the individual germs living in the cells." "It served as a conveyance of common nutrition, and also as transport for the new germs."

5. *Cellules.*—The individual cells, or cellules, are in intimate connection with the common canal, but each germ is individualised, and constructs a cell with solid walls for itself. These cellules are inclined to the axis at various angles, are

sometimes in contact for their whole length (*Gr. colonus*) or for part of their length (*Gr. priodon*), or they are quite free as in *Rastrites*.

In form, each of the cellules may be compared to " a small sac," the length of which is always greater than the width, and is either rectangular or circular in section. Sometimes they narrow towards their orifices, or they curve over in the form of a hook (*Gr. priodon*). Barrande considers that there is no doubt that the wall between two contiguous cellules is double, and he represents this in his figures of *Gr. priodon*.

Each cellule, according to Barrande, has two orifices :

(1) *The internal orifice*, communicating with the common canal, and generally subrectangular or round in shape ;

(2) *The external orifice*, which presents many variations in form.

He discusses the various forms of cell apertures in considerable detail, from the simple straight apertures of *Gr. bohemicus*, which open upwards, through the oblique and spinose ones of *Gr. testis*, etc., to the curved ones of *Gr. priodon*, which open downwards. The apertures of several species are provided with spines, and when two spines are present these are arranged either above and below the aperture (*Gr. chimæra*), or symmetrically on the lateral parts of the border (*Gr. testis*). Barrande records all these spines or " ornaments " as " simple appendages of the test."

6. *Nature and Ornaments of the Test.*—The solid test of the Graptolites is about ·1 mm. in thickness, and of a horny character. Barrande believes that it contains little, if any, carbonate of lime, and is inclined to think that it is largely carbonaceous ; and he points out that the black colour of the shales in which Graptolites occur seems to confirm this opinion.

As regards the markings and superficial ornamentation of the test, he observes oblique striations on some of the cells, but he does not appear to have realised their vital importance as " growth-lines." Except in *Gladiolites*, the Graptolite test is smooth and continuous. With respect to the genus *Gladiolites*, which he was the first to distinguish, Barrande recognises and describes with great accuracy the peculiar network of threads forming the test, and is of the opinion that the meshes were either open, or else closed by " a membrane chemically different from that of which the network is composed."

7. *State of Preservation.*—Barrande points out that in Bohemia the Graptolites are usually preserved as impressions, but occasionally (in the limestones) they occur in relief with the test preserved, while in certain of the higher beds internal casts alone are found.

The accuracy of the observations of Barrande, the breadth and lucidity of his conclusions and descriptions, and the beauty of his illustrations, make his work not only a classic, but also most valuable for reference even at the present day. But Barrande's work, important as it was, was only a stage in the progress of graptolitic research, and in the subjects treated of in the subsequent sections of his memoir, later observers have made considerable advances upon his original views.

8. *Mode of Growth of the Graptolites.*—Barrande's view of the method of growth of the Graptolites is very different from that accepted at the present day. Observing that the polypary increased in width in one (the distal) direction and decreased in the other (the proximal), he considered that the proximal part with the smaller cells was the younger, or, as he termed it, the *growing portion*, while the widest part (distal) was the *adult portion*. He concluded that " the growth of the polypary took place, therefore, by the successive appearance of new ' germs ' at the narrow end." These germs are at first very distant (as in *Gr. proteus*), while in the adult portion they are in contact. He was therefore of the opinion that the " elongation of the body of the polypary must have preceded the production of any new germ. Thus this body would seem to have served as a canal of propagation." Barrande, however, points out that there are at least two foreign species in which such a mode of growth is impossible, viz. *Gr. Murchisoni* and *Gr. geminus*. These consist of two branches united by a small terminal " stem." In these forms he seems clearly to have realised that the growth must be in the opposite direction to that advocated by himself for the Bohemian forms.

Again, the fact that many single- and double-celled forms show a naked axis prolonged beyond the so-called adult end, Barrande considers may be explained by the " perishing of the older cells, one after the other, as the younger cells are developed." The occasional prolongation of the axis also beyond the narrow end of the polypary can, according to Barrande, only be interpreted by the " accidental decomposition of the young cells after the death of the individual before fossilisation." Another difficulty, presented by the not infrequent decrease of the cells towards the adult end, is also explained by him in the same way. Considering the cautious deductions from facts observed in most of the other parts of this book, it might appear strange at first sight that Barrande should have adopted so definitely this theory as to the method of growth; but it must be remembered that at that time the complex Canadian forms, in which such a mode of growth is demonstrably impossible, had not been described.

9. *Mode of Existence.*—As respects the mode of existence of the Graptolites, Barrande only considers the single question whether the Graptolites were free or fixed ; and he agrees with M'Coy that, as far as the evidence of the Bohemian species goes, it is " very probable that they were entirely independent." There is no evidence of adherence of Graptolites to rocks, as is the case with modern Zoophytes and Molluscs. It would have been impossible for them to be fixed to the ground by the growing portion, on account of its slender and frequently curved character, and there is no evidence whatsoever to suggest that they were fixed by the adult end. With regard to such forms as *Gr. bicornis* and *Gr. Murchisoni*, however, Barrande considers that the invariable presence of a " point " or

" pivot " at one end seems to indicate with fair certainty a fixed mode of life. He therefore concludes that it is probable that "certain Graptolites lived attached, while others floated or swam freely in the Silurian seas."

10. *General Characters of the Family of the Graptolites.*—Barrande points out that since the name *Graptolithus* was founded on such forms as *Gr. sagittarius* and *Gr. scalaris*, it should be maintained for all those species showing some analogy to these; but he suggests that the genus *Graptolithus* should be separated into two sub-genera, viz. **Monoprion,** including those with one row of cells, and **Diprion,** embracing those with two parallel rows.

As regards the curious Y-shaped form figured by Hall as *Gr. ramosus*, Barrande is inclined to regard it as an accidental phenomenon produced by the splitting of a Diprion form. He considers that the name *Graptolithus* should be restricted to those forms having their cells in contact, and he suggests the name of **Rastrites** for those in which the cells are quite distinct from each other. (This last name has become generally adopted, but the titles *Monoprion* and *Diprion* have never come into general use as generic terms.) Another new generic name proposed by Barrande was that of **Gladiolites** for a Diprion form covered with a fine network of threads, " and having no axis." His alternative title for this genus, **Retiolites,** has been more generally adopted.

11. *Distribution and Range of Graptolites in Bohemia.*—Barrande deals next with the distribution of Graptolites, first in Bohemia, and afterwards in other parts of the world. In Bohemia the vast majority of Graptolites occur at the base of the (Upper) Silurian; some occur in the lowest parts of the inferior limestone, Stage E, but they become rarer in the higher beds, and eventually disappear altogether before Stage F. Regarding the lower limits of their range, there is no evidence of their presence in the beds below Stage D, and therefore it is certain that in Bohemia, at any rate, the Graptolites did not make their appearance until after the disappearance of the primordial fauna.

As respects the kind of rock in which Graptolites are found, Barrande calls attention to the fact that they are most abundant in shales, and rare in siliceous rocks; or, as Murchison had previously put it, " the region of Graptolites is a great region of mud."

Barrande notices especially the occurrence of Graptolites in his " colonies " of Stage D, and indeed employs the Graptolites as evidence of his famous " Theory of Colonies." The existence in the colonies of the same species as are found in the much later Stage E, in association with those characteristic of Stage D, seemed to him to afford a proof that the colonial organisms, including the Graptolites, were " derived from a centre of creation quite different from that where the fauna proper to Stage D took its origin; " and that collectively they were forerunners, so to speak, of a typical Upper Silurian fauna, which did not attain its maximum development until a much later date.

12. *Distribution of Graptolites in Foreign Countries.*—Barrande discusses the question of the general distribution of Graptolites at considerable length, and gives the result of the researches of all previous writers as to their range. Although he collected all the material obtainable on this subject, he contented himself with general remarks as to the range of the Graptolites in time and space. His conclusions may be summed up as follows:—Graptolites belong characteristically to the Silurian system; in England they made their appearance during the period marked by the primordial fauna of Bohemia, reaching their maximum development about the middle of the Silurian period,—that is to say, at the top of the Lower and base of the Upper Silurian; and dying out altogether in the Lower Ludlow beds.

Descriptive Sections.

After giving a historical summary of the literature of the subject, Barrande devoted the rest of his memoir to the description of the various Bohemian genera and species. Both the descriptions and the illustrative figures given by Barrande were far in advance of anything done previously by graptolithologists, and the descriptive letterpress of his work has remained a standard type for diagnosis down to the present day. Several of the beautiful figures upon his illustrative plates, however, are, in accordance with the practice of the time, either generalised from more than one specimen, or somewhat idealised in the matter of detail.

But the state of knowledge of specific characteristics was then naturally inferior to that of the present day, and some of his species included forms now acknowledged to be specifically distinct. Thus under the name *Graptolithus priodon* (Bronn) Barrande not only included Bronn's typical form (Pl. I, figs. 3, 5—9), but at least three other forms also, which are now usually regarded as belonging to distinct species, viz. *Cyrtograptus Murchisoni* (Carr.), *Monograptus spiralis*, var. *subconicus* (Tullb.), and *Mono. Jaekeli* (Perner, ' Études sur les Graptolites de Bohème,' part 3). Again, the illustrations of his own species *Gr. colonus* embrace those of three distinct forms, of which the most characteristic (Pl. II, figs. 2 and 3) is now by common consent regarded as that of the type species, while fig. 5 may probably be assigned to the form at present classed as *Mono. dubius* of Suess, and figs. 1 and 4 to *Mono. vomerinus* of Nicholson. Perhaps the most conspicuous case is that of his *Gr. Nilssoni*, which he illustrates by three figures. These are drawn as if the specimens represented occurred together upon one and the same slab of rock, but they are now known to have been obtained from different localities and from different horizons (Perner, *op. cit.*). Fig. 16 is now regarded as the type form of *Gr. Nilssoni;* fig. 17 as a specimen of *Cyrto. Lundgreni;* while fig. 18 belongs to *Cyrto. tubuliferus* of Perner.

The species of Graptolites identified, described, and figured by Barrande include the (1) *Mono. priodon* and (2) *Mono. spiralis* of earlier authors, and the following forms which were then new to science :—(3) *Gr.* **Nilssoni**; (4) *Gr.* **colonus**; (5) *Gr.* **bohemicus**; (6) *Gr.* **Roemeri**; (7) *Gr.* **turriculatus**; (8) *Gr.* **Becki**; (9) *Gr.* **proteus**; (10) *Gr.* **Halli**; (11) *Gr.* **testis**; (12) *Gr.* **palmeus**, var. lata and var. **tenuis**; (13) *Gr.* **ovatus**; (14) *Gr.* **chimæra**; (15) *Rastrites* **fugax**; (16) *R.* **peregrinus**; and (17) *R.* **gemmatus**.   Little or no doubt is left as to the specific characteristics of most of these forms, and they are easily recognisable from Barrande's descriptions and figures.

In the description of his (18) *Gr.* **tectus** and (19) *Gr.* **nuntius**, Barrande enters into a lengthy explanation of the various aspects that the cellules may assume according to the direction of compression.   Thus he shows that specimens of Graptolites apparently biserial may really be the compressed scalariform views of a uniserial form.   Here, however, Barrande's inference, correct as it was respecting Graptolites in general, led him occasionally astray, and he erroneously described a true biserial species, viz. his *Gr. tectus*, as a uniserial form.

Barrande described the structure of his remarkable new genus, *Gladiolites* (or *Retiolites*), and its type species (20) *G.* **Geinitzianus**, with special care and accuracy of detail.   The median filiform axis visible in some specimens he holds to be part of the external network, and not to represent a division between the two series of cellules, for it is often discontinuous.   As regards the test of *Retiolites*, Barrande considers it is very improbable that it was continuous as in other Graptolites, but the network might, perhaps, have been covered by a thin film.

Barrande's brilliant memoir acted as a great incentive to the study of individual Graptolite species, and since its publication papers on this subject have been numerous.   His clear presentation of the different specific criteria trained the eyes of those who immediately followed him, and for the next twenty to twenty-five years research was mainly in the direction of the collection and description of new species; and side by side with this some important additions were made to our knowledge of the structure, mode of life, and zoological position of the Graptolites.   The detailed study of their geological distribution was, however, somewhat neglected, and was not taken up until a much later date.   This was, no doubt, owing largely to the fact that graptolithologists, fascinated by the brilliance of Barrande as a palæontologist, naturally accepted also his ideas with respect to migration and colonies, unaware that stratigraphy in the Lower Palæozoic rocks had not yet advanced to that degree of detail by means of which it was possible to interpret correctly the complexities of the Bohemian succession.

1850.
Verneuil,
'Bull. Soc. Géol. de France,' vol. vii.

In 1850 Verneuil noted the occurrence of *Gr. colonus* and *Gr. testis* in the Ampelite schists of Neuvilette (Brittany). These are included in his " second fossiliferous stage."   In his " third fossiliferous stage," the representative of the Upper Silurian, he records *Gr. priodon* as being the most common species.

1851.
Suess,
"Ueber Böhmische
Graptolithen,"
'Naturwissensch. Ab-
handl. von Haidinger,'
Bd. iv, Abth. 4.

Immediately after the appearance of Barrande's memoir on the Bohemian Graptolites, Suess published the results of his own researches on this subject. His labours covered much the same ground as those of Barrande; but his paper, though long and detailed, did not materially advance the subject beyond the point to which Barrande's researches and conclusions had already brought it, and the figures upon the plates which illustrate his paper are indifferent. We may here notice those points respecting which Suess added to or differed from the results already arrived at by Barrande.

Mode of Existence.—In dealing with the character of the rock in which the Graptolites occur, Suess points out that the shales are often very carbonaceous, and frequently contain balls of anthraconite, and he suggests that these balls may have been formed by the rolling in the mud of the remains of marine plants. This is interesting in the light of later opinion concerning the mode of life of Graptolites, and the recent view that they were attached to seaweeds.

Classification.—As regards the classification of the Graptolites, that of Suess differs considerably from that proposed by Barrande. He suggests that there are only three genera, viz.—

1. *Retiolites* (in Barrande's sense);

2. *Petalolithus* (including the sub-genus *Diprion* of Barrande and the sub-genus *Diplograpsis* of M'Coy);

3. *Graptolithus* (embracing forms having only one row of cells). This genus is subdivided into three sub-groups, namely, those having—

(*a*) a strong axis bent into a fixed curve in one plane, and their cells always in contact (*Gr. bohemicus*);

(*b*) an axis at the older end bent into a definite conical screw line; at the younger end thread-like; cells in contact when fully developed, not in young stage (*Gr. proteus*);

(*c*) an axis thread-like and curved; cells not in contact (*Rastrites*).

He agrees with Barrande (and is even more consistent in this respect) in considering the narrowest end of the polypary as the younger, and therefore always figures it with that end uppermost.

Structure.—Suess described the genus *Retiolites* and its species *R. Geinitzianus* in much detail, but his explanation of its structure is markedly different from that of Barrande. According to Suess, the axis, which is "sometimes distinct, sometimes almost invisible, not rigid, but flexible," gives rise to alternating secondary branches at regular intervals. As the polypary increases in size, interspaces appear between the secondary branches, and these are formed by the splitting of the branches into two parts from the centre towards the outside. These interspaces are then divided by five or six strong transverse walls at right angles to the secondary branches. These smaller divisions constitute true "cells." In the widest part of the polypary the branches may split as often as four or five times. Suess admits that it is somewhat difficult to understand how the transverse walls bounding the "cells" were produced. The genus *Retiolites* is distinguished from *Petalolithus* (*Diprion*, Barr) by having its secondary branches united by a distinct cell system instead of by a membranous substance.

Description of Species.—In addition to (1) *Retiolites Geinitzianus* (which Suess regards as synonymous with *Fucoides dentatus*, Brong., *Gr. pristis*, *Gr. scalaris*, *Gr. foliaceus*, *Gr. secalinus*, and other biserial forms) Suess describes and figures a new form under the name (2) *Retiolites* **grandis**. The individuality of this last-named species was not recognised by subsequent writers until a comparatively recent date, when Tullberg made it the type of a new genus, *Stomatograptus*, but his interpretation of its structure is very different from that of Suess. Suess considered it to be distinct from *Ret. Geinitzianus* on account of its larger size, the peculiar form of the outer margin, and the small vertical distance apart of the secondary branches; while Tullberg distinguished it because of its median row of "pores."

Under the description of his new genus **Petalolithus** Suess notes the prolongation of the solid axis at both ends, and agrees with Barrande that it is very probable that the cells have fallen off these naked prolongations. He makes a great point of the alternate or opposite position of the secondary "branches," and thinks it is due to subsequent movement or disturbance. He considers that it is possible to recognise in *Petalolithus* something akin to the cell system in *Retiolites*, and suggests that in *Petalolithus* "those parts which take the place of the cell system in *Retiolites* consisted of a skin-like substance provided with stomata (?)."

He divides the forms assigned to *Petalolithus* into two groups :

(*a*) those in which there is a decrease in width at both ends of the polypary, producing a characteristic oval shape ;

(*b*) those in which the distal walls are parallel.

The following species of *Petalolithus* are described and figured :

(3) *P. palmeus;* (4) *P. ovatus;* and a new species, (5) *P.* **parallelo-costatus**.

The structure of the genus *Graptolithus* (*Monoprion*) is next discussed, and he considers that the axis in *Graptolithus* is tube-like, and not solid as Barrande had stated.

In his first group of the genus *Graptolithus* Suess includes forms like *Prionotus geminus*, *Gr. ramosus*, etc. (which, according to him, had the power of branching dichotomously), and describes (6) *Gr. priodon;* (7) *Gr. bohemicus;* (8) *Gr. serratus* (which he believes to be identical with *Gr. dentatus*, *Gr. Roemeri*, and *Gr. latus*); (9) *Gr. testis* (the structure of the aperture and spines of which he describes in great detail); (10) *Gr. lævis* (Hall) (Suess' figure of this is possibly a branched form—a *Cyrtograptus*,—but quite unidentifiable from his drawings); (11) *Gr. Sedgwickii;* (12) *Gr. tænius;* (13) *Gr. Becki;* (14) *Gr. Nilssoni;* (15) *Gr. convolutus* (?). Under the name (16) *Gr. colonus* (Barrande) he figures several specimens which probably represent various aspects of *Mono. vomerinus* (Nicholson). In addition to these forms, which had been already described by previous authors, Suess figures two new species in this group, viz. (17) *Gr.* **dubius** and (18) *Gr.* **falx**. Of these, however, only the first is identifiable at the present day as a distinct species.

In his Group 2 (*Graptolithus*) Suess describes (19) *Gr. turriculatus;* (20) *Gr. proteus*, and a new form (21) *Gr.* **armatus**.

In his Group 3 (*Rastrites*) he includes (22) *Gr. Linnæi;* (23) *Gr. fugax;* (24) *Gr. peregrinus;* and a new species, (25) *Gr.* **Barrandei**.

1851.
Richter,
" Ueber Thüringische
Graptolithen,"
' Zeit. d. Deutsch. geol.
Gesell.,' Bd. iii.

In 1851 Richter published a second paper, entitled " Ueber Thüringische Graptolithen." This paper is supplementary to the one previously published by him in 1850, and while the views of Barrande as regards the structure of the Graptolites are, to a large extent, incorporated in it by Richter, they are in many particulars amplified by the results and conclusions drawn from his own researches.

The paper contains a descriptive list of species obtained by him from the Alum shales of Thüringia, viz. (1) *Gladiolites Geinitzianus;* (2) *Diprion palmeus;* (3) *D. ovatus* (which he regards as probably only a variety of *D. palmeus* or *D. folium*); (4) *Gr. priodon;* (5) *Gr. colonus;* (6) *Gr. Becki;* (7) *Gr. nuntius;* (8) *Gr. Halli;* (9) *Gr. bohemicus;* (10) *Gr. testis;* (11) *Gr. chimæra;* (12) *Gr. proteus;* (13) *Gr. turriculatus;* (14) *Rastrites gemmatus;* (15) *R. Linnæi;* (16) *R. peregrinus.*

Richter confirms Barrande's views on almost all points of structure, and, in addition, he draws attention (simultaneously with Scharenberg) to the special organ now known as the " sicula " (Lapworth), which he calls the *fuss* (foot, base, or pedestal). His figures (1 to 3) represent it as being jointed. He regards it as a " prolongation of the axis," generally directed upwards, but occasionally downwards (*Diprion folium*). In *Gladiolites* and *Diprion* it is always bodkin-shaped; in *Rastrites* spindle-shaped. If this organ is actually a " foot," then he considers Barrande's and Suess' view that the thinner end of the polypary is the younger, and the wider part the older, must be erroneous. He believes that the skeleton of the polypary probably possessed little or no rigidity, and therefore that the form of the polypary, except in *Gr. turriculatus*, was not constant.

Mode of Occurrence.—Richter draws attention to many interesting facts bearing on the mode of occurrence of the Graptolites. He points out that they always lie on the *surface* of the rock, and are never found upright or traversing the shale; and that they are generally arranged quite irregularly—implying, apparently, that they did not live fixed in the mud where they are now found. According to

d

Richter, the occurrence of such numbers of specimens on these extremely thin layers of rock indicates that the Graptolites were very short-lived, and that the fact that they gradually disappear as the Alum shales thin out, shows that they were confined to deep water.

1851.

Scharenberg,
'Ueber Graptolithen, mit besonderer Berück-sichtigung der bei Christiania vorkom-menden Arten.'

A remarkable paper published by Scharenberg in 1851, which may be regarded in a sense as a direct outcome of Barrande's memoir, is of great importance, the author criticising fully the observations and views of Barrande in the light of his own researches among the Graptolites from the Christiania district of Norway.

Description of Species. — In this paper Scharenberg describes and figures six species: (1) *Gr. geminus*, (2) *Gr. virgulatus*, (3) *Gr. folium*, (4) *Gr. teretiusculus*, (5) *Gr.* **Barrandei**, and (6) *Gr.* **personatus**, of which the last two only are new. The figures are conscientiously drawn, every imperfection being represented, and are consequently most valuable. But the extreme caution by which not only the figures, but also all the details in his work are characterised, leads him to include under one specific name several forms now regarded as specifically distinct (although, on the other hand, his own two supposed new species cannot be sustained). Under the name of *Gr. teretiusculus* he included at least two distinct species. One of these is the character-istic Arenig-Llandeilo species of *Climacograptus* (subsequently named after him by Lapworth as *C. Scharenbergi*). The specimens described by him as *Gr. folium* are examples of *Phyllograptus*.

Specific Criteria.—Scharenberg prefaced the descriptive portion of his work by a careful analysis of the most important points to be considered in distinguishing different species. As these are of considerable interest, they may here be briefly considered.

1. *Length of Stipe.*—According to Scharenberg this character is not of much importance, owing to the fact that it is difficult to determine when the specimen is complete, and that even fully grown specimens of one and the same species may vary much in size.

2. *Width of Stipe.*—On the other hand, Scharenberg considers that there is a practically constant relation between the width of the cells and the common canal.

3. *General Form of the Polypary.*—Scharenberg regards this character as one necessitating extreme caution in its interpretation, concluding his remarks on the matter as follows:—" The determination of a species founded on the outer form of a Graptolite is only trustworthy when there is only one row of cells, and these cells are distant (*Rastrites*). Teeth, denticles, and appendages of the cell must be considered as distorted or falsely shaped cell apertures, until their normal condition has been proved with certainty in well-preserved specimens, in the same way as

Barrande has done in the case of *Gr. priodon* and *Gr. colonus*." So anxious was Scharenberg to avoid any confusion between normal and abnormal appearances, that it led him into the opposite extreme of misinterpreting several constant specific characters, and attributing them either to the results of deformation or to the special mode of preservation.

4. *Existence and Direction of Axis.*—Scharenberg disagrees emphatically with Barrande's view that the axis is never anything more than a single solid cylinder, the same in a *Diprion* as in a *Monoprion* form. He explains the curious double and vesicular appearance of the axis in some of the specimens of the *D. palmeus* as figured by Barrande, by the theory that it not only represents the true axis itself, but also a part of the common canal, from which the cells have been torn off before fossilisation. In the case of the two-branched forms, such as *Gr. geminus*, Scharenberg points out that the axis must have divided or branched during life. In all double Graptolites he thinks that the axis is invariably straight, while in single-rowed ones it may assume almost any form. He does not consider that the form of the polypary is constant in each species, except perhaps in the case of *Gr. turriculatus*.

5. *The Texture of the Graptolite Stipe.*—Scharenberg points out that this character would be of the greatest importance in determining species if it could be ascertained; but, unfortunately, it can very rarely be observed. In the case of *Gladiolites*, its peculiarities quite justify this form being made the type of a new genus. He observes the horizontal grooves and markings (growth-lines) on the cells in certain specimens of *Graptolithus* as " small thickened rings, like the texture of many snail shells;" but, like Barrande, he does not suggest that they are growth-lines.

6. *Number of Cell Rows, and the Angle which the Cells make with the Axis.*—Scharenberg does not regard either of these characters as of much systematic importance. He holds that Barrande's separation of the Graptolites into *Monoprion* and *Diprion* sections is also of little value, owing to the difficulty of determining in scalariform views to which division a specimen belongs.

7. *The Inclination of the Cells to the Common Canal, and their Direction.*—These characters are of great moment, especially in those forms in which the cells are in close contact.

8. *The Distance of the Cells apart from each other.*—This is also a reliable method of distinction of species; but Scharenberg does not agree with Barrande that it is of generic value, as in the case of *Rastrites*.

9. *The Form of the Cells and their Apertures.*—This is " undoubtedly the most important criterion for the distinction of species."

10. *The Breadth of the Common Canal in Relation to the Length of the Cells.*—This is a character which deserves more attention than it has hitherto received.

The most important part of Scharenberg's paper is the general preface, dealing with the classification, structure, and mode of life of the Graptolites, for in these particulars Scharenberg added very considerably to the observations made by Barrande on the subject.

Classification.—As regards the classification of the Graptolites, he considers that Barrande's suggested division into *Monoprion* and *Diprion* is quite inadequate, for it excludes all the branching forms, such as *Gr. geminus*, *Gr. sextans*, etc., of the existence of which he himself has no doubt whatsoever. Bronn's (? Richter's) classification into spiral, straight, double-row, and twinlike forms Scharenberg regards as better, but nevertheless not quite satisfactory, as many Graptolites are at first curved and then become straight. As *Gladiolites* has no distinct axis, and its structure is so peculiar, and without parallel, Scharenberg considers that it should be separated from the Graptolites altogether.

Structure.—From the structural point of view, Scharenberg's views as to the development of the *Diprion* forms are of interest. They were antagonistic to those previously held by the majority of graptolithologists, but they have subsequently been more or less confirmed by modern research. He points out that in *Diprion* the cells alternate without exception. " It may be concluded from this that in the development of these animals two cells never arose at the same time, and consequently there is no essential difference between the Graptolites which are distinguished as having one row and those which have two rows of cells."

As regards the nature of the skeleton of the Graptolites, Scharenberg agrees with Barrande that it was horn-like, and that it possessed a " high degree of flexibility ; " but owing to its easy destructibility, Scharenberg, as has been already mentioned, considered that many original characters, such as spines at the cell apertures, etc., were merely the result of decomposition.

Affinities.—With respect to the affinities of the Graptolites, Scharenberg held that they were most closely allied to the *Pennatulidæ*. He pointed out, however, as did Prout in the same year, that they differed from the *Pennatulidæ* in having an external and not an internal skeleton.

Mode of Growth.—Scharenberg strongly criticises Barrande's views as to the mode of growth of the Graptolites. He considers that many, especially the branched and double-rowed forms, were fixed in the mud by a kind of " stem," and that consequently the cell apertures opened upwards. If this were the case, then the narrow end, instead of being, as believed by Barrande, the youngest, must be the oldest part, and growth must have proceeded from below upwards. This special mode of growth would indeed be analogous to that in the *Pennatulidæ*, to which the Graptolites are closely related. He also points out, in support of this view, that the narrowness of the polypary does not necessarily imply that the smaller cells are the youngest, for in some species (*G. ovatus*) the polypary narrows at both ends alike.

1851.

Bronn,

'Lethæa Geognostica,'

vol. i.

Bronn, in the 1851 edition of his 'Lethæa Geognostica,' enumerates twenty-seven species of Graptolites belonging to the genus *Graptolithus,* of which he asserts that twelve are of Lower Silurian and fourteen of Upper Silurian age, while he records one species from the Bergkalk (Mountain Limestone). He gives figures of (1) *Gr. priodon,* (2) *Rastrites Linnæi,* and (3) *Retiolites Geinitzianus,* but these are all copied from Barrande.

1851.

Boeck,

'Bemaerkninger anga-aende Graptolitherne.'

During the same year that witnessed the appearance of Scharenberg's work (1851), a paper was published by Boeck, an author who, unlike Scharenberg, was not acquainted with Barrande's memoir. The illustrations of this paper are excellent, but the conclusions drawn from them are characterised by the same timidity of interpretation as that noticed in Scharenberg's paper; and, like many other writers at that time, Boeck attributed several characters, now known to be of specific value, to accidental deformation.

Classification.—This part of his paper is of little value. He considers that the *Prionotus scalaris* of Hisinger is the same as *D. pristis;* but none of the specimens figured by Boeck under these names are referable to either of these species as now defined. One of his forms, however, is easily identifiable from the excellent illustrations with *Climacog. Scharenbergi,* Lapw. (fig. 10). Under the title *Pr. sagittarius* he figures fragments of a form which may be a species of *Didymograptus.* Boeck's *Pr. folium* (fig. 27), which he considers to be identical with *Gr. foliaceus,* is possibly a *Phyllograptus.* *Gr. Murchisoni* (fig. 24) (or var. *geminus*) is well figured, but Boeck regards the branching in this species as an accidental structure, the result of splitting. One of the figures (fig. 30), which is supposed to represent *Gr. Murchisoni,* is apparently the species subsequently named by Lapworth *Bryograptus Kjerulfi.* Boeck considers that Schlotheim's *Orthoc. serratus* is identical with the *Pr. scalaris* of Hisinger.

Structure.—The structural features of the Graptolites are most carefully described and named by Boeck in this paper, though the connection and functions of these structures can hardly be said to have been fully realised by him. He considers that the skin or test of the Graptolite was partly elastic (or rather contractile). This test formed a kind of tube, which was prolonged into a narrow, pointed part—*apex.* The varying length of this apex was due to the fact that it was attached to an internal contractile organ, which could draw the apex back into the tube. The longitudinal dorsal groove ("*suture*" of Nicholson) he terms the *sulcus longitudinalis,* and this might be either straight, undulate, or angulate. The curious transverse grooves running from the angles from this sulcus in *C. Scharenbergi* he terms *sulci laterales,* while the grooves formed by the cell apertures themselves are named *sulci transversales.* The edges of these sulci (or costæ) are considered by him to be thickenings of the skin of the tube, and this thickening

e

increased "inwards inside the tube," and therefore was only discernible in compressed specimens. The thecal walls are noted and entitled *sulci obliqui*; but their continuations with the apertures were not observed by him. In addition he notices other structures, which he terms *septa longitudinales* and *transversales*.

Boeck did not consider that the outer skin had been anywhere provided with openings, but pointed out that the thickening observable at the base of the thecal walls had somewhat the appearance of such openings. He failed altogether to realise the meaning of his *sulci transversales*, which are the edges of the cell apertures. In various cross-sections of the polypary which he examined he recognised the fine thread-like part, *tendo* (virgula), "to which has been fastened contractile fibres, which could have effected the tube's shortening."

Affinities.—Boeck admits that practically nothing is known as respects the zoological affinities of the Graptolites. He naturally considers, however, that their structure, as above described, gives " reason for doubting the prevalent view of the time that the once living animals belonged to the family of Polyps, at least in so far as they have been placed among the living *Sertularia* and *Pennatula*," and he suggests that they might be " broken pieces of some tentacularly formed organ of a larger animal."

1851.
Harkness,
" Description of the Graptolites found in the Black Shales of Dumfriesshire," 'Quart. Journ. Geol. Soc.,' vol. vii.

Harkness in this paper gives a complete list of Graptolites obtained up to that date from the Moffat shales of South Scotland.

Twelve species are recorded, of which five are described as new forms. He arranges them under three generic types, adopting the classification of Barrande and M'Coy. 1. *Graptolites* (*Monoprion*). 2. *Diplograpsis* (*Diprion*). 3. *Rastrites*. All the species are described and figured. The previously recorded species included (1) *Rastrites peregrinus*, (2) *Gr. Sedgwickii*, (3) *Gr. Becki*, (4) *Gr. Nilssoni* (not the true *Mono. Nilssoni*, Barrande), (5) *Diplog. rectangularis*, (6) *D. folium*, (7) *D. foliaceus*. Five new species are described by Harkness, viz. (8) *Rastrites* **triangulatus**, (9) *Gr.* **Nicoli**, (10) *Gr.* **incisus** (11) *D.* **nodosus**, and (12) *D.* **pennatus**, but only the first two are identifiable with certainty at the present day.

As regards the zoological position of the Graptolites, Harkness considers that the *Diprions* are " not far removed from *Pennatula* and *Virgularia*," possessing in common with them a solid central axis, and a linear arrangement of cells. He asserts that the " existing analogies of *Graptolites* proper and *Rastrites* are not so well ascertained; but on the whole they appear not far removed from the modern genus *Sertularia*, and partake rather of the character of Hydroid than of Asteroid zoophytes."

1851.
Salter,
" Silurian Rocks of
Scotland," ' Quart.
Journ. Geol. Soc.,'
vol. vii.

In the same year Salter, in an appendix to a paper by Murchison on " The Silurian Rocks of Scotland," described and figured two species of Graptolites, one of which was new, from the Girvan district, viz. (1) *Gr. tenuis* and (2) *Diplog.* **bullatus.** The figures, however, are not good, and it is difficult, if not impossible, to identify the species.

1851.
Prout,
" Description of a New
Graptolite found in the
Lower Silurian Rocks
near the Falls of St.
Croix River," ' Amer.
Journ. of Science,'
vol. ii, ser. 2.

Two papers on Graptolites appeared in America during this year. Prout described and figured a new form of Graptolite under the name *Gr.* **Hallianus.** The drawings are poor, and it is difficult to identify the Graptolite, but it clearly belongs to the Dendroidea rather than to the Rhabdophora, and is probably a *Dendrograptus.*

Prout also discusses the zoological affinities of the Graptolites in some detail, and " with due deference to Beck " thinks that this new species, and all those with a hollow central tube and tubular cells, are allied rather to *Sertularia* than to *Virgularia.* For in the " Virgulariæ and Gorgonidæ the tubes of the polypi have no regular connection with the central axis, except through the medium of a fleshy or coriaceous envelope, and it would be difficult to account for the tubular character of both the polyparium and denticles of Graptolites, unless we suppose that the central or semi-calcareous stem was entirely destroyed, while the fleshy envelope remained in a perfect state of preservation, and this is extremely improbable." In other words, Prout considers that the skeleton of Graptolites was external, as in the Sertularidæ, rather than internal, as in the Gorgonidæ and Pennatulidæ. In his new form *Gr. Hallianus*, Prout recognises that there is no trace of a central axis, and this being very indestructible when present, probably therefore never existed. He believes that the " cup-like denticles " are " unilateral," " vaginated on their external sides," and " inserted into a common connecting tube." Prout considers that probably both the " Sertularidæ and Pennatulidæ existed at the same time in the ancient seas," and that some of the " differences of opinion on the origin of the Graptolites may be found more seeming than real."

1852.
Geinitz,
" Die Versteinerungen
der Grauwacke-
formation in Sachsen
und den angrenzenden
Länder Abtheilungen,"
Heft 1, ' Die Silurische
Formation.' " Die
Graptolithen."

The first edition of a general work by Geinitz appeared in 1852, entitled " Die Graptolithen." It resembles Barrande's memoir in its comprehensive character, but Geinitz's views on the structure, development, and zoological affinities of the Graptolites are little more than a repetition of those formulated by Barrande himself. Indeed, the work is critical and historical rather than original or suggestive. Geinitz differs from the views of Barrande, Suess, and Boeck, as to the direction of growth of the Graptolites, and consequently as to the proper position for figuring them. He regards the narrow end as the " beginning of the animal," the apertures of the cell thus opening upwards, " facing the light." Some forms, such as *Monograpsus* and *Nereograpsus*, he holds to be free-swimming; while others, like *Diplograpsus* and *Cladograpsus*, were probably attached to the mud by their narrow end.

Affinities.—He adopts the generally accepted view as to the Sertularian affinities of the Graptolites, and disagrees with Boeck's views that they may be part of a larger animal, and therefore not complete in themselves. Geinitz attributes some of the various aspects under which the species present themselves to their flexibility and contractibility, and considers that *Pr. teretiusculus* possessed a greater elasticity and contractibility than any other known form.

Classification.—The classificatory part of Geinitz's work is perhaps the most valuable. He recognises five genera :

1. *Diplograpsus* (*Diprion* and *Petalolithus*). Geinitz holds that Barrande's name *Diprion*, being already in use for an insect, must be dropped, and as *Petalolithus* is identical with *Diprion*, that must be abandoned also. Further, M'Coy's name of *Diplograpsus*, having the priority, should be adopted for the genus.

2. **Nereograpsus** (two-rowed Graptolites without or with a very slender axis). Geinitz emphatically states that these forms are not worm-markings, and that he has been able to recognise the openings of the polyp cells. Geinitz's views of these forms have, however, not been admitted by other palæontologists.

3. **Cladograpsus** (" two-armed or forked Graptolites "). Under this name Geinitz includes those species which are now grouped in the genera *Dicellograptus* (Hopkinson) and *Dicranograptus* (Hall).

4. **Monograpsus** (single-rowed Graptolites with solid axis). This name was proposed by Geinitz to include both the *Monoprion* and *Rastrites* of Barrande, and was suggested so as to be analogous to the *Diplograpsus* of M'Coy. It has since been universally adopted.

5. *Retiolites* (double-rowed Graptolites, which are covered on their surface by a net-like skin, and possess a superficial central axis).

Geinitz points out that the name *Graptolithus*, or *Graptolithes*, can no longer be used for any single division of the Graptolites, any more than the family names Trilobite or Ammonite for a single genus of either of these families.

Description of Species.—The following species were described and figured by Geinitz, for the most part with great accuracy :—(1) *Diplograpsus ovatus* ; (2) *D. folium* ; (3) *D. palmeus* (some of the figures represent *C. rectangularis*) ; (4) *D. pristis*, His. (which he believes to be identical with *Pr. scalaris*, His.) ; (5) *D. dentatus* (?) (fig. 25 undoubtedly represents the genus now known as *Dimorphograptus*, Lapworth) ; (6) *D. bicornis* (?) ; (7) *D. rectangularis* (?) ; (8) *D. bullatus* (?) (these three last forms Geinitz does not regard as true species) ; (9) *D. foliaceus* ; (10) *D.* **cometa** (a new form subsequently chosen as the type of the sub-genus

Cephalograptus, Hopkinson); and (11) *D. teretiusculus*. Geinitz regards the three Graptolite forms *D. secalinus*, *D. laxus*, and *D. distichus* as doubtful species.

His genus *Cladograpsus* is divided by him into two groups, for which, however, he did not suggest sub-generic titles. Those constituting his first group (*a*) are at present united in the genus *Dicranograptus*, Hall; those in his second group (*b*) are divided between *Dicellograptus*, Hopk., and *Didymograptus*, Hall. His first group contained Hall's species of (12) *Cladograptus ramosus* and (13) *C. furcatus*. In the second group are included a new species, (14) *C.* **Forchammeri**, and the previously described forms (15) *C. Murchisoni*, (16) *C. serra*, (17) *C. sextans*, and (18) *C. serratulus*.

Under *Monograpsus* the following are figured and described :—(19) *Monograpsus sagittarius*, (20) *M. Barrandei*, (21) *M. nuntius* (under this name are included some scalariform views of a *Climacograptus*), (22) *M. tectus*, (23) *M. Nilssoni*, (24) *M. bohemicus*, (25) *M. incisus*, (26) *M. virgulatus*, (27) *M. colonus*, (28) *M. latus* (= *M. Roemeri*), (29) *M. chimæra*, (30) *M. testis*, (31) *M. Sedgwickii*, (32) *M. Halli*, (33) *M. distans*, (34) *M. Becki*, (35) *M. Nicoli*, (36) *M. priodon*, (37) *M. millepeda*, (38) *M. proteus*, (39) *M. convolutus*, (40) *M. turriculatus*, (41) *M. triangulatus*, (42) *M. gemmatus*, (43) *M. peregrinus*, (44) *M. Linnæi*, and two new forms, (45) *M.* **Heubneri** and (46) *M.* **Salteri.**

Geinitz enters fully into the discussion of the structure of *Retiolites*, and gives his support to the views of Barrande rather than to those of Suess.

Distribution.—An important section of Geinitz's work is devoted to the general distribution and range of the Graptolites, but little or no advance is made on the facts and conclusions previously given by Barrande and others.

1852.
Salter,
" Description of some Graptolites from the Silurian of Scotland," 'Quart. Journ. Geol. Soc.,' vol. viii.

A short paper by Salter appeared in the same year. In this a few species of Graptolites obtained by Harkness from the South of Scotland are figured and described. Under the name (1) *D. teretiusculus*, some specimens of *Climacograptus* are figured, one of these being certainly referable to *C. Scharenbergi* (Lapw.); (2) *Gr.* **Flemingii** is described as a new species, and a form referred to (3) *Gr. sagittarius* is figured; but *Gr. laxus*, *Gr. Nicoli* and *Gr. tænia*, Salter, are not considered by the writer to be distinct species.

As regards the range of the Graptolites, Salter says that "there is no evidence of a double Graptolite being found above the Caradoc Sandstone" ("Llandovery" of modern geologists).

1852.
Hall,
'Palæontology of New York,' vol. ii.

Hall, in the second volume of his 'Palæontology of New York,' figures and describes the two species (1) *Gr. clintonensis* and (2) *Gr. venosus* from the Clinton beds, previously named by him in 1843 and 1849. The former is a *Monograptus*, allied to *M. riccartonensis*, the latter a *Retiolites*, similar to or identical with *R. Geinitzianus*.   He also gives a diagnosis of the new genus **Dictyonema,** from which it will be seen that he was as yet uncertain as to the zoological affinities of this form.   He writes, " The general structure of this coral is very similar to *Fenestella*.   The branches consist of a black film enveloping a semi-calcareous or corneous interior, and they have the appearance and texture of Graptolites, to which they are doubtless closely allied."   It has very little, if any, true relation with *Gorgonia*, to which it was previously referred; it " possesses no positive characters by which it can be identified, either as Bryozoa or true corals."   Two new species of this genus are described and figured by him—(3) *D.* **gracilis** and (4) *D.* **retiformis**.   He found also a specimen of a new genus **Inocaulis,** which is, however, probably not a Graptolite.

1852.
Barrande,
" Einige Bemerkungen über die Abhandlung des Suess," 'Jahrb. der k. k. Geol Reichanstalt.'

In the same year (1852) Barrande replied to the observations and conclusions made by Suess on " The Bohemian Graptolites."

As regards *Retiolites*, Barrande refuses to accept Suess' interpretation of its structure, believing that the apparent facts observed were due to the bad state of preservation of the specimens.   He argues strongly against the validity of Suess' genus *Petalolithus*, and his species *Ret. grandis*, *Petalog. parallelo-costatus*, *Gr. ferrugineus*, *Gr. dubius*, *Gr. lævis*, *Gr. falx*, *Gr. armatus*, and *Gr. Barrandei*.   Subsequent investigators agree with Barrande in most of these contentions.

1853.
Salter,
" Description of a New Species of Graptolite," 'Quart. Journ. Geol. Soc.' vol. ix.

A new Graptolite species was described in 1853 by Salter, from the Hudson River group at Lauzon Precipice, under the name *Gr.* **caduceus**.   Two species, however, are actually figured by him under this name, the one being *Tetragraptus serra* of Brongniart, and the other the *Didymog. gibberulus* of Nicholson.

1853.
Ribeiro,
" On the Carboniferous and Silurian Formations of the Neighbourhood of Bussaco," 'Quart. Journ. Geol. Soc.,' vol. ix.

In 1853 Ribeiro recorded *Gr. ludensis* of Murchison from the upper division of the Silurian of Portugal, probably from the rocks corresponding with those of the Wenlock formation of England.

1853.
Richter,
" Thüringische Graptolithen," 'Zeit. d. Deutsch. Geol. Gesell.' Bd. v.

A third paper by Richter on " Thüringische Graptolithen " was published in 1853.   In this paper he proposes the following classification of the Graptolites, which is considerably more detailed than the one previously given by him in 1850.

(A) Graptolites with many axes. *Cladograpsus.*

    (A) Many-celled. ? (1) [1] *Cl. nereitarum.*

    (B) One-celled. ? *Lophoclenium comosum.*

(B) Graptolites with one axis.

    (A) Many-celled.

        (1) Axis weak. *Nereograpsus* (Geinitz).

            (2) *N. Sedgwicki ;* (3) *N.* **Beyrichi,** *N. MacLeayi.*

        (2) Axis strong.

            (*a*) Skeleton net-like. *Retiolites,* Barr. ; (4) *R.* **rete** (?).

            (*b*) Skeleton complete. *Diplograpsus.*

                (*aa*) Cells overlapping.

                    D. ovatus, D. folium, (5) *D. palmeus, D. pristis, D. dentatus,*

                    (6) *D. teretiusculus.*

                (*bb*) Cells free.

                    (7) *D.* **birastrites** (= *D. acuminatus*), (8) *D. cometa.*

    (B) One-celled. *Monograpsus,* Geinitz.

        (1) Cells in contact.

            M. testis, (9) *M. nuntius* (? *M. Roemeri*), *M. colonus,* (10) *M. sagit-tarius,* (11) *M. Nilssoni, M. Halli, M. Sedgwickii, M. Heubneri,* (12) *M. priodon,* (13) *M. Becki, M. convolutus,* (14) *M. turriculatus, M. proteus, M. millepeda,* (15) *M.* **pectinatus.**

        (2) Cells free.

            M. triangulatus, (16) *M. peregrinus,* (17) *M.* **spina** (?), (18) *M.* **urceolus** (?), *M. Linnæi,* (19) *M. gemmatus.*

It will be seen, therefore, that Richter adopted the various genera recognised by Geinitz, even including those now referred to worm burrows. Most of the species of Graptolites mentioned are described and figured by Richter, and the drawings are fair ; but hardly any of his five new forms, viz. *Retiolites rete, D. birastrites, M. pectinatus, M. urceolus,* and *M. spina,* are recognised as distinct species at the present day.

Structure.—Richter did not advance our knowledge much as regards the structure of the Graptolites, but he pointed out more clearly than had previously been done those characters of the cells which are of specific value, such as (1) the general form, (2) the relative distance, (3) the shape, (4) the inclination, etc. Not only so, but in this memoir Richter makes many original suggestions of considerable importance respecting the development of the Graptolites. He was the first to suggest that the "foot" might have had some connection with the development of the Graptolite ; but he does not enter into any details. He points out that the greatest amount of growth is in the direction of length ; the circumference once attained does not increase, and the lowest cells are always the shortest (except in

---

[1] NOTE.—In this and all subsequent papers only the species figured are numbered.

Diplog. cometa). The distance apart of the cells is the same throughout, but the shorter and thinner lower cells appear to be more distant, because they do not overlap so much. In his drawings of *Diplog. palmeus*, Richter figures two aspects of the polypary ("dorsal and ventral"), and shows that the medium septum is incomplete. A similar structure is visible in his figures of *Diplog. cometa*.

Another most important point first brought out by Richter in this paper is that development among the Graptolites is in the direction of simplification of the general forms. Thus the many-branched forms of the earlier formations gave place in the newer formations to those with two rows of cells, and these in their turn to those with one row of overlapping cells, and eventually to those with free cells.

Mode of Existence.—As regards their occurrence and mode of existence, Richter appeared to be of the opinion that the Graptolites lived attached by the "foot" to the floor of the ocean. Owing to their large size, they were probably not short-lived, and they must have lived in enormous colonies to account for the numbers found in the rocks. They are seldom found in association with other animals. Richter observes that the forms with overlapping cells occur isolated, or in association with other species, whereas the free-celled forms are found in groups of their own kind. He suggests that this might be due to their greater or smaller power of moving from place to place. All the Graptolites mentioned by Richter in his paper occur in the Alum and Siliceous shales of Thüringia.

1854.
Fournet,
"Sur les terrains-anciens de Neffiez, Languedoc," 'Bull. de la Soc. Géol. de France,' ser. 2, vol. xi.

The papers which appeared during the next few years are of little importance, and are mainly stratigraphical in their bearing. Fournet simply states that Graptolites occur in limestones and shales associated with *Cardiola interrupta*, but records no species.

1854.
Beyrich,
"Ueber das Vorkommen von Graptolithen im Schlesischen Gebirge," 'Zeit. d. Deut. geol. Gesell.,' vol. vi.

Beyrich, in his paper "Ueber das Vorkommen von Graptolithen im Schlesischen Gebirge," mentions no definite Graptolite species.

1854.
M'Coy,
'Description of the British Palæozoic Fossils in the Geological Museum of Cambridge.'

In his 'Description of the British Palæozoic Fossils in the Geological Museum of Cambridge,' M'Coy gives a description of a few species of Graptolites, some of which also are figured. *Gr. convolutus*, (1) *Gr. latus*, (2) *Gr. lobiferus*, *Gr. ludensis* and var. **minor** (a new variety), (3) *Gr. millepeda*, *Gr. Murchisoni*, *Gr. sagittarius*,(4) *Gr. Sedgwickii*, (5) *Gr. tenuis*, *Diplog. foliaceus*, *D. folium*, *D. mucronatus*, *D. pristis*, var. *B.*, *D. ramosus*, (6) *D. rectangularis*, *Diplograpsus* (?) *sextans*.

Although M'Coy places the dichotomously branching forms provisionally in the genus *Diplograpsus*, he points out that they constitute a " peculiar little group ; " and he suggests that if necessary they might be called **Didymograpsus** = twin Graptolites. This generic name is now in general use for a certain section of these two-branched forms, namely, for those which have their simple, sub-cylindrical, tube-like cells on the inner margin.

1854.
Murchison,
' Siluria.'

In the second edition of Murchison's ' Siluria,' published in 1854, a brief description of four genera of Graptolites, viz. *Diplograpsus*, *Graptolites*, *Rastrites*, and *Didymograpsus*, is given, with figures of a few of the most characteristic species in illustration.

Murchison does not commit himself to any opinion as to the zoological affinities of the Graptolites, but merely points out that they have been supposed by some to be nearly allied to Virgularia, the Corallines or Sertularia, and he considers that they " grew on the fine mud at the bottom of the sea."

As regards their geological range, he asserts that they are exclusively Silurian, and that in Britain neither *Diplograpsus* nor *Didymograpsus* are " ever met with above the horizon of the Caradoc sandstone ; " in Bohemia, however, the double-celled *Diplograpsus* occurs in association with the single-celled form *Graptolites*.

1855.
Harkness,
" Anthracitic Schists
and Fucoidal Remains
in the Lower Silurian
Rocks of the South of
Scotland," ' Quart.
Journ. Geol. Soc.,'
vol. xi.

In the year 1855 Professor R. Harkness described but did not figure a new species of Graptolite under the name **Rastrites Barrandi,** from the Glenkiln shales of the Moffat country. Judging from his description, this may have been a form of *Thamnograptus*. Many Graptolite species are recorded by him in this paper from various localities in South Scotland, including *Gr. sagittarius*, *Diplog. pristis*, *D. ramosus*, *D. mucro-natus*, and *D. bicornis*, *D. folium*, *Gr. Sedgwickii*, *R. peregrinus*, and *R. Linnæi*.

In the light of recent opinion as to the mode of life of the Graptolites, it is interesting to note that Harkness makes the same suggestion as that already made by Suess (1851) ; viz. that the carbonaceous matter of the anthracite shales has not been derived from the Graptolites themselves, but is " due to the existence of *seaweeds* during the earlier portion of the Lower Silurian epoch."

1855.
Cassiano de Prado,
" Sur la géologie
d'Almaden d'une
partie de la Sierra
Morena and des
Montagnes de Toledo,"
' Bull. de la Soc. géol.
de France,' vol. xii,

In the same year Cassiano de Prado recorded *Gr. spiralis*, *Gr. Halli*, *Gr. priodon*, and *Gr. palmeus* from the Silurian rocks of Almada, the Sierra Morena and Toledo.

1855.

Roemer,

" Graptolithen am
Harz," ' Neues
Jahrbuch,' vol. vii.

The German palæontologist, Roemer, described and figured eight species of *Monograpsus* from the Harz region, namely, the (1) *M. priodon*, (2) *M. latus*, (3) *M. sagittarius*, (4) *M. proteus* of earlier observers, and four new forms, viz. (5) *M.* **Jüngsti**, (6) *M.* **polyodonta**, (7) *M.* **oblique-truncatus**, and (8) *M.* **sub-dentatus**; but his drawings are so poor that it is impossible to identify the species. His form *M. sub-dentatus* is, however, probably the same as the *M. dubius* of Suess.

1855.

Emmons,

' American Geology,'
vol. i.

A number of American species of Graptolites and some new genera were described and figured by Emmons in his ' American Geology,' in 1855. These include (1) *Diplograpsus secalinus* (= *Fucoides simplex*, Emm.), (2) *D.* **rugosus**, (3) *D.* **dissimilaris** (= *Climacograptus* fragment), (4) *D.* **ciliatus**, (5) *D.* **obliquus**, (6) *Monograptus* **elegans** (= probably a fragment of a Dichograptid), and (7) *M.* **rectus**. The first new generic name proposed by him is **Cladograptus,** for Graptolites bearing " serrations or cells arranged on the outer edges of a branching stipe, axis none." This genus is held to embrace two new species, viz. (8) *C.* **dissimilaris**, and (9) *C.* **inequalis**. These species, however, certainly belong to the genus *Dicranograptus;* and as the generic name *Cladograptus* had already been employed by Geinitz, Emmons' title never came into use.

The second new genus proposed by Emmons is, however, generally accepted at the present day. This is named **Glossograptus,** with his *G.* **ciliatus** as its type. He suggests that probably his own species *Diplog. ciliatus* and *D. crinitus* may belong to this genus. Emmons lays great stress on the existence of the free solid axis extended beyond the body of the stipe in *Glossog. ciliatus*, and believes that it may have " served to attach the Graptolite to other objects ; " or else the animal may have floated freely in the sea. The first of these views of Emmons has received much support from the recent discoveries of Ruedemann.

Emmons' third genus, **Staurograptus,** with its type species (11) *S.* **dichotomus**, is evidently founded upon the proximal part of a *Dichograptid*, with only four branches showing clearly, and the generic name has never come into use.

His fourth genus, **Nemagraptus,** is suggested for forms having an " axis elongated and thread-like, simple or compound branches round at the base, and flattened at their extremities, with cells which appear to be arranged on the flattened part of the axis instead of the margins." Of the two examples, viz. (12) *N.* **elegans** and (13) *N.* **capillaris**, described by Emmons, the former is clearly a fragment of the form subsequently described and figured by Hall as his *Graptolithus gracilis*, and afterwards made by him the type of his genus *Cœnograptus*.

Emmons' descriptions of all his species are meagre in the extreme, and his figures so poor that it is very difficult to identify the forms. All the species described by him came from the so-called Taconic beds of New Hampshire, from strata now known to be of Ordovician age.

1856.
Bronn,
' Lethæa Geognostica.'

The only work that appeared in 1856 in which Graptolites were referred to was Bronn's ' Lethæa Geognostica.' The reference to these fossils in this book is almost entirely a *résumé* of the work of previous observers, and contains little or nothing that is original. He accepts Barrande's three genera of *Monoprion, Rastrites* and *Retiolites*, but rejects Geinitz's *Cladograpsus* and *Nereograpsus*. These he considers to be trails or footprints of annelids.

As regards the affinities of the Graptolites, Bronn is opposed to Barrande's view of their relationship to the Pennatulidæ, on account of their very different habit, and also from the fact that the family of the Pennatulidæ is not represented in any of the Palæozoic formations.

1857.
Naumann,
' Lehrbuch der
Geognosie,' 2nd edit.

A few Graptolites were figured in 1857, in the second edition of Naumann's ' Lehrbuch der Geognosie ;' but the author adds nothing new, and the few figures that are given are mainly copies of those of Barrande.

1857.
McCrady,
"Zoological Affinities
of the Graptolites,"
' Proc. Eliot Soc. of
Nat. Hist. of Charleston,' vol. i.

In 1857 McCrady published a paper on the "Zoological Affinities of Graptolites," in which he endeavoured to show the similarity of the graptolitic forms to the Echinoderm larvæ. (We have not, however, been able to obtain access to this paper.)

1857.
Meneghini,
' Palæontologie de
l'Isle de Sardaigne.'

Descriptions and figures of the Graptolites of the Island of Sardinia were published by Meneghini in 1857. Nine species are named, and of these seven are described as new. Unfortunately the accompanying illustrations are so poor that it is impossible to identify any of the specific forms with certainty ; and in spite of Meneghini's lengthy descriptions, his suggested names have never been adopted. It is possible, however, that his figures are attempts to represent well-known species. His (1) *Gr. (Monograpsus)* **antennularis** is of the *M. Becki* type, and his (2) *Gr.* **Larmarmoræ** is possibly identical with Salter's *M. Flemingii.* The form entitled (3) *Gr. colonus*, Barrande, is wrongly identified ; but (4) *Gr.* **belophorus** may be that species. His (5) *Gr.* **hemipristis** is represented by views partially scalariform, and is therefore unidentifiable. His (6) *Gr.* **Gonii** belongs clearly to the *M. vomerinus* group. His (7) *Gr.* **falcatus** resembles the *M. limatulus* of Törnquist in general form and in the long proximal prolongation of the stipe. His (8) *Gr.* **mutuliferus** is of the type of *M. Flemingii ;* while his (9) *Gr. priodon* is perhaps correctly referred to Bronn's species, but even this is uncertain. His list is closed by two unnamed species of *Diplograpsus*, which are

not specifically recognisable.   But while his figures do not admit of the recognition
of the individual species, the general habit is fairly well given, and it is evident
that nearly all the forms are of Wenlock age.

1857.
Salter,
Assoc. of Amer.
Naturalists, Montreal.

In the same year Salter described a new genus of Graptolite, which he called **Graptopora,** and he considered this to be
intermediate in character between a Graptolite and a *Fenestella*.   The species which he regarded as typical of the
genus was his own *Graptopora (Fenestella) socialis*.   This
genus had already been described by Hall under the title of *Dictyonema*.

1858-9.
Hall,
" Notes upon the Genus
Graptolithus and a
description of some
new forms from the
Hudson River Group,"
' 12th Report on the
State Cabinet, Albany.'

In a paper entitled " Notes upon the Genus *Graptolithus*,"
in 1859—a part of which paper had been communicated to
Sir William Logan in 1857 (' Report of Progress, Geol.
Survey of Canada '),—Hall criticises to some extent the views
expressed by Barrande in his ' Graptolites de Bohême,' and
by other European graptolithologists.   It is interesting to
note that the two greatest palæontologists of that day, living
on opposite sides of the Atlantic, had totally different graptolitic material to work upon.   Hall's examples of Graptolites
were practically all branching forms from Lower Silurian (Ordovician beds), while
Barrande's were all simple forms of Upper Silurian age.   This was a disadvantage
to them in the interpretation of their results at the time, but it was nevertheless
extremely fortunate for the progress of science, for owing to this circumstance
both the Lower and Upper Silurian Graptolitic faunas were worked out independently by these two great palæontologists, neither being biassed by the views
of the other.

Hall was unwilling to accept the European genera, *Monograpsus, Diplograpsus,*
and *Cladograpsus*, believing that there was not sufficient reason at that time for
separating branching forms from the unbranched, and regarding all those with a
single series of serratures as having been originally composed of two, four, or more
branches.   He also rejects Geinitz's genus *Nereograpsus*.   He states that the
Canadian specimens " sustain the opinion already expressed that *Dictyonema* will
form a new genus of Graptolites, the serratures being on the inner side;" while as
regards the form of the polypary in this genus, he concludes that while its mode of
growth was " probably flabelliform in some species, it is clearly funnel-shaped in
D. retiformis."

Many new species are described by Hall in this work and partly figured, the
majority being grouped under the single generic name of *Graptolithus*.   These
embrace (1) *Gr.* **logani** and (2) *Gr.* **abnormis** (now placed in the genus *Loganograptus*), **and** (3) *Gr.* **flexilis** (*Clonograptus*).   Hall's specimens of *Loganograptus
Logani* show the presence of a disc for the first time in the history of graptolitic
research.

The remaining forms, all of which were new to science, are (4) *Gr.* **octo-brachiatus**, (5) *Gr.* **octonarius** (*Dichograptus*); (6) *Gr.* **quadribrachiatus**, (7) *Gr.* **crucifer**, (8) *Gr.* **bryonoides**, (9) *Gr.* **Headi**, (10) *Gr.* **alatus**, (11) *Gr.* **fruticosus**, (12) *Gr.* **denticulatus** (*Tetragraptus*); (13) *Gr.* **indentus**, (14) *Gr.* **nitidus**, (15) *Gr.* **bifidus**, (16) *Gr.* **patulus**, (17) *Gr.* **similis**, (18) *Gr.* **extensus** (*Didymograptus*); (19) *Gr.* **pristiniformis** (*Diplograptus*); (20) *Gr.* **ensiformis** (*Retiolites*, Hall); (21) *Gr.* **tentaculatus** (*Retiograptus*, Hall).

In addition to these, Hall describes three species of a remarkable four-rowed type, viz. (22) *P.* **typus**, (23) *P.* **ilicifolius**, and (24) *P.* **angustifolius**. For these he suggests the new generic name of **Phyllograptus**. He explains the structure of these forms, and shows that they consist of "four semi-elliptical parts joined at their straight sides."

Hall was well aware that the exact number of branches did not constitute either a generic or specific character in some of his new forms, for he notices a specimen of *Gr. octobrachiatus* with only seven branches, and a *Gr. Logani* with nine.

The genera *Thamnograptus* and *Dendrograptus* [1] are briefly referred to, and he also describes a new genus which he calls **Plumulina**, of which he gives two species, (25) *P.* **plumaria** and (26) *P.* **gracilis**.

Hall also figures some Graptolite stipes with reproductive cells. As the question of the reproduction of the Graptolites is entered into in greater detail by Hall in his 'Graptolites of the Quebec Group' (1865), the discussion of his results will be referred to later.

He disagrees with Barrande's view that the narrower end is the newest part of the polypary, and shows that the origin of the stipe in some of the Canadian forms renders this untenable. In other respects he does not add much to our knowledge of the structure of the Graptolites. Referring to those forms which have cells on two sides (*Diplograpsus*), he is "disposed to believe that they may have been simple from the base," though the bifurcated appearance of *D. bicornis* offers objection to this view.

As regards the mode of life of the Graptolites, Hall is somewhat in favour of the theory of their having been free floating animals; for, as he points out, "in many specimens there is no evidence of a point of attachment or radix, and they have much the appearance of bodies floating free in the ocean."

He notes how strongly the Graptolites simulate the Palæozoic Bryozoa in mode of growth, though they differ essentially from all of them, not only in form, but in the arrangement of their cellules, in the nature of their substance, and in the structure of their skeleton.

---

[1] The actual description of *Thamnograptus* was not published until 1859 (Pal. New York, vol. iii, Suppl.). *Dendrograptus*, though named in 1857, does not seem to have been fully described until 1865 (Graps. of Quebec Group).

1858–9.
Carruthers,
" On the Graptolites
from the Silurian
Shales of Dumfries-
shire, with a Descrip-
tion of three new
Species," ' Ann. Mag.
Nat. Hist.,' ser. 3,
vol. iii.

Carruthers' first paper on the Graptolites appeared in 1858 in the 'Transactions of the Royal Physical Society of Edinburgh.' It was, however, republished in 1859 in the 'Annals and Magazine of Natural History.' It contains a description of three new species from the Silurian shales of Dumfriesshire, viz. (1) **Cladograpsus** (*Pleurograptus*) **linearis,** (2) *Diplograpsus* (*Cryptograptus*) **tricornis,** and (3) *Didymograpsus* (*Dicellograptus*) **Moffatensis.** Carruthers suggests the generic title of *Cladograpsus* for forms " having two main stems each supporting the cells on their upper sides," and with " branches given off at irregular distances." This name, as we have already seen, had been employed by Geinitz and Emmons for forms generically distinct from those for which it was proposed by Carruthers.

1859.
Giebel,
" Die Silurische Fauna
des Unterharzes,"
'Zeit.für die Gesammte.
Natur. Wiss,' vol. xi.

In the year 1859 also Giebel recorded *Monog. sagittarius* from the Silurian beds of the Lower Harz.

1859.
Hall,
Supplement to vol. i,
' Palæontology of New
York,' vol. iii.

In 1859, in a supplement to vol. i of his ' Palæontology of New York' (which supplement was, however, actually published simultaneously with vol. iii), the preceding " Notes " were reproduced by Hall, and some additional new forms from the Hudson River group were figured and described.

These include (1) *Gr.* **multifasciatus** (a many branched form, *Amphigraptus*) ; (2) *Gr.* **divaricatus** (*Dicellograptus*) ; (3) *Gr.* **gracilis** (*Cœnograptus*, Hall) ; (4) *Gr.* **marcidus** (*Cryptograptus*) ; (5) *Gr.* **Whitfieldi,** (6) *Gr.* **angustifolius** (*Diplograptus*) ; (7) *Gr.* **spinulosus** (*Glossograptus*) ; (8) **Retiograptus Geinitzianus** (*Clathrograptus*) ; (9) **Thamnograptus typus** and (10) *T.* **capillaris,** (11) *Rastrites* **Barrandi** (? *Thamnograptus*).

As regards the points of structure brought out in the examination and description of these species, Hall remarks that in *Thamnograptus* the cellules are at present unknown. In *Gr. divaricatus* he calls attention to a row of small *nodes* placed obliquely to the direction of the axis. He figures a large specimen of *Gr. gracilis*, and traces the stages in development from the earliest, in which there are no lateral branches, through the form now referred to *Nemag.* (*Cœnog.*) *surcularis*, up to the large and typical form. He notices the radicle or axillary bar from which the main stipes diverge, and writes that " it is barely possible that the apparent central radicle may be the remains of two other stipes, corresponding to the two usually preserved." He argues that it is " still possible that these small bifurcate fronds are but the separated offshoots from a rhizoma, which extended along the muddy bottom of the sea, giving off the ascending stipes in pairs, which in their progress become branched, as before shown; and in this case the little transverse bar in the bending of the frond is a part of the broken rhizoma."

1859.
Goeppert,
'Die Fossile Flora der Silurischen, Devonischen, und unteren Kohlen-formation des sogenannten Uebergangsgebirges.'

Although the Hydroid nature of the Graptolites would seem to have been well established by this time, Goeppert, in 1859, rejected the generally accepted view, and asserted that they were algæ. He figures part of a Graptolite branch, apparently bearing a fruit like that in *Callithamnum*, as a direct proof of their true algal nature. Goeppert figures many forms of true Dendroid Graptolites, to most of which he gives names corresponding to those usually given to algæ.

Thus his (1) **Sphærococcites Scharyanus** is a *Callograptus?*; his (2) **Calithamnites Reussianus** is a good *Ptilograptus*; while (3) **Chondrites fruticulosus** and (4) var. **articulatus** are probably *Dendrograpti*. He retains the generic name *Dictyonema* of Hall, but diagnoses it as an alga, and substitutes the specific name (5) *Dictyonema* **Hisingeri** as a common title for *Gorgonia flabelliformis*, Eich., *Fenestella socialis*, Salt., and *Dicty. fenestratum*, Hall, which he held to be the same species. He agrees with Brongniart in regarding *Fucoides dentatus* and *F. serra* as algæ, though he admits that the former also resembles Graptolites; but he changes the generic name of both to **Amansites.** Goeppert's drawings are good for the most part, and the paper is of value on this account.

1860.
Hall,
13th Report of the State Cabinet.

Two papers on Graptolites were published in America during the year 1860, one by Hall and one by Dawson. The paper by Hall was merely a repetition of that published by him in the previous year in the 'Palæontology of New York.'

1860.
Dawson,
"Note on the Silurian and Devonian Rocks of Nova Scotia," 'Canad. Naturalist and Geologist,' vol. v.

Dawson's paper was essentially stratigraphical, but in it he records *Gr. clintonensis* from the Lower Arisaig series of Nova Scotia, and gives a brief preliminary description and figure of a new species of *Dictyonema*, *D.* **Websteri,** from the *Dictyonema* shales near New Canaan.

1860.
Eichwald,
'Lethæa Rossica,'
edit. 2.

A second edition of Eichwald's 'Lethæa Rossica' was published in 1860. In this edition five Graptolite species were described and figured, three being new. The new forms are (1) *Diplog.* **pennula** (? *Petalograptus*); (2) *D.* **paradoxus,** and (3) *D.* **tumidus.** The old species are the (4) *D. distichus* of Eichwald, and (5) *Monoprion serratus* of Schlotheim. The figures of the new species are very poor, and it is impossible to identify any of them with certainty at the present day. Eichwald re-describes his own *Gorgonia (Dictyonema) flabelliforme* under the new generic name of **Rhabdinopora.**

1860.
Michel,
" Coupe du terrain Silurian aux environs de Domfront.," ' Bull. Soc. Géol. de France,' ser. 2, vol. xvii.

In this year Michel recorded the existence of Silurian shales, near Domfront, containing *Gr. colonus*, characteristic of stage E. (A collection of Graptolites from this locality was subsequently sent to Barrande, and he identified in addition *Gr. bohemicus*.)

1861.
Salter,
" New Fossils from the Skiddaw Slates," ' Geologist,' vol. iv.

In 1861 Salter figured from the Skiddaw Slates a new Graptolite forming the type of a new genus. This is a *Clonograptus*, showing, however, only ten stipes. Salter observes that it certainly does not belong to the genus *Graptolithus*, which includes forms which are simple and perfect from end to end, and concludes, " I shall shortly, I hope, describe the new branched dichotomous form under the name of **Dichograpsus**."

1861.
Billings,
" On the Occurrence of Graptolites in the Base of the Lower Silurian," ' Canad. Naturalist and Geologist,' vol. vi.

In 1861 Billings reviewed the work of Schmidt (' Silurische formation von Estland,' 1858), and compared the geological appearance of Graptolites in Europe and America, with the intention of showing that the "occurrence of Graptolites in rocks so ancient as those of the Quebec group is not inconsistent with what we know of their geological range in other countries," and that therefore they need not be of the age of the Hudson River group. He concludes, however, that his investigations tend to prove that "Graptolites cannot always be relied upon to show that exposures of rock widely separated from each other are either of a different or of the same age."

1861.
Dalimier,
" Stratigraphie des terrains primaires dans la presqu'île du Cotentin," ' Bull. Soc. Géol. de France,' ser. 2, vol. xviii.

In the same year Dalimier recorded the occurrence of Graptolites of the second Silurian fauna from the shales and associated grits above the Fucoid grit in the Peninsula of Cotentin.

1861-2.
M'Coy,
" Note on the Ancient and Recent Natural History of Victoria," Ann. Mag. Nat. Hist.,' vol. ix.

In 1861 M'Coy had noted the occurrence of several species of Graptolites from the Palæozoic rocks of Victoria in a pamphlet published for the Intercolonial Exhibition. The year following, 1862, the main results of his study of these forms were published in England. He records the following species:—*Diplograpsus pristis, D. mucronatus, D. rectangularis, D. ramosus, D. folium,* and *D. bicornis; Phyllograptus typus; Didymograpsus serratulus, D. caduceus, D. furcatus; Graptolites gracilis, Gr. Logani, Gr. quadribrachiatus, Gr. octobrachiatus, Gr. ludensis, Gr. tenuis, Gr. latus,* and *Gr. sagittarius.*

The remarkable similarity between the graptolitic fauna of Australia, America, and that of the Northern Hemisphere appears to have greatly astonished him; and he lays especial stress on the "extraordinary fact of the specific identity of this marine fauna over the whole world during the most ancient Palæozoic periods."

1861.
Baily,
"Graptolites of Meath, etc.," 'Geol. Soc. of Dublin.'

A paper on the Graptolites of Ireland was published in 1861 by Baily. It is mainly stratigraphical in its bearing, but the various Graptolite species mentioned are briefly described and figured.

Didymog. Murchisoni is recorded from Bellewstown, co. Meath; and he erroneously asserted that with this form were associated *Gr. Sedgwickii* and *Gr. Nilssoni.* The species (1) *Diplog. pristis,* (2) var. *scalariformis,* (3) *D. mucronatus,* (4) *Gr. gracilis,* and (5) *Cladog. Forchammeri* are recorded and figured by him from "beds of Llandeilo age" in county Clare. As regards the first two of these species, he points out that the specific names given by him are merely provisional. His *D. pristis* var. *scalariformis* appears to be a *Climacograptus.* The example shown on fig. 2 *b,* is probably *Climacog. tubuliferus,* Lapw. His *Diplog. mucronatus* is the *D. bimucronatus,* Nich. The structure of the cells in his *Cladog. Forchammeri* are well represented.

From co. Tipperary he records *Gr. priodon* and a new species, *Gr.* **hamatus** (*Cyrtograptus*).

Referring to the range of the Graptolites in general, Baily concludes that " all the double forms are confined to the Lower Silurian division, being most abundant in the lowest series of beds, the equivalents of the Llandeilo Flags; one species only, the *Gr. priodon,* a single form, ranging through the series."

1863.
Salter,
" Note on Skiddaw Slate Fossils," 'Quart. Journ. Geol. Soc.,' vol. xix.

In 1863 Salter, in an appendix to Harkness' paper on " The Skiddaw Slate Series," recorded and figured several species of Graptolites. They include the (1) *Phyllog. angustifolius* of Hall, and two species of *Tetragrapsus.* One of these is clearly (2) *T. serra,* and the other a presumed new species named by Salter (3) *Gr.* **crucialis.** This last, however, is identical with the *T. quadribrachiatus* of Hall.

Two new species of *Dichograpsus* are named, (4) *D.* **aranea** (*D. octobrachiatus*) and (5) *D.* **Sedgwickii,** and a form is figured with a disc. A figured fragment (fig. 12) assigned by Salter to *Dichograpsus* is really an example of *Bryograptus Kjerulfi.* Four species of *Didymograpsus* are figured :—(6) *D. caduceus,* (7) *D. geminus,* (8) *D.* v. **fractus,** and (9) *D.* **hirundo.** The last two are new forms.

Salter agrees with Huxley's suggestion that the disc in *Dichograpsus* is analogous to the basal plate of *Defrancia,* and he regards it as of great systematic importance. He accepts Hall's genera *Dendrograptus* and *Dictyonema,* and points out the strong external resemblance of *Dictyonema* to *Fenestella*. He adds, "The presence of the projecting Graptolite cells, and horny texture, however, prevents its being confounded with that genus; but the resemblance is very close, and I think we have here a real affinity."

He discusses the geological age of the Skiddaw slates as evidenced by their included Graptolite species, and compares them with the beds of the Quebec group.

1863–4.
Logan,
" Geology of Canada,"
and " Graptolites of
the Quebec Group,"
' Reports of Progress.'

In the two reports of Logan on the " Geology of Canada," which appeared in 1863–4, Graptolites are referred to, but nothing is added in them to our previous knowledge of the group in general.

They are stratigraphical in their bearings, and Logan distinctly recognises the probable identity in age of the Norman's Kill shales and the Utica and Hudson River groups on the ground of the similarity of their Graptolites.

1863.
Dewalque,
" Notes sur les Fossils
Siluriens de Grand
Manil," ' Bull. de la
Soc. Géol. de France,'
ser. 2, vol. xx.

Dewalque recorded in 1863 the occurrence of certain " scalariform impressions " of Graptolites (identified by Barrande) in the Silurian shales of Grand Manil, Belgium, but no names were given.

1863.
Billings,
" Parallelism of the
Quebec Group, etc.,"
' Geol. Survey of
Canada.'

In the same year Billings again discussed the question of the age of the Quebec group, as evidenced by its included Graptolites. He concludes that it must lie between the middle of the Calciferous group, and the bottom of the Black River limestone. As regards the identity in age of the Chazy and the Quebec groups, he considers (from the dissimilarity in the species) that in a " portion of the Quebec group we have a set of strata representing those which are absent (elsewhere) in Canada." " The remainder may possibly be of the age of the Chazy."

1864–5.
Törnquist,
" Om Fågelsangs-
traktens Undersiluriska
Lager," ' Lunds Univ.
Årsskrift,' tom. ii.

In 1864–5 Törnquist described and figured six species from the Swedish Lower Silurian beds of Fågelsang. These are (1) *Diplograpsus teretiusculus* (probably including also *Climacog. Scharenbergi*), (2) *Phyllograptus typus*, (3) *Didymograpsus Murchisoni*, (4) *Dendrograpsus gracilis* (considered by Holm to be a *Pterograptus*), and a new species (5) *Didymograpsus* **virgulatus**. Finally he gives examples of (6) *Dictyonema flabelliforme*, showing the structure of the branches.

1865.

Kjerulf,

" Veiviser ved geologiske excursioner i Christiania omegn," 'Universitets-program for andet Halvaar. Christiania.'

In the catalogue of fossils from the Christiania district, prepared by Kjerulf in 1865, figures are given of species of Graptolites in Etage 2. These are (1) *Dictyonema norwegicum,* Eich., and (2) *Dictyonema* **graptolithinum** (*D. flabelliforme*). He also records *Gr. gracilis,* Hall (*Pterograptus,* cf. Holm). Under the name (3) *Gr. tenuis,* Portlock, Kjerulf figures the branching forms subsequently named and described by Lapworth as *Bryograptus Kjerulfi.* A few species are also recorded from Etage 3. From Etage 8, under the name of (4) *Gr. ludensis,* he figures *M. priodon,* a second species of *Monograptus,* and a form which is probably a *Cyrtograptus.*

1865.

Malaise,

" Notes sur quelques fossiles du massif Silurien du Brabant," 'Bull. de l'Acad. R. de Bruxelles,' tom. xx.

In 1865 Malaise recorded the discovery of Graptolites in the Silurian shales of Brabant, but did not adduce the names of any of the specific forms obtained by himself.

1865.

Peck,

" Graptolithen schiefer bei Lauban," 'Neues Jahrb.'

In the same year four species of Graptolites, identified by Geinitz as *M. sagittarius, M. priodon, M. colonus,* and *M. Sedgwickii,* were found by Peck in the Graptolite shales near Lauban.

1865.

Hall,

" Graptolites of the Quebec Group," 'Geol. Surv. of Canada,' Dec. 2.

In 1865 the valuable results of Hall's long-continued researches on the Graptolites of America, worked out by him during the previous ten years, and already partly laid before the public in various papers, were embodied in a collective and exhaustive monograph entitled 'The Graptolites of the Quebec Group.' It will be well to give here a general summary of the whole, so as to realise fully the views of this great palæontologist on the Graptolites.

The work is in the main descriptive, almost every species being illustrated by several drawings of great excellence. Hall acknowledges only one inclusive family —*Graptolitidæ,* and gives the following table of the component genera :

Classification.

Family GRAPTOLITIDÆ.

I.

Species consisting of stipes or fronds, with a bilateral arrangement of the parts ; a solid axis with a common canal extending along each series of cellules.

1. The successive buds developed in tubular cellules which are usually in contact for a greater or less proportion of their length, and inclined towards the axis.

(*a*) Cellules in single series along one side of a common solid axis. Stipes, two or more, from a common origin, with or without a central disc. Sub-genera *Monoprion, Didymograptus, Monograptus, Tetragraptus,* etc.

Graptolithus.

(*b*) Cellules on one side of slender branches, which are developed on one or two sides of a long slender axis or rachis, the free extremities of which are likewise celluliferous. Ex. *Gr. gracilis* and *Gr. divergens.*

(*c*) Cellules developed in parallel arrangement on two sides of a common solid axis. Stipes narrow, elongate. Sub-genus *Diprion,* = *Diplograptus.*

Phyllograptus.

(*d*) Cellules developed in a cruciform arrangement on the four sides of a common or coalescent axis. Stipes elliptical or sub-elliptical.

2. Cell-apertures excavated in the margins of the stipes, without tubular or cup-form extension; the cell-apertures upon one or both sides of the stipe. *Graptolithus bicornis* and others.

Climacograptus.

3. Solid axis eccentric or subexterior, with cellules developed in parallel ranges on opposite sides of the stipe, and in contact throughout their entire length.

Retiolites.

(*a*) Known only as separate stipes, with reticulate test.

Retiograptus.

(*b*) Occurring as single stipes, and as compound fronds; test smooth.

II.

Species having a common trunk, or growing in sessile groups of stipes from a common origin, without distinct bilateral arrangement of the parts. Cellules in single series on one side of the stipes or branches, and arranged along a common canal or axis.

1. Branches free (*i. e.* not connected by transverse bars); cellules in contact or closely arranged.

Dendrograptus.

2. Stipes and branches more or less regularly united in a reticulate front, without elongate stem.

Dictyonema.

3. Branches unfrequently and irregularly connected by transverse processes.

Callograptus.

4. Stipes round or flattened, growing in groups, and bifurcating above; margins denticulate; surface rough and scaly.

Inocaulis.

III.

Rastrites. Slender cylindrical branches, with tubular cellules arranged in single (or in double) series. Cellules not in contact in any part of their length.

IV.

Thamnograptus. Species having a common axis or rachis, with slender lateral alternating branchlets. Cellules unknown.

V.

Ptilograptus. Species having a common axis, more or less frequently bifurcating, with pinnulæ closely and alternately arranged on the opposite sides; cell-apertures on one face of the pinnulæ.

VI.

Buthograptus. A simple flexuous rachis, with slender flattened pinnulæ, arranged in alternating order at close and regular intervals on the two sides. Cell-apertures unknown or circular.

VII.

Oldhamia. Strong stems, which are numerously branched. Branches and branchlets slender, arranged in whorls. Cellules undetermined.

Nomenclature.—It will be seen from the above synopsis that Hall still maintains his original view that all the simple stiped forms described by Barrande and others are really only isolated branches of a more complex form, and he therefore rejects all such generic names as *Monograpsus*, *Didymo-*, *Tetra-*, and *Dichograpsus*. He writes, "These subdivisions may be of some value when the entire frond and all its appendages are preserved, but unfortunately this is rarely so; and when we have but fragments of the stipes or branches there is no force or value in the application of these terms; we are thus reduced to the necessity of adopting the old term *Graptolithus*." In criticising M'Coy's genus *Didymograpsus* he points out that the name had been used for two distinct groups, namely, those of the type of *Graptolithus patulus*, or the true *Didymograpti* (as we know them at the present day), and those of the type of *Gr. divaricatus* (*Dicellograpti*), and he shows how valueless are the genera founded on the number of stipes by citing the case of *Didymog. caduceus* (Salter), which he believes to be a four-stiped form, though it so closely simulates one with two stipes.

Hall points out that the corneous disc in Graptolites is not a character of generic value, as it occurs in some 4-, 8-, and 16-stiped forms, while other species with the same number of stipes do not possess one.   He does not accept Geinitz's genus *Cladograpsus*, and agrees with Bronn in rejecting Geinitz's *Nereograpsus* and Emmon's *Nemapodia*, which are probably worm tracks.   Emmon's genera *Glossograpsus* and *Nemagrapsus* are rejected by Hall, but he observes that Emmon's *Staurograpsus*, "if accurately represented in the figure, merits generic distinction."

He retains the genus *Diplograpsus* of M'Coy, not on account of its form, but on account of the shape and arrangement of the cellules, which are the same as those in the various species of *Graptolithus* (*Didymograptus*) described by himself.   From this genus *Diplograpsus*, however, as previously accepted, Hall separates under the name *Climacograptus* those double forms, the cell-apertures of which are "excavated in the margins of the stipes, without tubular or cup-form extension."   He gives a very careful and detailed description of the structure of this new genus, illustrated by excellent figures.   He regards his *C. bicornis* as the type, but includes also *Pr. teretiusculus*, His., and *D. rectangularis*, M'Coy, in the same genus; and he conceives that many if not all of the scalariform specimens figured by previous authors belong to species of this character.

Hall also recognises the fact that the cells in such forms as *Gr. ramosus* and *Gr. Forchammeri* are similar in structure to those of *Climacograptus*, and he proposes the sub-generic title of *Dicranograptus* to include all such species.

Barrande's genera *Rastrites* and *Retiolites* are retained, but Hall differs from Barrande in believing that in the latter genus the axes are two in number—on one side a straight cylindrical solid axis, and on the opposite side an undulating or zigzag filiform axis.   The structure of *Retiolites*, so far as Hall was able to work it out from the material at his disposal, is given in illustrative figures.

As regards his own genus *Retiograptus*, Hall supposes that "the two sides of the stipes are very unlike each other in form and external characters, as is the case in *Retiolites*."

The four genera of Dendroid forms, viz. *Dendrograptus*, *Callograptus*, *Dictyonema*, and *Ptilograptus*, are placed by him in the family of the Graptolitidæ, but he suggests that when further information has been obtained as to their structure, it may be necessary to separate them from this family.   In *Dictyonema* he is unable to recognise a solid axis, but he believes that it existed in *Ptilograptus*.   The peculiar genera *Inocaulis*, *Buthograptus*, and *Oldhamia* he holds can only be doubtfully classed among the Graptolitidæ.

It may be noted that Hall here first adopts the termination " graptus " instead of " grapsus " for all his new genera, " since the latter is used in description of Crustacea."   This modification of the earlier terminology may be said to be universally adopted at the present day.

Description of Species.—The following large number of species are described and figured by Hall in this work (the generic names in brackets are those employed at the present day) :—

(*Didymograptus*) .   (1) *Gr. nitidus*, (2) *Gr. patulus*, (3) *Gr. bifidus*, (4) *Gr. indentus*, (5) *Gr.* **extenuatus**, (6) *Gr. constrictus*, (7) *Gr. similis*, (8) *Gr.* **arcuatus**, (9) *Gr. extensus*, (10) *Gr.* **pennatulus.**

(*Tetragraptus*)   .   (11) *Gr. byronoides*, (12) *Gr. denticulatus*, (13) *Gr. quadribrachiatus*, (14) *Gr. fruticosus*, (15) *Gr. crucifer*, (16) *Gr. Headi*, (17) *Gr. alatus*, (18) *Gr. Bigsbyi.*

(*Dichograptus*)   .   (19) *Gr. octobrachiatus*, (20) *Gr. octonarius.*

(*Loganograptus*) .   (21) *Gr. Logani* and a variety.

(*Clonograptus*)   .   (22) *Gr. flexilis*, (23) *Gr.* **rigidus**, (24) *Gr. abnormis*, (25) *Gr.* **Richardsoni**, (26) *Gr.* **ramulus.**

Diplograptus   .   (27) *Gr. pristiniformis*, (28) *Gr.* **inutilis**, (29) *Gr.* **putillus**, (30) *Gr.* **quadrimucronatus.**

Climacograptus .   (31) *C.* **antennarius**, (32) *C. bicornis*, (33) *C.* **typicalis.**

Dicranograptus .   (34) *D. ramosus.*

(*Leptograptus*)   .   (35) *Gr.* **flaccidus.**

Retiolites   .   .   (36) *R. ensiformis*, (37) *R. venosus.*

Retiograptus   .   (38) *R. tentaculatus*, (39) *R. eucharis.*

Phyllograptus   .   (40) *P. typus*, (41) *P. ilicifolius*, (42) *P.* **anna**, (43) *P. angustifolius.*

Dendrograptus .   (44) *D.* **flexuosus**, (45) *D.* **divergens**, (46) *D.* **striatus**, (47) *D.* **erectus**, (48) *D.* **diffusus**, (49) *D.* **gracilis.**

Callograptus   .   (50) *C.* **elegans**, (51) *C.* **Salteri.**

Dictyonema   .   (52) *D.* **irregularis**, (53) *D.* **robusta**, (54) *D.* **Murrayi**, (55) *D.* **quadrangularis.**

Ptilograptus   .   (56) *P.* **plumosus**, (57) *P.* **Geinitzianus.**

Thamnograptus .   (58) *T.* **anna.**

(*Monograptus*)   .   (59) *Gr. Clintonensis.*

In his introductory chapters Hall devotes various sections of his work to the consideration of such matters as the structure, method of reproduction, and mode of existence of the Graptolites. Such of his views as were new to the science of the time may here be briefly summarised.

Structure.

The Radicle.—This was the name employed by Hall for the basal spine, initial spine, or "initial point" of the Graptolite. He admits that in those species with a single row of cellules it may have served as a temporary organ of attachment in the earlier period of its growth, "though all the evidence is opposed to this view." In *Retiograptus,* however, it " is only a broken process of attachment of the individual stipe, which existed as one of the members of the entire frond, the true initial point of which would be in the centre of the whole." It is clear from Hall's observations that the term was not in all cases given by him to one and the same structure, sometimes being identical with what is now known as the sicula (*Cœnograptus*), sometimes to the apical part of the sicula (*Didymograptus*), and sometimes to the " apertural spine" of the sicula (*Diplograptus*). Hall emphasises, however, one important fact, now known to be true of the sicula, viz. that the radicle passes into the " commencement of the solid axis."

Funicle.—Hall believes that in the Graptolites with four stipes (*Tetragraptus*) " the condition appears like that of two individuals of the two-stiped forms conjoined by a straight connecting process of greater or less extent, destitute of cellules," and this he calls the " funicle." The greater the number of stipes, the greater are the number of the divisions of this non-celluliferous funicle. (It has subsequently been shown by more recent observers that the non-celluliferous character of the dividing and subdividing primary and secondary stipes is only apparent, being merely due to their mode of preservation. Consequently Hall's funicle as originally defined by him is non-existent.)

Central Disc.—This curious structure, first recognised and described by Hall, " appears to be composed of two laminæ which, at least in the central parts, are not conjoined, and the space is probably occupied by some softer portion of the animal body." According to Hall, the functions of this disc are to " give strength and support to the bases of the stipes," but beyond this " it probably serves other purposes of the animal economy," and he seems to hint that reproduction was one of these (*loc. cit.*, p. 35).

Solid Axis.—Hall adds but little in this work to our knowledge of the " solid axis," merely corroborating Barrande's observations ; but he recognises its importance in the Graptolite structure as a whole, regarding it as the " foundation on which the other parts are erected." Although believing the axis to be solid, he distinctly acknowledges in a note that " the aspect of the axis, when marked by a longitudinal groove, is precisely that which a *hollow* cylindrical body would have if extremely compressed" (p. 22).

Common Canal.—While recognising the fact that the common canal gives rise to the cellules in most cases, Hall points out that in such forms as *Gr. gracilis* it must also give origin to " simple small stipes with solid axis, common canal, and cellules." He lays great stress on this double function of the common canal, but unfortunately his view of this question is largely due to his erroneous idea of the non-celluliferous character of the main stipe in *Gr. gracilis*. In the case of *Diplograptus*, Hall considers that there may be either two common canals separated by the " axis becoming a flattened plate," or else a single canal with only a filiform axis. He did not therefore distinguish between the septal walls between which the axis runs, and the axis itself. In *Phyllograptus* he infers that there are four slender common canals, which "may or may not communicate with one another."

Calycles or Cellules.—The structure and form of the various kinds of cellules are discussed by Hall in considerable detail, and excellent figures are given illustrating the longitudinal and transverse sections of various species, especially of those of *Climacograptus*. Hall agrees with Barrande that the cell walls are double. He considers that the cellule is limited by the cell walls, and he states that he has not discovered "evidence of such cell diaphragms" as were described by M'Coy; but he observes that there is "sometimes a swelling of the test of the common body below the cellule, indicating an enlargement of the parts at the bases of the buds," with occasionally an "undulation of the axis corresponding to this enlargement." The structure of the cellules of *Climacograptus* is minutely discussed, and Hall concludes that the cell partitions originate from the solid axis, and "appear to consist of triangular plates, so that there is an apparent double communication with the common body, giving not only the usual bilateral arrangement of the parts generally, but a bilateral arrangement of the parts in the individual alveoles."

The various kinds of cell-apertures are described, and the different ornaments of the test—such as spines, striæ, and pustules—in the several species, but few additions are made by him to the observations of Barrande on the same subject.

Mode of Reproduction.—One of the most important parts of Hall's work is the section devoted to the Reproduction of the Graptolites. In it he brings to light many novel facts, and by several valuable suggestions directs attention to the need for further research in this direction. In 1858 he had described certain "elongated sacs with swollen extremities," which he supposed contained the ovules or germs. These sacs "have scarcely any apparent substance," but are supported by numerous fibres, and in one case at any rate there is "conclusive evidence that they are connected with the solid axis of the parent stipe." The various figures given by Hall in illustration, both in the former and in the present work, show well the shape and arrangement of these "sacs." In his fig. 9 they are attached to what appears to be an example of *D. Whitfieldii*. In other cases there is apparently no ordinary cell structure on the stipes bearing the so-called reproductive sacs. These forms have since been separated under the provisional title of *Hallograptus* by Carruthers.

Several so-called germs or "young Graptolites of extremely minute proportions" are also described and figured by Hall in this work. Although they have never been actually found inside any of the "sacs," he has no doubt that they were derived from them. Hall's figures of these "germs" will best illustrate his views on their structure. In some of his typical instances, his so-called little "sac" containing the germ of the zoophyte is practically synonymous with the body afterwards distinguished by Lapworth as the sicula, though Hall considers the "radicle

h

and lateral spines to be distinct from the sac." This sac "extends itself as the common body in its canal along the axis, and gives origin to the budding which develops the successive cellules." He calls attention to the long axis and to an "extension of the common body along the axis above the incipient cells."

Although these " germs " were observed by Hall only in Diplograptid forms, he considers that they probably differed but little in other species; in the branching forms the only difference would be that the common body would divide into 2, 4, 8, etc., divisions, each one bearing its axis and common canal. As respects the development of the cellules in general, Hall calls attention to their invariable "lesser development towards the base of the stipe," although the same is often the case at the distal end as well.

Affinities.—He points out that the method of reproduction in the Graptolites thus observed "shows much analogy with the Hydroidea, and would indicate the Sertularians as their nearest analogues."

Mode of Existence.—In the case of all those Graptolite species having a single row of cellules, and also in the cases of the two- and four-rowed forms, such as *Retiolites*, *Retiograptus*, and *Phyllograptus*, Hall believes that they were " free floating bodies in the Silurian seas."

With respect to the Dendroid or tree-like forms, however, he holds that there is some evidence indicative of a different mode of existence, and he infers that these were fixed to the sea bottom by a root or bulb-like expansion at their base.

Distribution of the Graptolites.—The short section dealing with the geological and geographical distribution of the Graptolites must also be briefly referred to in this place. Hall considers that Graptolites came into existence at the time of the deposition of the Potsdam Sandstone, and attained their greatest development at the epoch of the Quebec group. " Several genera are known in the Trenton formation, and a greater development occurs at the period of the Hudson River formation." " In the Clinton group there is one species of *Graptolithus* and a *Retiolites*, while *Dictyonema* and *Inocaulis* occur in the Niagara beds. *Dictyonema* also ranges up into the Upper Helderberg and Hamilton formations." The wide geographical distribution of Graptolites in America is pointed out by Hall, but no attempt is made by him to fix the geological range of the individual species.

CHAPTER III.

1866 to 1880.

Previous to the year 1866 most of the work among the Graptolites had been done by geologists living on the Continent of Europe or in America; but at this date what might be called the British period of investigation began, and for the next twenty years the great majority of papers that appeared emanated from British investigators—Salter, Carruthers, Nicholson, Hopkinson, and Lapworth.

1866.
Nicholson,
" Ovarian Vesicles or
' Grapto-gonophores,' "
' Geol. Mag.,' vol. iii.

Nicholson's first paper was devoted to the description of certain bodies occurring with Graptolites at Garple Linn, near Moffat, which he suggested might be " gonophores " or " ovarian vesicles," and for which he proposed the name " Grapto-gonophores." These bodies he describes as being " corneous and bell-shaped," " provided at one extremity with a prominent spine or mucro, the other terminating in a gentle curved or nearly straight margin." Generally they are found free; but in the case of *Gr. Sedgwickii* they occur in such close juxtaposition as to " justify the belief that the connection was organic, and not simply accidental." They appear to spring from the common canal or cœnosarc with the mucro at the free end.

Nicholson points out that, if this interpretation be correct, the Graptolites must be " finally referred to the Hydrozoa, and would find their nearest analogues in the Sertularidæ," " from which, however, they would always be separated by sufficiently distinctive and definite characters."

1866.
Salter,
Memoir Geol. Survey,
vol. iii.

In H.M. Geological Survey Memoir on North Wales (published in 1866) Salter described and figured several forms of Graptolites.

Of new species, he figures two—(1) *Diplog.* **barbatulus** (which it is at present impossible to identify), and (2) *Dendrog.* **furcatula.** Of species already named by previous observers, he refigures (3) *Gr. sagittarius* (which he considers to be identical with *Gr. virgulatus* and *Gr. Barrandei* of Scharenberg, and which is probably a fragment of a Dichograptid); (4) *Diplog. teretiusculus;* (5) *D. mucronatus;* (6) *D. ramosus;* (7) *D. bicornis* (his two figures do not show the characteristic features of that species); (8) *Didymog. geminus;* (9) *D. hirundo;* and (10) *Dictyonema sociale.* The question of the structure and the affinities of this last mentioned form (which had first been figured by Salter in ' Siluria,' Edit. 2, 1854) is entered into in much detail. Salter somewhat reluctantly accepts Hall's view of the identity of Eichwald's *Gorgonia* (*flabelliformis*), Angelin's *Phyllograpsus,* and his own *Graptopora,* with Hall's genus *Dictyonema.* While he considers the genus to have a " true relation with the Graptolite group," he regards it as a link connecting Graptolites with the Fenestellidæ among the Bryozoa, under which class, indeed, following Huxley, he groups the Graptolites in general.

Salter combats Hall's view that all the single-stiped forms of Graptolites are broken fragments of branched species, holding that the " evidence against it lies both in the mode of occurrence of these bodies and even more in the very complete series of forms which can be furnished by our cabinets." " It is, moreover, certain that the Graptolites, occurring in great shoals in muddy deposits, probably of a deep sea, often unaccompanied by other fossils, and tranquilly laid down on the soft carbonaceous floor, would be less likely to be broken up than most fossils." He also asserts that *Diplograpsus* " could never have consisted of two single Graptolites, the line of junction being quite soldered up," and he refers to such forms as *D. ramosus* in support of his opinion. He points out that in the genus *Didymograpsus* the " radix " is sometimes " lengthened out into a long acute point, and the branches reflexed," from which it appears that he did not accept Hall's assertion that *D. caduceus* (Salter) was really a *Tetragraptus*.

1866.
Geinitz,
" Bemerkungen über Hall's 'Graptolites of the Quebec Group,'"
' Neues Jahrb. f. Min.'

In a review of Hall's Memoir on the " Graptolites of the Quebec Group," Geinitz criticises Hall's generic list, and points out that there must be four distinct genera included in Hall's single genus *Graptolithus*, viz. *Monograptus*, *Didymograptus*, *Tetragraptus* and *Dichograptus*. He considers his own *Cladograpsus* to be identical with Hall's *Dicranograptus*, but he accepts Hall's genera *Phyllograptus* and *Retiograptus*. He again emphasises the close relationship of the genera *Monograptus* and *Rastrites*, and suggests a name, **Birastrites,** for those *Rastrites*-like forms having apparently two rows of thecæ, to which genus he thinks *Thamnograptus* and *Buthograptus* might also possibly belong. He considers that *Oldhamia* is an alga, but on the other hand he argues strongly in favour of *Nereograptus* being a true Graptolite. He proposes the new generic name of **Stephanograptus** for Hall's species *Gr. gracilis*, which, however, Hall himself had already acknowledged as probably identical with Emmon's *Nemagraptus* (*elegans*).

1867.
Carruthers,
" Graptolites : their Structure and Systematic Position,"
' Intell. Observer,'
xi (4) and xi (5)
Nos. 64 and 65.

A general paper on Graptolites was published by Carruthers in 1867 in the pages of a scientific magazine known as the ' Intellectual Observer.' In this he discusses their structure, systematic position, and classification. He divides them into four sections—

Section I.—" Species with a single series of cells," including—1. *Rastrites*, 2. *Graptolithus*, 3. *Cyrtograpsus*, 4. *Didymograpsus* (including *Tetragrapsus* and *Dicellograpsus*), 5. *Dichograpsus*, 6. *Cladograpsus*, 7. *Dendrograptus*.

Section II.—" Species with two series of cells." 8. *Diplograpsus*, 9. *Climacograptus*, 10. *Retiolites*.

Section III.—" Species with single and double series of cells on different parts of the same polypary." 11. *Dicranograptus*.

Section IV.—" Species with four series of cells." 12. *Phyllograptus*.

He figures a large number of previously named species, viz.: (1) *Gr. priodon*, (2) *Gr. convolutus*, (3) *Gr. Roemeri*, (4) *Gr. Sedgwickii*, (5) *Gr. Halli*; (6) *Rastrites Linnæi*, (7) *R. capillaris*; (8) *Diplog. pristis*, (9) *D. tricornis*, (10) *D. cometa*,

(11) *D. folium ;* (12) *Climacog. scalaris ;* (13) *Retiolites Geinitzianus ;* (14) *Didymog. Murchisoni,* (15) *D. crucialis ;* (16) *Dichog. aranea ;* (17) *Phyllog. ilicifolius ;* (18) *Cladog. linearis* ; and three new forms, viz. : (19) *Gr.* **Hisingeri** (a name which he suggests for Hisinger's *Prionotus sagittarius* to " prevent further confusion "), (20) *Gr.* **Clingani,** and (21) *Didymog. (Dicellog.)* **elegans.** These, however, are not described.

In the diagnoses of the various forms of Graptolites, Carruthers employs for the first time in scientific literature the same nomenclature as that already proposed by Huxley and Allman for the Hydrozoa in general, using such terms as " polypary," " cœnosarc," " hydrotheca," etc., in his descriptions. This plan has since been followed by the majority of palæontologists.

He agrees generally with Barrande's account of the structure of a typical Graptolite, such as *Monog. Roemeri* and *M. priodon*, but he notices " what seems to be a septum at the base of each hydrotheca," in *Gr. sagittarius* and *Gr. latus.* He calls attention to certain specimens of *Diplog. pristis* which he found on one slab, in which the " naked axes met." This circumstance suggested to him the " possibility of the supposed perfect specimens of *Diplograpsus* being only fragments of more complex forms " (as in *Retiograptus*). He thus, like Emmons (p. xxxviii), anticipated in theory Ruedemann's subsequent discoveries. Carruthers, however, finally rejects this idea as being " anomalous and improbable."

He also discusses what is known or surmised about the development of the Graptolites, and figures certain young forms with their " radicle," and also a specimen of a form closely resembling those examples of *Diplograpsus* bearing " reproductive sacs " figured by Hall, but in this case only the interlacing fibres are seen, not the sacs themselves.

As regards the affinities of the Graptolites, he considers that they are more closely allied to the Hydrozoa than to the Polyzoa or any other group of animals, the polyps rising directly from the cœnosarc.

1867.
Carruthers,
Murchison's ' Siluria,'
4th edit.

In an appendix to the fourth edition of Murchison's ' Siluria,' which is dated 1867, Carruthers gave a second and briefer account of the Graptolites in general, and also a classification. This classification is essentially the same as that given by him in his paper in the ' Intellectual Observer ' already cited ; the only difference is that he subdivides his Section I into those with their—(*a*) polyparies simple and (*b*) polyparies compound ; his Section II into those—(*a*) with a slender solid axis and (*b*) without an axis (*Retiolites*).

He figures (1) *Retiolites Geinitzianus,* (2) *Dichog. Sedgwickii,* (3) *Dicranog. ramosus,* (4) *Phyllog. angustifolius,* (5) *Cladog. linearis* ; and also three new species — (6) *Cyrtog.* **Murchisoni,** (7) *Rastrites* **maximus,** (8) *Dendrog.* **lentus.** (The last form in reality belongs to the genus *Clonograptus.*) He again in this article argues in support of the Hydrozoal affinities of the Graptolites.

In 1867 Carruthers criticised Nicholson's views respecting the so-called " Grapto-gonophores." He denies their ovarian character, believing them to be Brachiopods, *i.e. Siphonotreta micula*, etc. He thinks that the supposed attachment is only a case of accidental juxtaposition, as it would be more natural for them to be attached by the mucro end, and he draws attention to the fact that no living Hydrozoon has " corneous " gonophores that become free swimming zooids. Moreover, no scars of attachment have been observed. Nor, he points out, do they bear any resemblance to the young Graptolite forms figured by Hall, the various stages of growth of which Carruthers himself had traced in the development of his own *Diplog. tricornis.*

As regards the affinities of the Graptolites themselves, Carruthers now inclines to the opinion that " although they resemble the Hydrozoa in general aspect, they are nevertheless more closely allied to the Polyzoa in the following characters :

1. There is no distinct common canal. Sometimes the polyps rise from a common substance which extends along the whole of the celluliferous portion of the organism, but there is no constriction or septum at the base of the cells. In other species the walls of the cells are continued to the solid axis.

2. The mouths of the cells are furnished with spines.

Graptolites, however, differ from all living zoophytes in possessing—(1) a solid axis, (2) free polypidoms.

In this paper Carruthers places his species *Cladograptus linearis* in the genus *Dendrograptus.*

The same year Nicholson described some Graptolites from the Lower Silurian beds of the South of Scotland, including three which were new to science, viz. (1) *Diplog.* **tubulari-formis**, (2) *D.* **acuminatus**, (3) *Didymog.* **anceps ;** and three species named by earlier observers—(4) *Didymog. flaccidus*, (5) *Diplog. quadrimucronatus*, (6) *D. Whitfieldi.* He also describes a new genus, **Corynoides**, typified by the species (7) *C.* **calicularis.** As regards this genus, the name of which was suggested by Harkness, Nicholson defines it as a " simple hollow tube, probably corneous, provided with a single or double radicle or mucro, and developed distally into a cup-like hydrotheca." The single polypite is " closely analogous to some of the Corynidæ or Tubularidæ," especially resembling *Coryomorpha.* He holds that it was " undoubtedly a free floating and independent organism."

In the same paper Nicholson also describes and figures three types of Hall's "germs" of Graptolites, belonging apparently to " *Diplog. pristis.*" He figures three stages of growth in this form, also the early stage of a uniserial species. In this last example it will be seen that the true sicula is well represented as forming a constituent part of the " germ," but the solid axis, instead of being shown as a continuation of the apex of the sicula, is figured as continued into, and is confused with, the apertural spine of the sicula.

1867.
Nicholson,
' Geol. Mag.,' vol. iv.

In a letter written during this year Nicholson replies to Carruthers' criticism of his " Grapto-gonophores," and he suggests that possibly the pustuliform elevations at the bases of the cells in *Didymog. nitidus* and *D. anceps* " may be the cicatrices of ovarian capsules."

1867.
Carruthers,
' Geol. Mag.,' vol. iv.

A subsequent letter from Carruthers called forth a reply from Nicholson, but no new facts were given by either writer.

1867.
Nicholson,
" On a New Genus of Graptolites, with Notes on Reproductive Bodies," ' Geol. Mag.,' vol. iv.

A month later Nicholson published another paper, in which he suggested the generic name **Pleurograptus** for the form typified by the *Cladograptus linearis* of Carruthers. Nicholson re-describes the species, showing how it differs from *Dendrograptus*, and from all known genera, in having no 'funicle,' the " primitive parent stem being itself celluliferous." He considers that *Gr. gracilis*, with its marked ' funicle,' is unique in its character, and " should form the type of a new genus."

Nicholson also figures in this paper a stipe " studded with small rounded tubercles," " apparently springing from the common canal on either side," and he suggests that this may be an " instance of ovarian vesicles in their young condition," which may either remain permanently attached, or may possibly become free at a later stage.

He also figures several more " gonophores " of *G. Sedgwickii*, and states that he believes that he has " made out with certainty that these capsules are reproductive in function," while their resemblance to orbicular Brachiopods when compressed is " purely mimetic and illusory." Associated with examples of *Gr. sagittarius* the capsules occur in the greatest confusion, but he " failed to detect any organic connection between them and the cells." This, he suggests, may be due to the fact that they were thrown off when extremely minute, attaining their full development subsequently; or they " were attached to the sides of the polypites, or to ' gonoblastidea,' as in many living Hydrozoa."

He points out the fact that no ovarian sacs are found among the Graptolites at Hartfell—where there are no forms of *Graptolithus* (*Monograptus*); but he explains this by suggesting that the sacs belonging to the genera there represented had possibly no corneous envelope, and that therefore they have not been preserved. Some additional young forms or germs are figured by him, differing but little from those previously described, except as respects their greater size.

Nicholson strongly upholds in this paper the Hydrozoal affinities of the Graptolites, on the following grounds :

(1) The true Graptolites (except *Dictyonema* and possibly *Dendrograptus* and *Callograptus*) are all free-swimming forms, whilst the Bryozoa are invariably fixed.

(2) The undoubted presence of a common canal " in many, if not all."

(3) The mode of growth and the nature of the embryonic forms.

(4) The existence of forms like *Corynoides*.

He points out that the Graptolites differ from the whole sub-class of the Hydroida in the fact that the polypidom was free, and not fixed by a hydro-rhiza, and he is disposed to place them in a " new sub-class, intermediate between the fixed and oceanic Hydrozoa."

1867.
Carruthers,
" Note on Systematic
Position, etc., of
Graptolites," ' Geol.
Mag.,' vol. iv.

The same year Carruthers replied to these statements and opinions of Nicholson in much detail, but his paper contained no new facts or theories.

1867.
Törnquist.
" Om Lagerföljden
i Dalarnes Under-
Siluriska Bildningar."

In the same year also a short stratigraphical paper was published by Törnquist dealing with the appearance of Graptolites in Central Sweden. He considers that *Prionotus sagittarius* (His.) and *Didymog. virgulatus* are identical with *D. ludensis ;* while *Petalog. folium* (His.) is really a *Retiolites.*
Pr. scalaris is founded partly upon *Diplog. pristis* and partly upon *D. teretiusculus.* The species discussed, together with *Rastrites? convolutus*, are recorded by him from the Lower Silurian beds of Dalarne.

1868.
Nicholson,
" The Graptolites of
the Skiddaw Series,"
' Quart. Journ. Geol.
Soc.,' vol. xxiv.

In the following year (1868) no less than six papers on Graptolites were written by Nicholson. The first of these, " On the Graptolites of the Skiddaw Slates," is mainly descriptive in character, but the works of previous observers are carefully reviewed, and several new forms of Graptolites are named and figured. The following genera and species are recorded by him from these rocks :

Didymograpsus. Salter's *Didymog. caduceus* (which he considers to be non-existent as a distinct species—at any rate in the Skiddaw Slates—and to be probably identical with *Tetrag. bryonoides* (Hall)), *D. v-fractus, D. sextans* (which he regards as "somewhat peculiar among the *Didymograpsi*"), *D. patulus* (= *D. hirundo*, Salter), *D. nitidus, D. bifidus,* and *D. serratulus* are described, and (1) *D. geminus* is figured.

Phyllograpsus. Phyllog. angustifolius, and (2) *P. typus.*

Tetragrapsus (Nicholson defends this genus against Hall). *T. Headi, T. quadribrachiatus, T. bryonoides, T. crucifer.*

Dichograpsus. Nicholson proposes the retention of Salter's name for this genus, to include such forms as possess a "variable number (always more than four) of simple stipes, united centrally at the base by a non-celluliferous stem or funicle." He distinguishes two groups within the limits of the genus: (*a*) Those typified by *D. Logani* and *D. octobrachiatus*, in which the celluliferous stipes are never divided at all; (*b*) Those like *D. flexilis*, *D. rigidus* and *D. multiplex*, in which the celluliferous stipes themselves branch and rebranch repeatedly. Like Hall, he does not regard the disc in the Graptolite as of generic value, and is inclined to believe that its "homologue is to be found in the 'float' or 'pneumatophore' of the Physophoridæ." The disc seen in *Climacog. bicornis* and others is probably of the same character, and it "may have been developed only at certain stages of growth, in certain individuals of the species, and probably for certain definite purposes." The species of *Dichograpsus* described are *D. Logani*, (3) *D. octobrachiatus* and two new forms, (4) *D.* **multiplex** (*Temnograptus*) and (5) *D.* **reticulatus** (*Schizograptus*).

In discussing the genus *Diplograpsus*, it is noteworthy that Nicholson lays great stress on the importance of observing the character of the base for determining the various species, "forming as it does the most valuable aid to a correct diagnosis." The suggestion thus made has subsequently proved to be of especial value in the discrimination of both genera and species. He divides the Skiddaw Slate species of *Diplograpsus* by their basal characters into three classes, viz. those having (*a*) a median radicle, flanked by two lateral processes, which spring from primary cellules on each side (*D. bicornis*, etc.); (*b*) two primary cellules, greatly elongated, forming with the solid axis a broad tapering "radicle" (*D. cometa*, etc.); (*c*) the base formed by a basal extension of the solid axis beyond the proximal extremity of the frond (*D. pristiniformis*, etc.). He re-describes *Diplog. mucronatus* and *D. antennarius*, and re-figures (6) *D. teretiusculus* and (7) *D. pristiniformis*. For *D. antennarius* and *D. teretiusculus* he considers that it would be advisable in future to accept Hall's title of *Climacograpsus*.

As regards the genus "*Graptolites*" or "*Graptolithus*," as then understood (the *Monograptus* of later authors) Nicholson states that he is "inclined to think that the genus is not represented" in the Skiddaw Slates, and that the forms ascribed to it, such as *Gr. sagittarius*, *Gr. tenuis*, and *Gr. Nilssoni* are in reality fragments only of compound species. This view has been fully justified by subsequent research. A fragmentary branching form is referred by him to (8) *Dendrograptus Hallianus*, which species he considers is probably identical with *D. furcatula*, Salter.

A new form (9) *Pleurograpsus* (?) **vagans** is also described and figured. This is not a *Pleurograptus*, but belongs to the Dichograptidæ; owing, however, to its fragmentary condition it is impossible to refer it with certainty to any known genus of that family.

i

1868.
Nicholson,
" On the Graptolites
of the Coniston Flags;
with Notes on the
British Species of the
Genus *Graptolites,"*
Quart. Journ. Geol.
Soc.,' vol. xxiv.

Nicholson's second paper was entitled " On the Graptolites of the Coniston Flags." This also is almost entirely descriptive, twenty-four species of Graptolites being described, of which five are new.

The following are the genera and species noticed in this paper: (1) *Diplograpsus folium,* (2) *D. palmeus,* (3) *D. angustifolius,* (4) *D. putillus, D. vesiculosus, D. pristis,* and two new *Diplograpti* (5) *D.* **tamariscus** and (6) *D.* **confertus** ; (7) *Climacograpsus teretiusculus ;* (8) *Retiolites Geinitzianus,* and a new form (9) *R.* **perlatus** ; (10) *Rastrites peregrinus,* and (11) *R. Linnæi ;* (12) *Graptolithus lobiferus,* (13) var. *Nicoli* and (14) var. **exiguus,** (15) *Gr. Sedgwickii,* (16) var. *triangulatus* and (17) var. **spinigerus,** (18) *Gr.* **fimbriatus,** (19) *Gr. Nilssoni,* (20) var. **major** and (21) var. **minor,** (22) *Gr. tenuis,* (23) *Gr. bohemicus,* (24) *Gr. priodon,* (25) *Gr. colonus,* (26) *Gr. sagittarius,* (27) *Gr. turriculatus,* and a new form, (28) *Gr.* **discretus,** of which Nicholson remarks that " the long sub-mucronate extremities of the cellules are often furnished with little ovoid, or triangular, vesicular bodies depending from their apices."

It is impossible to discuss the identification of each species in detail, but in the light of our present-day knowledge we are aware that several of the forms assigned by him to species already named must be regarded as incorrectly referred. The paper added very greatly to the number of British Graptolites and to our knowledge of the Graptolite species occurring in the higher beds of the Lake District.

As regards the age of the Coniston Flags, Nicholson considers them to be Lower Silurian, and the term as applied by him included all the beds between the summit of the Coniston Limestone proper and the base of the Coniston Grits.

1868.
Nicholson,
" On the Nature and
Zoological Position of
the Graptolitidæ,"
' Ann. and Mag. of
Nat. Hist.,' ser. 4,
vol. i.

The four remaining papers by Nicholson were published in 1868 in the 'Annals and Magazine of Natural History.' The third, entitled " On the Nature and Zoological Position of the Graptolitidæ," gives a clear account of the general state of knowledge at that time with respect to the morphology, zoological affinities, etc., of the Group.

Treating of the morphology of the Graptolites, Nicholson discusses in turn the " three factors, structurally and developmentally distinct," of which each single linear stipe is composed—*i.e.* (1) the solid axis, (2) the common canal, and (3) the cellules.

Solid axis.—In Monoprions this is a solid cylindrical rod, but in Diprions it is " certainly a corneous plate dividing the frond into two vertical compartments, apparently composed of two laminæ, with a median cylindrical rod and perhaps *including a central canal.*" The proximal extension of the axis is probably present in all true Graptolites, and constitutes the " radicle " or " initial point " of

Hall; the "funicle" is regarded as being composed of the proximal extensions of the axis, together with, probably, the common canal. The distal extension of the axis is only seen in *Diplograpsus*. It may consist of the solid axis only, or of a "bladder-like body, more or less elliptical in form, with a distinct filiform margin, and of uncertain function." "This dilatation," as seen in the new species *D. vesiculosus*, "seems always to be a direct expansion of the axis, which would thus appear to be tubular."

The non-solid character of the axis, at any rate in some forms, though hinted at by previous authors (Suess, etc.), was thus definitely stated by Nicholson for the first time. As regards the homologies of this axis, Nicholson thinks it is "probably related (but by analogy only) to the horny or calcareous 'sclerobasis' of the Gorgonidæ and Pennatulidæ." Its chief function was to give support, and its radicle was not used for purposes of attachment; therefore there is no close parallel between it and the foot stalk of the Sertularidæ.

The common canal.—He considers that the common canal is an individual structure, "giving origin to the cellules" and conveying a "soft connecting substance uniting the various polypites into an organic whole." He considers it to be homologous with the cœnosarcal canal of zoophytes generally.

Cellules.—Nothing new is added by Nicholson in this paper concerning the structure of the cellules. He points out their resemblance to the hydrothecæ of the Sertularidæ, but is opposed to the view that they were cut off from the common canal by a diaphragm.

As regards the development of the Graptolites, Nicholson inclines to the opinion that the "germs" at present discovered are not the earliest forms of the embryo; these probably had no corneous test. He agrees by implication with Barrande's view that the youngest cells are at the proximal end of the polypary, and in consonance with this he expresses the opinion that the secondary cellules appear to be intercalated between the radicle and the primordial cellules, so that the youngest cellules are proximal, the oldest distal in position. This mode of development "corresponds with that observed in the Calycophoridæ and Physophoridæ."

Nicholson's previously published views on the reproductive organs of the Graptolites are summarised in this paper. He thinks that when the capsules dropped off, probably minute, ciliated, free-swimming organisms (? planulæ) were liberated, which, at a later stage, developed a corneous covering. He suggests that the vesicle of *D.* **vesiculosus** (which was here figured for the first time) was in some way connected with the process of reproduction.

As to the mode of existence of the Graptolites, Nicholson says "there can be no question that by far the greater number were free-swimming or free-floating organisms." Some had floats; others were very probably provided with "necto-calyces," or swimming-bells, but these would not be preserved. The Dendroid forms, which most closely resemble the Sertularidæ, may have been fixed.

He regards the Graptolites generally as the "primitive stock" from which the various existing sections of the living Hydrozoa originally diverged.

1868.
Nicholson,
"On the Occurrence of the Genus *Ptilograpsus* in Britain, with Notes on the Ludlow Graptolites," 'Ann. and Mag. of Nat. Hist.,' ser. 4, vol. i.

In his fourth paper Nicholson recorded the occurrence of *Ptilograpsus* in rocks of Ludlow age in Britain, and described a new species (*P.* **anglicus**.) Nicholson agrees with Hall that the genus *Ptilograpsus* is closely related to *Plumularia*, and that it was probably an attached form. He here modifies his previous view as to the invariable presence of the axis in all Graptolites, and admits that "the axis is not so constantly present as has generally been thought," that it "is certainly absent" in all the Dendroidea, and "probably in other families."

In addition to *P. anglicus*, Nicholson records several other forms of Graptolites from the Ludlow rocks, viz. *Gr. priodon*, *Gr. colonus*, and *Gr. Nilssoni*.

1868.
Nicholson,
"On *Helicograpsus*," 'Ann. and Mag. of Nat. Hist.,' ser. iv, vol. ii.

This paper was followed in June by a fifth, in which Nicholson proposed the new generic name of **Helicograpsus** for the species *Gr. gracilis* of Hall. The essential difference, according to him, between this and his own genus *Pleurograpsus* (Carruther's *Cladograpsus*) consists in the presence of a distinct "funicle" and regular branching in the former; whereas in the latter there is no funicle, and the branching is quite irregular.

1868.
Nicholson,
"On the Geological Distribution of Graptolites," 'Ann. and Mag. of Nat. Hist.,' ser. 4, vol. ii.

The sixth paper by Nicholson published in 1868 dealt with the "Distribution in time of the British Species and Genera of Graptolites." It may be here briefly summarised as giving an excellent idea of the general state of opinion on this subject at that date.

(1) The Graptolites as a whole are characteristically Silurian, and fourteen out of seventeen genera are exclusively confined to the Lower Silurian, the Upper Silurian only possessing two peculiar species.

(2) In the Tremadoc Slates (= Upper Cambrian) *Dictyonema* occurs.

(3) To the Lower Llandeilo (Skiddaw Slates) the genera *Dichograpsus*, *Tetragrapsus*, and *Phyllograptus*, etc., are strictly confined. They occur in association with species of *Didymograpsus* and *Diplograpsus*.

(4) The Upper Llandeilo Rocks (which include all the graptolitic beds of Scotland) contain the genera *Diplograpsus*, *Climacograpsus*, *Graptolites*, *Rastrites*, and *Dicranograpsus*.

(5) The Caradoc beds do not as a rule yield Graptolites, but in Ireland they afford *Diplog. pristis*, *Didymog. sextans*, *Helicog. gracilis*, *Gr. Nilssoni*, *Gr. Sedgwickii*, *Callog. elegans*, etc.

(6) In the Lower Llandovery one Graptolite only has been found—*Climacog. teretiusculus*,

(7) In the Wenlock *Gr. Flemingii* is characteristic; *Retiolites Geinitzianus* occurs here, but also in the Lower Silurian and the Ludlow; *Gr. priodon* and *Gr. colonus* occur in the Wenlock and in the Lower and Upper Ludlow; *Ptilograpsus* is peculiar to the Lower Ludlow.

In spite of the apparently wide range in time of nearly all the species and genera cited, Nicholson remarks that they afford " very reliable and valuable data whereby formations in different parts of the world may be correlated with one another, or the exact position held by any group of beds in the stratified series may be more or less exactly ascertained," an assertion which, however slightly founded at that time, has been shown by subsequent research to be practically correct.

1868.
Carruthers,
" Revision of the British Graptolites, with Description of New Species and Notes on their Affinities," ' Geol. Mag.,' vol. v.

During the same year, 1868, and previous to the publication of some of Nicholson's papers mentioned above, Carruthers brought out his ' Revision of the British Graptolites,' with descriptions of several new species and notes on their affinities.

In the classificatory part of this paper a large number of species are described, and the new ones (some of them previously mentioned by him in his appendix to Murchison's 'Siluria,' Edit. 4) are figured and described.

The genus *Rastrites*, as acknowledged by him, includes four species: *R. peregrinus*, (1) *R. Linnæi*, (2) *R. maximus*, and a new form (3) *R.* **capillaris.** He holds that *R. triangulatus* (Hark) was founded on the proximal part of *Gr. convolutus* (Monograptus), and he gives a figure showing that this species of Graptolite " really terminates proximally in a polypary which cannot be distinguished from that of *Rastrites*," thus throwing doubt on the stability of the genus *Rastrites* itself. He points out that *R. Barrandei* (Hark) was founded on fragments of *Cladograpsus (Cænograptus) gracilis.*

The genus *Graptolithus* he restricts in the same manner as other palæontologists of the time to forms now classed as *Monograptus*. He considers that this generic term ought properly to be applied to double forms like *Gr. scalaris* (Linn), for which it was first employed by its founder, but that it would create too much confusion to make the correction now. *Graptolithus* is represented by twelve species in his list: *Gr. Nilssoni, Gr. Flemingii, Gr. tenuis, Gr. Salteri, Gr. Hisingeri,* (4) *Gr. convolutus, Gr. Sedgwickii, Gr. priodon, Gr. Halli, Gr. Becki,* (5) *Gr. Clingani* and a new species, (6) *Gr.* **intermedius.** His own new genus, *Cyrtograpsus* (previously named in ' Siluria'), is described, and his species (7) *C. Murchisoni* is re-figured. He shows that *C. hamatus* (Baily) also belongs to this genus.

He does not regard the number of branches in allied forms of Graptolites as a generic distinction, and therefore includes under *Didymograpsus* species of *Tetragrapsus*, as well as forms of *Dicellograpsus*. He considers that " the possession of an obvious branching hydrocaulus stipe " might be a good reliable generic character, but points out that unfortunately there are no materials in Britain to enable one to determine this.

Many species of the genus *Didymograpsus* as thus enlarged are referred to, viz. *D. hirundo, D. Murchisoni, D. v-fractus, D. sextans, D. Forchammeri,* (8) *D. elegans, D. moffatensis, D. caduceus, D. bryonoides, D. quadribrachiatus.*

The forms of *Dichograpsus* noticed by him are : *D. octobrachiatus* and *D. Sedgwickii.*

He reinstates the genus *Cladograpsus* for his own *C. linearis*, refusing to accept Nicholson's generic name of *Pleurograpsus*, and includes in the genus *C. linearis,* (9) *C. capillaris*, and *C. gracilis.*

The forms of *Diplograpsus* noted and figured are : (10) *D. pristis* (= *D. vesiculosus* and *D. physophora*), (11) *D. minimus, D. angustifolius*, (12) *D. Whitfieldi* (including *D. quadrimucronatus*, Nich.), (13) *D. tricornis*, (14) *D. cometa* (including *D. tubulariformis*, Nich.). This last named form he thinks "should perhaps be made the type of a new genus." As regards (15) *D. mucronatus*, Carruthers suggests that those forms with "several branching and apparently anastomosing processes from the cell mouth," which Hall considered to be the marginal fibres of the reproductive sacs, may really prove to be a distinct species, for which he proposes the name *D. Bailyi.* *Diplog. persculptus* is referred to, but not described or figured.

Two forms of *Dendrograptus* are noticed : *D. furcatulus* and (16) *D. lentus*, Carr.

Under *Climacograptus* Carruthers gives figures of (17) *C. scalaris* and a new species (18) *C. minutus.*

Under the genus *Dicranograptus* he includes *D. ramosus* and a new species (19) *D. Clingani,* thus for the first time restricting this genus to those forms with a biserial proximal portion and uniserial distal stipes.

The genus *Retiolites*, according to Carruthers, possesses no axis or septum. The forms recognised are : *R. Geinitzianus* and *R. venosus.*

Only one form of *Phyllograptus* is noticed from Britain, viz. the *P. angustifolius* of Hall.

Carruthers' paper is prefaced by a general description of the structure of a typical Graptolite, but this contains nothing new. He, however (as in his paper in the 'Intellectual Observer'), strongly recommends the adoption of the nomenclature already in use for the Hydrozoa, and he consistently employs it throughout this memoir.

The affinities of the Graptolites are discussed by Carruthers at great length. He considers that the general form of the polypary, its free or attached nature, its chitinous character, are of no systematic value, whereas the presence or absence of a common canal is of very great importance. The affinities of the Graptolites to the Polyzoa are fully considered, but the absence of a common canal in the Cheilostomata, and the fact that the cells are in communication only through a perforated septum in the Ctenostomata, "distinguish them at once." The various characters of the six groups of living Hydrozoa are given, but the only two which he acknowledges have any affinities with the Graptolites are the Corynidæ and the Sertularidæ, and as there are no cells in the former, Carruthers considers that the nearest allies of the Graptolites are the latter, although they have no axis.

As respects the mode of life of the Graptolites, Carruthers is inclined to the opinion that they were attached, and points to the long proximal extension of the radicle in *C. scalaris* as an example. He rejects Nicholson's idea of "floats" and "swimming bells," and also his reproductive sacs and gonophores.

1868.
Hall,
" Introduction to the Study of the Graptolitidæ,"
'Twentieth Annual Report of the State Cabinet.'

The 20th Annual Report of the State Cabinet of New York, published in the same year, contains a paper by Hall, entitled "An Introduction to the Study of the Graptolitidæ." This is in the main a reprint of selections from his previous memoir on the "Graptolites of the Quebec Group," but he gives some "Supplementary Notes" on the genera *Didymograptus, Cladograptus, Dicranograptus,* and *Cœnograptus.* He points out that both M'Coy and Geinitz included under each of the first two generic names two distinct types, and he suggests the employment of *Didymograptus* (M'Coy) for such forms as *D. Murchisoni,* and *Cladograpsus* (Geinitz) for forms of the type of *D. ramosus,* etc., thus relinquishing for the time his own genus *Dicranograptus* in favour of the older title suggested by Geinitz. As regards the *Cladograpsus* of Carruthers, Hall points out that there is little doubt that it is similar to his own *Cœnograptus,* which may, again, be identical with Emmon's *Nemagraptus.*

Hall still asserts that it is generally impossible to distinguish between *Didymograptus, Tetragraptus,* and *Dichograptus;* but if the last of these names is to be used, he suggests that it be restricted to such forms as *Gr. Sedgwickii* and *Gr. aranea;* and he proposes the name **Loganograptus** for "those forms with central corneous discs, while those which are repeatedly dichotomous, like *Gr. flexilis,* will constitute a third genus."

Hall also makes some additional remarks on *Phyllograptus,* and on the presence of a common body in this genus.

1869.
Heidenhain,
" Ueber Graptolithen-führende Diluvial Geschiebe der Norddeutschen Ebene,"
'Zeit. d. Deutsch. Geol. Gesell.,' Bd. xxi.

In 1869 Heidenhain gave descriptions and some good figures of a few species from the Graptolite-bearing boulders of the Drift of Northern Germany. Descriptions only are given of *M. priodon, M. sagittarius, M. colonus,* and *M. testis,* while there are figures of a new species—(1) *M.* **distans** (Heidenhain's species is identical with the subsequently described *M. scanicus,* Tullb.), (2) *M. Nilssoni,* (3) *M. Salteri?,* (4) *M. Bohemicus,* (5) *M. Roemeri* (which, according to Heidenhain, occurs without associates in a harder dark limestone), and (6) *M. sp.* (This last was afterwards named by Jaekel *M. micropoma.*) The age of the beds from which these erratics must have been derived is from this record of Graptolites now known to be Lower Ludlow.

He records from rocks somewhat different in character, and therefore probably from another geological horizon, *D. palmeus,* var. *tenuis* and *D. pristis* ?, and gives descriptions of them.

1869.
Nicholson,
" On Some New Species of Graptolites," ' Ann. Mag. of Nat. Hist.,' ser. 4, vol. iv.

During the same year Nicholson described and figured several new species, and one new genus of Graptolites from the Lake District. The new genus, which he names **Trigonograptus** (1) (*T. ensiformis* type), he regards as intermediate between *Retiolites* and *Diplograptus.* A form which he here denominates (2) *Dichograpsus* **fragilis** was afterwards made by himself the type of a new genus *Trichograptus.* Other new species described and figured by him in this paper are (3) *Dichograptus?* **annulatus,** (4) *Diplog.* **Hopkinsoni** (= *Cryptograptus*), (5) *D.* **armatus** (? *Glossograptus*), (6) *D.* **Hughesi,** (7) *D.* **sinuatus,** (8) *D.* **bimucronatus,** (9) *D.* **insectiformis,** (10) *Climacog.* **innotatus,** (11) *C.* **tuberculatus,** (12) *Gr.* **argenteus,** (13) *Didymog.* **affinis,** and (14) *D.* **fasciculatus.** He also re-figures (15) *D. vesiculosus.*

1869.
Linnarsson,
" Om Vestergötlands Cambriska och Siluriska Aflagringer."

The same year Linnarsson recognised two distinct graptolite horizons in Sweden: (1) the Lower Graptolite Shales, with *Phyllograptus, Didymograpsus,* etc. (the equivalents of the Skiddaw Slates), and (2) the Upper Graptolite Shales, with *Graptolithus, Diplograpsus, Rastrites,* and *Retiolites,* containing fossils similar to those in South Scotland. As the latter occur above the Brachiopod Shales (Caradoc) he thinks that the Llandeilo age of the Scotch beds (Murchison) is probably incorrect.

1869.
Hopkinson,
" On British Graptolites," ' Journ. Quekett Microscopical Club,' vol. i.

In a paper read before the Quekett Microscopical Club Hopkinson gave a generalised account of the British Graptolites. The history of research among the Graptolites is briefly dealt with, and their structure is described in some detail. He adopts throughout the Hydrozoal nomenclature first employed by Carruthers, each term being carefully defined.

He accepts Carruthers' classification for the most part, but he places all true Graptolites in the single order of the Graptolitidæ, and regards Carruthers' four classificatory sections as " sub-orders " or " families," which he names respectively Monoprionidæ, Diprionidæ, Monodiprionidæ, and Tetraprionidæ.

In the family Diprionidæ, Hopkinson suggests a new genus, **Cephalograptus,** to include the single species *D. cometa,* Gein. He figures (1) *Rastrites peregrinus,* (2) *Gr. priodon,* (3) *Gr. Hisingeri,* (4) *Gr. Sedgwickii,* (5) *Cyrtog. Murchisoni,* (6) *Didymog. Murchisoni,* (7) *Tetrag. bryonoides,* (8) *Dichog. octobrachiatus.* He describes and figures a new species of *Diplograpsus*—(9) *D.* **penna**. The genus *Retiolites* he does not consider to be a true Graptolite. He regards *Dendrograptus* as forming a connecting-link between his true Graptolites (the Rhabdophora of the later works of Allman and others) and the genera *Callograptus*, *Dictyonema*, etc.

He also treats in brief of the reproduction and development of Graptolites, and he concludes that in their mode of reproduction " Graptolites are nearly allied to Sertularian Hydrozoa."

1870.
Nicholson,
" On the British
Species of *Didymograpsus*," ' Ann. and
Mag. of Nat. Hist.,'
ser. 4, vol. v.

A revision of the genus *Didymograpsus* and its British species was made by Nicholson in 1870. He groups the species which he assigns to this genus into three sections, according to the " angle of divergence "; and he carefully distinguishes, therefore, between what he terms the angle of divergence and the " radicular angle " of the stipes, and the position of the cells with reference to the " radicle."

The distinguishing characters of his three groups are as follows :—

(*a*) Radicle on the inferior aspect, and cells on the superior aspect, angle of divergence not greater than 180°—*D. Murchisoni, D. affinis, D. patulus.* (*b*) Radicle as in group (*a*), but the angle of divergence more than 180°—*D. flaccidus* and *D. anceps.* (*c*) Situation of cells reversed, on the inferior aspect, on the same side as radicle—*D. sextans* and *D. divaricatus.*

It will be seen from this classification that Nicholson had not yet recognised, even to the extent to which Hall had done previously, the systematic difference between the true genus *Didymograptus* and *Dicellograptus* (*Dicranograptus*—pars of Hall), nor yet the distinction between the sicula proper and its apertural spine. The species described and figured by him are (1) *D. patulus*; (2) *D. v-fractus*; (3) *D. extensus;* (4) *D. nitidus ;* (5) *D. affinis ;* (6) *D. serratulus* (= *D. Nicholsoni*)*;* (7) *D. fasciculatus ;* (8) *D. geminus;* (9) *D. bifidus;* (10) *D. divaricatus* (= *Dicello. elegans,* in part)*;* (11) *D. anceps* (*Dicellograptus*) *;* (12) *D. flaccidus* (*Leptograptus*)*;* and (13) *D. sextans* (*Dicellograptus*).

1870.
Nicholson,
" Revision of the Genus
Climacograpsus, with
Notes on the British
Species of the Genus,"
' Ann. and Mag. of
Nat. Hist.,' ser. 4, vol. vi.

Nicholson followed up this paper by a corresponding " Revision of the genus *Climacograpsus*," of which genus he took *C. teretiusculus* (= *C. scalaris* or *rectangularis*) as his type. His diagnosis is as follows: " Composed of two simple unicellular stipes placed back to back, their internal walls coalescing to form a single vertical septum, along the centre of which runs a delicate solid axis in the form of a fibrous, filiform rod." This rod is always prolonged distally, and generally proximally. Nicholson doubts Hall's statement that in *Cl. typicalis* " there seems to be no septum, but the solid axis runs up the centre of a tube common to both series of cellules." He agrees with Hall, however, that the cell

k

partitions are attached to the solid axis, and that the only way in which communication could take place was by assuming that the cell partitions are triangular plates, their apices attached to the axis, having "an unequally arched or convex upper surface, and a concave lower surface." Nicholson states that in those examples of *Climacograpsus* studied by himself there exists a distinct common canal, and the figures of the various "aspects" given by him illustrate well his views of the structure of the genus. The term "suture," which was in this connection here first suggested by Nicholson for the median groove or line formed at the surface by the septum, has subsequently been generally adopted.

He discusses the true character of Linnæus' *Gr. scalaris*, and gives a brief historical sketch of the species. He is inclined to think that it is not a Graptolite at all, "at any rate it is impossible to say whether it is the scalariform impression of a mono-prionidian or di-prionidian form."

He describes four species in addition to (1) *C. teretiusculus*, viz. (2) *C. innotatus*, (3) *C. tuberculatus*, (4) *C. antennarius*, and (5) *C. bicornis*. His figures of this last named species include both the *peltifer* and *tridentatus* varieties of later authors. He states that he has "little or no hesitation in comparing the basal disc or cup in *C. bicornis* with the disc of *Dichograpsus*," &c.

1870.
Hopkinson,
" On the Structure
and Affinities of the
Genus *Dicranograptus*,"
' Geol. Mag.,' vol. vii.

In 1870 Hopkinson published a paper on the genus *Dicranograptus*, Hall. He regards *Dicranograptus* as a distinct genus (not a sub-genus as Hall believed), and agrees with Carruthers in restricting it to those Graptolite forms in which the proximal portion is di-prionidian and the distal mono-prionidian. He differs, however, from Carruthers in believing it to be more nearly allied to *Climacograptus* than to *Diplograptus*. Although he describes the proximal extremity of the polypary as "composed of two series of thecæ, each having its own common canal," he somewhat modifies this assertion by saying in a footnote that "I am by no means certain that the two series are thus isolated." "*Climacograptus* and *Dicranograptus* alike differ from *Diplograptus* in the fact that the separation of the hydrothecæ is only occasionally seen, and very seldom extends to the common periderm, and their apertures are in a hollow which appears to be excavated out of the polypary. *Dicranograptus* only differs from *Climacograptus* in that its thecæ are usually, but by no means always, more or less prolonged distally."

Hopkinson describes and figures the following species in this paper: (1) *D. ramosus*, (2) *D. Clingani*, (3) *Dicranog. sextans* (doubtfully referred to this genus), (4) *D.* **Nicholsoni** (the web which seems to unite the branches for a short distance after bifurcation, he suggests, may be possibly analagous to the central disc of *Dichograptus*), and (5) *D.* **formosus**.

1871.

Hopkinson,

" On *Dicellograptus*, a
New Genus of Grapto-
lites," ' Geol. Mag.,'
vol. viii.

In 1871 Hopkinson published a paper on **Dicellograptus,** a new genus which he proposed for those simple bifurcating forms which had been previously included along with others by Hall in his genus *Dicranograptus*, but had been retained by Carruthers, Nicholson, and others in the genus of *Didymograptus.* In the forms assigned to this new genus the solid axis bifurcates in the "axil" of the branches; in one species the polypary is slightly enlarged at the axil, in others we get a spine of variable length, while in another the branches are connected by a membrane very like the corneous disc of *Dichograpsus.* The thecæ are the same as in *Climacograptus*, "undistinguishable from each other for the greater portion of their length."

Hopkinson discusses the nature of the so-called "axillary spine," which is especially conspicuous in this genus, and which had been regarded by Nicholson as the true "radicle," and also by Carruthers as the true "initial process," while Hopkinson claims that the "proximal spine" (which is usually flanked by two lateral spines) is the true "radicle," and that the "axillary spine" is, so far as we know at present, "an organ without its analogue in any other genus." This was the first recognition of the distinction between what is now known to be the "apertural spine" of the sicula and the apex of the sicula itself, a distinction which has proved to be of first-class systematic importance in this diagnosis.

Hopkinson also points out the unavoidable confusion in measuring the angle of divergence of the branches resulting from this failure to distinguish between the "initial spine" and the "axillary spine;" and he shows that it is not always possible to measure the angle of divergence along the polypiferous margin, as suggested by Nicholson.

He believes that in *Dicellograpsus, Dicranograptus,* and *Climacograptus* the branches are organically connected where in juxtaposition, "there being no septum observable."

The following species are included by him in his genus *Dicellograpsus*: (1) *D. Forchammeri,* (2) *D. elegans,* (3) *D. moffatensis,* (4) *D. anceps,* and a new species,(5) *D.* **Morrisi.**

He gives the range of *Dicellograpsus* as "exclusively Lower Silurian," and he states that it is eminently characteristic of the Llandeilo formation.

1871.

Hopkinson,

" On a Specimen of
Diplograpsus pristis
with Reproductive
Capsules," ' Ann. and
Mag. of Nat. Hist.,'
ser. 4, vol. vii.

Hopkinson's second paper contained a description of an interesting specimen of *Diplog. pristis* collected by the Geological Survey of Scotland from Leadhills, bearing "reproductive capsules." These reproductive organs, which he considers "represent the gonothecæ of the recent Sertularian Zoophytes," appear to have budded from the periderm at right angles to the thecæ. They are pear-shaped and "bounded by a single marginal fibre slightly thickened at its edges." These fibres, he suggests, may have been slender tubes. One specimen appears to indicate that the capsule may have been composed of two "membranes joined together at their edges through which the fibre has run."

He also figures two young forms of *Diplograpsus*, apparently lying within one of the capsules, but points out that they are large enough to have "entered on an independent existence."

He remarks on the agreement of this specimen with the capsules figured by Hall, but not with those figured by Nicholson, of the existence of which he seems to doubt. Their possible bearing on the affinities of the Graptolites is discussed, and Hopkinson considers that they confirm the near alliance of the Graptolites to the Sertularina, "though all the characters of their reproductive organs are not found in any one genus of the Hydroida."

1871.
Lapworth,
" On the Graptolites of the Gala Group,"
' Brit. Ass. Report.'

To the Meeting of the British Association of 1871 Lapworth communicated a list of the characteristic Gala Graptolites of South Scotland, and described but did not figure two new species : (1) *Retiolites* **obesus** and (2) *Graptolithus* **socialis** (afterwards identified with *M. exiguus*, Nich.).

1871.
Baily, W. H.,
' Memoir of the Geological Survey of Ireland.'

In 1871 W. H. Baily contributed some palæontological notes on the Silurian rocks of the country round Downpatrick and the shores of Dundrum Bay and Strangford Lough. In dark-grey slates near Downpatrick and Portaferry he recognises three Graptolite species, one *Gr. priodon* and two new forms which he named (1) *Gr.* **plumosus** and (2) *Gr.* **gradatus.** *Gr. plumosus* is certainly the *M. exiguus* of Nicholson, while *Gr. gradatus* is allied to *M. communis* (Lapw.). Both species are described and figured.

1871.
Richter,
" Aus dem Thüringschen Schiefergebirge,"
' Zeit. d. Deutsch. Geol. Gesell.,' Bd. xxiii, Heft 1.

In the same year Richter published a paper giving the results he had arrived at in the continuation of his work on the Thüringian Graptolites. He gives a general revision of his views on their structure, development, affinities, etc., and the new points brought out by him may here be briefly summarised.

He founds a new genus which he names **Triplograptus,** with a single species (1) *T.* **Nereitarum.** (This, however, is now known not to be a Graptolite and need not, therefore, be discussed.) He also describes and figures Thüringian examples of older species, viz. (1) *Diplog. pristis*, (2) *D. teretiusculus*, (3) *Monog. cf. sagittarius*, (4) *M. priodon*, (5) *M. gemmatus*, (6) *M. peregrinus*, (7) *Phyllograptus*, and in addition three new forms : (8) *Diplog.* **pennatulus,** (9) *Monog.* **crenatus,** and (10) *M.* **chorda.** Most of the species are readily recognisable from his illustrations.

The greater part of Richter's paper deals with the structure of the Graptolites. He states for the first time in the history of graptolitic research, that the skeleton, which possessed considerable flexibility, consists of two layers : (1) an *inner* layer of considerable thickness, generally marked by transverse lines which meet on the upper and lower surfaces of the cells so as to form a zigzag line, but in a few cases quite smooth (as in *M. Sedgwickii*, *M. convolutus*, etc.) ; and (2) an *upper* layer consisting of two thin lamellæ, so thin that the ornamentation of the inner layer is seen through. This outer layer is rarely preserved. The same two layers occur in *Retiolites*, but the double character of the upper layer is not yet determined.

In this paper also Richter was the first definitely to describe and figure as an independent structure in the genus *Monograptus* the initial Graptolite body now known as the "sicula." He calls it the *foot* or *haft-organ*. The shape of this "foot" —pointed, cone-like, rounded at the lower end—is admirably given in his figure of *M. priodon*, and he recognises the fact that the canal arises from this "foot." Richter also observes that the presence of the "foot" at the base of single-branched forms shows clearly that they are independent forms, and not merely branches of forms belonging to compound genera.

Richter considers that the axis of Graptolites is solid and of a fibrous structure, lies in a groove of the inner skeletal layer, is covered by the outer double one, and consists of as many parts as there are rows of cells. Even when the groove is sinuous, as in *D. teretiusculus*, the axis is quite straight. In *Diplograptus* he holds that there is only one canal and one axis, and he does not agree with Hall that in *Phyllograptus* there are four of each. The peculiar structure of the cells in *D. teretiusculus*, as interpreted by him, is described in detail, and illustrated by figures. Examples of the varieties of apertural ornamentations are given, especially in the case of *Rastrites*. These, he considers, are formed of the double upper layer. He believes that in *Rastrites* the apertures of the cells, which scarcely project beyond the common canal, are situated in the axil of the spine, the "cells" of previous authors being merely ornamental processes.

As regards the development of the Graptolites, he observes that nothing like the reproductive capsules described by Nicholson and Hall has been found in Germany; but he notices the occurrence of isolated "feet" with long, lash-like threads at their pointed ends. In *Monograptus* the first cell arises on the ventral side of the "foot"; in *Diplograptus* a cell arises from each side of the broader end of the "foot," but at different levels. He has noticed, in association with Graptolites, great numbers of spherical bodies, apparently surrounded by a double membrane, but he does not offer any suggeston as to the nature of these bodies, or the cause of this association.

His views of the mode of life of the Graptolites are remarkable. He thinks that the "foot" was movable, and could be turned upwards or downwards. When turned downwards it could be used to support the polypary vertically, by sticking into the mud. Some species, perhaps, could move freely from place to place, which would account for the fact that the full-grown forms do not occur in such groups as do the small individuals, but are scattered among other species.

The affinities of the Graptolites with the Pennatulidæ, Sertularidæ and Polyzoa are discussed, and Richter concludes that they are most closely allied to the last named. He adduces the rod of Rhabdopleura in support of his views, and compares the thin laminæ of the Graptolites to those of the cells and endocysts of a Polyzoon.

Other points are touched upon and discussed by Richter, such as the food of the Graptolites, their length of life, existence of enemies, etc., but nothing new in these directions is added to the views expressed by him in previous papers.

1871.

Linnarsson,

" Om nôgra Försten-
ingar från Sveriges och
NorgesPrimordialzone,"
' Kongl. Vet. Akad.
Förh.,' no. 6.

In 1871, also, Linnarsson described and figured a few fragments of a new species—(1) *Dichograptus* **tenellus** (*Clonograptus*), from the upper part of the Olenus shales in Westrogothia. This was the first instance of the occurrence of true Graptolites (excluding *Dictyonema*) in beds then regarded as being of true Cambrian (Primordial) age.

1871.

Lapworth, C.,

" On the Silurian Rocks
of the Counties of
Roxburgh and Selkirk,"
' Brit. Assoc. Reports,'
and ' Geol. Mag.,' vol.
viii.

In a paper read at the Edinburgh Meeting of the British Association and afterwards published in the ' Geological Magazine,' Lapworth and Wilson separated the Silurian strata of Roxburgh into five groups : the Hawick Rocks, the Selkirk Rocks, the Moffat Series, the Gala Group, and the Riccarton Beds, the last three of which are characterised by special graptolitic faunas. The rich graptolitic black shales of the Uplands are all referred to a single Moffat band which occurs not only in the country between Selkirk and Melrose, but re-appears in the Moorfoot Hills to the north.

1872.

Hopkinson,

" On Callograptus
radicans, a new
Dendroid Graptolite,"
' Geol. Mag.,' vol. ix.

In the following year, 1872, three papers were published by Hopkinson.

In the earliest of these he described a new species of *Callograptus* (*C.* **radicans**) from the Arenig rocks of the St. David's district. He gives a full diagnosis of the genus.

He discusses the affinities of the Dendroidea and considers that while *Dendrograptus* and *Ptilograptus* fall naturally into already accepted families of the Thecaphora (or Sertularina), *Callograptus* and *Dictyonema* (the latter being more Polyzoan than Hydrozoan in its affinities) should form a new family.

1872.

Hopkinson,

" On the Occurrence of
a Remarkable Group
of Graptolites in the
Arenig Rocks of St.
David's, South Wales,"
' Geol. Mag.,' vol. ix.

In Hopkinson's second paper he gives a list of graptolites from the Arenig rocks of St. David's. This list is preliminary to the paper by himself and Lapworth, which appeared three years later, when the various species were described and figured.

Hopkinson's third paper contained a description of several new species from the South of Scotland.

The first two, according to him, do not belong to the Graptolitidæ proper, but are nearly allied forms. Order, Hydroida, Sub-order, Athecata? Family, Corynidæ.—(1) *Corynoides* **gracilis**, (2) *Dendrograptus* **ramulus**. Sub-order, Rhabdophora, Fam. Monoprionidæ, (3) *Gr.* **attenuatus**, (4) *Gr.* **acutus**, (5) *Diplogr.* **Etheridgii**, (6) *D.* **penna**, (7) *D.* **pinguis**, (8) *D.* **fimbriatus**, (9) *D.* **Hincksi**, (10) *Dicranog.* **rectus**.

Hopkinson, in discussing the age of the Moffat shales, acknowledges it as probable that but one band of black graptolitic shale runs through the Llandeilo rocks of the South of Scotland, there being in this band "several distinct zones, each marked by a different assemblage of fossils, but with many species in common."

1872.
Hopkinson,
" On some Species of Graptolites from the South of Scotland," ' Geol. Mag.,' vol. ix.

A paper by Nicholson in this year " On the Migrations of the Graptolites " is mentioned here because of its historical value. His conclusions may be thus summarised :—The Skiddaw fauna was the oldest in Britain and migrated into Wales, Ireland, and America, four species only migrating northward into the Moffat area. The South of Scotland became a second centre of dispersion at the end of the Upper Llandeilo period, one migration proceeding southwards into the Lake district, founding the fauna of the Coniston group, and another going westward through Ireland to America and originating the fauna of the Hudson River group and that of the Utica Slates, while a third travelled in a south-east direction into the Silurian seas of Saxony and Bohemia.

1872.
Nicholson,
" On the Migrations of the Graptolites," ' Quart. Journ. Geol. Soc.,' vol. xxviii.

In a short note published in the ' Geological Magazine,' towards the end of 1872, Lapworth summarised his views on the age and stratigraphical relations of the Moffat Shales, as partly given in a paper read by him at the beginning of the year before the Geological Society of Glasgow (subsequently published, with additions, in the ' Transactions ' of that Society, vol. iv, p. 164). He points out that there are three main divisions, " lithologically and palæontologically separable "; which " naturally subdivide into several distinct zones, each characterised either by the exclusive possession of some well-marked species, or by the constant possession of some peculiar group of species."

1872.
Lapworth,
" Note on the Results of some Recent Researches among the Graptolitic Black Shales of the South of Scotland," ' Geol. Mag.,' vol. ix.

During the same year Nicholson published the first part (the " General Introduction ") of his ' Monograph of British Graptolitidæ,' a work unfortunately never completed. It is of especial historical interest, as giving a complete summary

1872.
Nicholson,
' A Monograph of the British Graptolitidæ.'

of the views then held by himself and others as to the nature, structure, reproduction, classification, etc., of the Graptolites. The Introduction is divided into nine chapters.

CHAPTER 1.—*Historical Notices*, summarising the progress of research among the Graptolites from the time of Linnæus to 1872.

CHAPTER 2.—*Form and Mode of Preservation of Graptolites.*—In discussing the mode of nomenclature to be employed in describing a graptolite, Nicholson here first suggests the title of " polypary " for the whole of the graptolite skeleton. He adheres to the old term " calycles," or " cellules," declining to adopt the term " hydrothecæ."

The various states of preservation in which Graptolites occur are described. The carbonaceous material of the rock in which they are generally preserved is considered to be of "animal origin," and can " hardly be ascribed to anything else except the Graptolites themselves."

CHAPTER 3.—*General Morphology of Graptolites.*—In this chapter Nicholson describes in considerable detail the structure of a typical Monoprionidian and a Diprionidian form, and the various aspects—" profile," " axial," and " scalariform " —under which they may be seen.

CHAPTER 4.—*Special Morphology of Graptolites.*—(1) *Solid Axis.* The hollow character of this structure previously suggested by him is again asserted, and this time with more certainty. The curious tube-like rod in the Rhabdopleura he considers " lends great support to this view." ⸳This also explains the apparent ability of the axis to grow independently of the rest of the polypary, and to prolong itself distally as in species of *Diplograpsus.* The distal extension of the axis is very rare in *Monograpsus*, and never exists in *Didymograpsus, Cœnograpsus*, etc. He considers that the axis is never present in the shape of a "thin, flat, apparently double plate " as believed by Barrande (' Grapt. de Bohême,' p. 4), and Hall (' Grapt. Quebec Group,' p. 22).

The character and position of the axis in *Phyllograpsus, Retiolites,* and *Trigonograpsus* are discussed ; the fact that the axis is *inside* and not outside the polypary in the two latter genera is insisted upon, while the absence of any axis whatever in the Dendroidea is remarked.

(2) *Surface Markings and Ornamentation of the Polypary.*—The striæ observed running parallel with the aperture of the cells are for the first time described as " growth lines," and good figures are given of the " pustules " visible at the base of the cell walls, especially in *M. vomerinus.*

CHAPTER 5.—*Nature and Function of the Base in the Graptolites.*—(1) *Radicle.* The form of the so-called " radicle " or " initial point " as then accepted is more clearly represented in this work than in previous papers, and it is distinguished from the " radicle " as referred to by Hall, by which he meant merely the commencement of the solid axis. Nicholson, however, did not realise the invariable presence of the initial cone-shaped body (the sicula) in all Graptolites, and he uses the term "radicle" to "signify the basal median process," "whether this consists of the solid axis alone, or of the solid axis along with the common body, or simply of the investing envelope of the latter."

(2) *Funicle* and (3) *Central Disc.*—Both these structures are described practically from Hall's point of view, and the former opinions held by Nicholson as to the "float"-like character of the latter are adhered to.

CHAPTER 6.—*Reproduction and Development.*—The various "ovarian vesicles" previously described by himself are re-described, and he changes his previous title of "Grapto-gonophores" to the more precise hydroidal title of "gonothecæ." The peculiar sac-like bodies figured by Hall (1865) and Hopkinson (1871) in specimens of *D. Whitfieldii* and *D. pristis* are considered by Nicholson to be also reproductive in function.

CHAPTER 7.—*Zoological Position of the Graptolites.*—(1) *Mode of Existence.* The free-floating habit of the true Graptolite is strongly emphasised, and consequently the systematic separation of the fixed Dendroidea from the true Graptolites is considered inevitable.

(2) *Systematic Position and Affinities.*—These questions are entered into very fully, and the various points of resemblance and difference between the Graptolites and (1) the Actinozoa, (2) the Polyzoa, and (3) the Hydrozoa are discussed in much detail. The general conclusion reached is, that they "find their *nearest* living allies in the Sertularians," and are "truly referable to the Hydrozoa," though they "cannot be placed in any living group of the Hydrozoa."

CHAPTER 8.—*Geological Distribution.*—The author's general views of the range and distribution of the Graptolites differ in no essential particulars from those expressed in previous papers, and need not here be referred to.

CHAPTER 9.—*Genera.*—The final chapter is devoted to a classification of the Graptolites, and to a description of the various genera, which are thus arranged:

Class, Hydrozoa, Sub-class Graptolitidæ.

Section A.—Monoprionidæ, Hopk.: (1) *Graptolites* or *Graptolithus;* (2) *Didymograpsus* (including *Dicellograpsus*); (3) *Tetragrapsus;* (4) *Dichograpsus;* (5) *Loganograpsus;* (6) *Pleurograpsus;* (7) *Cœnograpsus,* Hall (= *Helicograpsus,* Nich.); (8) *Cyrtograpsus;* (9) *Rastrites.*

Section B.—Diprionidæ: (10) *Diplograpsus;* (11) *Climacograpsus;* (12) *Dicranograpsus;* (13) *Retiolites;* (14) *Trigonograpsus;* (15) *Retiograpsus.*

Section C.—Tetraprionidæ: (16) *Phyllograpsus.*

Section D.—Dendroidea: (17) *Ptilograpsus;* (18) *Dendrograpsus;* (19) *Callograpsus;* (20) *Dictyonema.*

Section E.—Incertæ Sedis: (21) *Thamnograpsus:* (22) *Buthograpsus;* (23) *Inocaulis;* (24) *Corynoides.*

1872.
Allman,
'A Monograph of
the Gymnoblastic
or Tubularian
Hydroids.'

The year 1872 was also marked by the appearance of Allman's classical monograph on the 'Gymnoblastic or Tubularian Hydroids,' in which a considerable section is devoted to the discussion of the affinities of the Graptolites. Accepting without criticism the views already current among palæontologists as to the structure of these fossils, Allman draws from them some highly original and suggestive conclusions as to the homologies of the various organs. He considers that it is doubtful whether such anomalous forms as *Retiolites* and *Phyllograptus* should be included among the Graptolitidæ, while forms like *Corynoides, Dendrograptus,* and especially *Dictyonema,* are "almost certainly not Graptolites."

Acknowledging that the resemblance of the polypary in the Graptolites to the trophosome of a calyptoblastic Hydroid—Sertularian or Plumularian—is "sufficiently obvious," Allman considers that while their affinities with the Hydroida "are too decided to justify their omission from any complete exposition of the palæontological history of this group of the animal kingdom," yet their peculiar characters "necessitate the establishment for them of a separate sub-order of Hydroida." For this he proposes the name **Rhabdophora** (Rhabdos = rod), from the presence of the characteristic solid axis or virgula. The Polyzoan affinities of the Graptolites are very briefly discussed, but he admits that " were it not for the discovery of the graptolite gonosome, we should have nearly as much to say for this view as for that which would refer them to the Hydroida."

Allman discusses at considerable length the homologies of the most characteristic structure of the Graptolite, viz. the virgula or axis, "the presence of which can hardly be regarded as offering an insurmountable obstacle to the admission of the Graptolites into immediate relation with the Hydroida." He regards it, like the perisarc, as "an excretion from the cœnosarc." The distal and probably the proximal prolongation of a naked axis beyond the celluliferous part of the polypary he considers to be probably only an apparent phenomenon. He says that there is " reason to believe that the cœnosarc invested by a proper perisarc was originally continued " along the rod, but this perisarc on account of its delicacy has not been preserved.

Denticles.—Perhaps the most original part of this work is that in which Allman suggests that the structures in the living Hydroida homologous with the denticles of the Graptolites are not the hydrothecæ but the nematophores, such as those of *Aglaophenia,* which contain simple protoplasm and not true hydranths. He points out that it is not only in general form that the nematophores resemble the Graptolite-cells, but also in their method of communication with the common canal, for the continuous and open communication of the calycles of Graptolites with the main tube is very different from the constricted communication (often associated with an imperfect diaphragm) which exists between the hydrothecæ and the perisarcal tube in the Hydroids. If this suggestion is correct, then the Graptolites would be "morphologically Plumularidans in which the development of hydrothecæ had been suppressed by the great development of the nematophores," "while on the other hand, the existing Plumularidian" "would present in its nematophores the last traces of the structure of its ancient representative, the Graptolite."

Reproductive Organs.—Allman accepts the "gonosomal" nature of the reproductive sacs described by Hall and Hopkinson in *Diplo. pristis*, but rejects the "ovarian vesicles" of Nicholson, regarding their connection with the Graptolite as probably "purely accidental." Although admitting the probability that the appendages observed by Hall belong to the generative system, he is "unable to satisfy" himself that they are the remains of gonangia; indeed he thinks that they are not capsules at all but "hollow laminæ." He finds an analogue to these in the leaflets which compose the corbulæ in *Aglaophenia.* An explanation of the scarcity of these "corbulæ" in the case of the Graptolites is hinted at "in their free if not floating habit," for while no specimen of *Sargassum* in fructification has been discovered in the Sargasso Sea, the fructification of closely allied species which grow attached to rocks, etc., is not at all uncommon.

Allman in a subsequent note to this work briefly refers to Richter's views on the structure of Graptolites as given in his paper in 1871. He expresses considerable doubt as to the presence of two laminæ in the test, and of the development of the common canal from the "foot," and says that he is "unable to find in Richter's arguments any grounds for accepting the Polyzoal affinities of Graptolites," although the striæ observed by him in the Graptolites compare well with those in *Rhabdopleura.*

1873.
Hopkinson,
" On some Graptolites from the Upper Arenig rocks of Ramsay Island, St. David's.,"
' Geol. Mag.,' vol. x.

Some further facts regarding the occurrence of Graptolites in the St. David's district were furnished by Hopkinson in 1873. He gives a list of seven species from the Upper Arenig Rocks of Ramsay Island.

1873.
Hopkinson,
" On the Occurrence of Numerous Species of Graptolites in the Ludlow Rocks of Shropshire,"
' Geol. Mag.,' vol x.

The same year Hopkinson wrote a note "On the Occurrence of Numerous Species of Graptolites in the Ludlow Rocks of Shropshire," adding very considerably to our knowledge of the fauna of these rocks. Six species of *Monograptus* and two of *Ptilograptus* were new forms, viz., *M.* **capula**, *M.* **clavicula**, *M.* **incurvus**, *M.* **leintwardinensis**, *M.* **Salweyi**, *M.* **serra** (the three last alone are now identifiable), *Ptilog.* **elegans**, and *P.* **Nicholsoni**.

1873.

Nicholson,

" On some Fossils from
the Quebec Group of
Point Levis."

' Ann. and Mag. Nat.
Hist.,' ser. 4, vol. xi.

In 1873 Nicholson described some new Graptolite species collected by himself from the shales of Point Levis, *viz.,* *Dictyonema* **grandis**, *Tetragraptus* **approximatus**. In this paper the generic name **Clonograptus** (for such forms as *Gr. flexilis* and *Gr. rigidus*) was first used. This name was given by Hall at Nicholson's request. The three genera *Clonograptus*, *Loganograptus* and *Dichograptus* are clearly defined and distinguished.

Nicholson had by no means given up his idea of the ovarian character of the small capsules found in conjunction with certain species of *Monograptus,* and in this paper he names them, for the sake of convenience, **Dawsonia**. Of these he considers there are four distinct species : *D.* **acuminata**, *D.* **rotunda**, *D.* **tenuistriata**, *D.* **campanulata**.

1873.

Dames,

" Beitrag zur Kenntniss
der Gattung
Dictyonema, Hall,"

' Zeit. d. deutsch. geol.
Gesell.' bd. 25.

Fresh light was thrown on the structure of the genus *Dictyonema* by Dames in 1873. In the majority of his observations he agrees with Hall, and he considers that there is no doubt as to its graptolitic nature. In some well-preserved specimens from the Silurian Limestones in Prussia, the cross threads have broken, thus setting free the branches, and the characteristic graptolite cells with long threads coming from their apertures are easily recognisable. Such specimens demonstrate that the cells are in one row only, not in alternating rows on *both* sides as Nicholson had stated.

Dames suggests that *Dictyonema* is allied most nearly to *Dichograptus* and *Dendrograptus,* the branches of which are spread out and are not united by cross threads.

1873.

Stache,

" Die Graptolithen-
schiefer am Osternig-
Berge in Kärnten," etc.,

' Jahrb. der. k. k.
geol. Reichsanstalt,'
bd. 23.

In 1873 Stache recorded and gave brief descriptions of some species of Graptolites from Osternig in Carinthia. These include *Diplograpsus folium, D. pristis, Graptolithus (Monograpsus) proteus, G. triangulatus, G. millepeda, G. Nilssoni, Rastrites,* sp. *Retiolites, cfr. Geinitzianus.* From these fossils he concludes that the beds correspond to the Silurian Strata of the Fichtelgebirge and the Thüringian-Saxon Schiefergebirge, and also to the base of Barrande's Stage E., and the Coniston Flags of England.

1873.

Erdmann,

" Graptolith delvis
omsluten af en svafvel-
kisboll," ' Geol. Fören.
i Stockh. Förh.' bd. 1 ,
no. 11.

Erdmann figured a Graptolite which he refers to *Gr. sagittarius,* partly enclosed in a nodule of iron pyrites.

1873.

Malaise,

" Note sur la descrip-
tion du terrain Silurien
du centre de la
Belgique," ' Ann. Soc.
Malacologique de la
Belgique,' t. 8.

In the same year Malaise recorded the existence of the second Silurian Fauna, containing *Climaco. scalaris* and *Graptolithus priodon* in the Upper part of the Silurian " terrain " in the centre of Belgium; the fossiliferous horizons representing the Upper Caradoc and Lower Llandovery.

1873.

Lapworth,

" Notes on the British
Graptolites and their
Allies. On an
Improved Classification
of the Rhabdophora,"
' Geol. Mag.' vol. x.

In 1873 Lapworth published a comprehensive paper " On an Improved Classification of the Rhabdophora."

Classification.—Accepting Allman's term Rhabdophora as the collective term for all the presumed virgula-bearing Graptolites, Lapworth separates them into two sections, namely :

Section I. — Graptolitidæ (or **Graptoloidea**), in which the polypary is developed from a true sicula, the cœnosarc originates a single series of thecæ only, and the virgula is dorsal and on the exterior of the periderm.

Section II.—**Retioloidea,** in which the polypary is not developed from a siculiform " germ," the cœnosarc originates a double series of thecæ, and the epiderm is more or less supported by a framework of chitinous filaments.

The GRAPTOLOIDEA are divided into six families :

1. **Monograptidæ,** distinguished by the nucleated arrangement of the parts. Genera : 1, *Rastrites ;* 2, *Monograptus ;* 3, *Cyrtograptus.*

2. **Nemagraptidæ** (Hopk. MS.), slender forms, with ornate thecæ and irregular branches. Gen. : 4, **Leptograptus;** 5, **Amphigraptus**; 6, *Nemagraptus ;* 7, *Cœnograptus.*

3. **Dichograptidæ,** regularly branched, with prismatic thecæ. Gen. : 9, *Didymograptus ;* 10, *Tetragraptus ;* 11, *Dichograptus ;* 12, *Loganograptus ;* 13, *Clonograptus ;* 14, 15, 16, 17 (unnamed).

4. **Dicranograptidæ,** two-branched, with incurved thecæ. Gen. : 18, *Dicellograptus ;* 19, *Dicranograptus.*

5. **Diplograptidæ,** polypary duplicate. Gen. : 20, *Climacograptus ;* 21, *Diplograptus.* (Subgenera : **Glyptograptus,** *Petalograptus, Cephalograptus,* **Orthograptus.**)

6. **Phyllograptidæ,** polypary composite, quadribrachiate. Gen. : 22, *Phyllograptus.*

The RETIOLOIDEA are divided into two families :

7. **Glossograptidæ,** virgulæ coalescent, central. Gen.: 23, *Glossograptus ;* 24, *Retiograptus ;* 25, **Lasiograptus.**

8. **Retiolitidæ,** virgulæ separate, lateral. Gen.: 26, **Clathrograptus;** 27, *Trigonograptus ;* 28, *Retiolites.*

This classification is elaborated in a detailed Analytical Table.

Development.—In this paper Lapworth points out for the first time that in all the bilateral genera included in the families assigned to the Graptoloidea, the Graptolite polypary first becomes visible as a small, pointed, triangular, or rather dagger-like " germ," which he names the *sicula* (already recognised by Richter in the case of *Monograptus* and *Diplograptus* only, and denominated by him the " foot "). In the majority of bilateral examples studied a solid axis is developed in the outer wall of the sicula extending along its entire length, and a small protuberance or " bud " makes its appearance usually in the neighbourhood of the sicula, and becomes moulded into a theca: a similar bud is given off from the opposite margin, and from these primordial buds the two main branches of the polypary are evolved by a process of continuous gemmation.

But while the sicula appears always to have been present, the place of origin of the primordial bud or buds is somewhat different in the different genera, and in some forms there still remains a doubt whether the polypary is not a direct outgrowth of the major (*Dichograptus,* etc.) or of the minor (*Monograptus*) extremity of the sicula. The sicula normally ceases to grow after the primordial buds have been given off and may occasionally become imbedded, absorbed, or obsolete ; but in the great majority of cases it permanently retains its shape and form. " It is simply this persistent sicula which constitutes the axillary spine in *Dicellograptus,*" the " radicular bar " in *Cœnograptus,* and the " radicle " in *Didymograptus, Phyllograptus,*" etc.

The sicular or dorsal angle of the two main branches of the bilateral Graptolites is adopted by Lapworth as the " angle of divergence," and he points out that it ranges throughout the complete circle. It may be 0°, in which case the branches grow parallel with each other distally along the line of the sicula and coalesce by their dorsal surfaces either for the whole of their length (*Diplograptus*) or for a portion of it only (*Dicranograptus*). It may be less than 180° (*Dicellograptus*) or may exceed 180° (*Didymograptus*), or it may be as high as 360°, when the branches again coalesce (*Phyllograptus*).

Structure.—As respects the structure of the monoprionidian Graptolite Lapworth asserts that " the common portion of the polypary preserved in relief appears to be composed of the conjoined bases of the successive thecæ. These bud from each other in a single linear series. The budding orifices remain permanently open and form together a continuous tube or canal of communication for the conveyance of the common body." He restricts the application of the term *theca* to the " exterior and separable portion of the chamber—in other words, to that which is capable of being broken off from the common portion." He points out that "the line of junction of the thecæ is thickened and projects into the common canal as a rounded shelf or flange."

As respects the Diplograptidæ he states that "the polypary is merely composed of two of these monoprionidian polyparies placed back to back," their dorsal walls coalescing into a median septum, "between the two laminæ of which the duplicate virgula is imbedded." While he regards this as the normal mode of growth, he states in the sequel that in some forms of the Diplograptidæ the facts go to show that "the common canal was only partially divided, thus in effect communicating with both rows of thecæ as in *Retiolites*."

The peculiar characters of the diprionidian forms grouped as Retioloidea leads Lapworth to the view that they form a very distinct group from that of the typical Diplograptide.

The genus *Glossograptus*, originally suggested by Emmons for forms like *D. spinulosus*, Lapworth considers should be retained, as, in addition to the two long thecal spines, there occur "two opposite longitudinal rows of gigantic, isolated spurs developed along the median line of the periderm at right angles with the thecæ." A new genus, **Lasiograptus**, is proposed for such diprionidian forms as have "a connected network" of minute, inosculating threads, "almost completely surrounding the polypary." The type species (**Lasiograptus costatus**) is described but not figured in this paper.

Three other new genera are described, *i. e.*, **Clathrograptus** (type **C. cuneiformis**), **Leptograptus** (*L. flaccidus*), and **Amphigraptus** (*A. divergens*, Hall).

Lapworth regards the grouping of the first four families as given in his Analytical Table as the natural one, but that of the other families as "temporary and provisional," especially that of the Diplograptidæ.

1873.
Salter, J. W.,
'A Catalogue of the Collection of Cambrian and Silurian Fossils in the Geological Museum of the University of Cambridge.'

In 1873 Salter published an exhaustive Catalogue of the Cambrian and Silurian Fossils in the Woodwardian Museum at Cambridge, illustrated by some figures. The following species of Graptolites are figured and briefly described : (1) *Dictyonema sociale*, (2) *Graptolithus Hisingeri* (*sagittarius*), (3) *Diplograpsus mucronatus*, (4) *Phyllograptus angustifolius*, (5) *Didymograpsus geminus*, (6) *Tetragrapsus bryonoides*, (7) *Dichograpsus*, sp. *Loganograptus*, (8) *Dendrograpsus furcatule*, (9) *Graptolithus Sedgwickii*, (10) *Rastrites* (*Graptolithus*) convolutus, (11) *Rastrites peregrinus*, (12) *Diplog. folium*, (13) *Diplog. pristis*, (14) *Gr. ludensis* (*priodon*), (15) *Retiolites Geinitzianus*.

1874.
Miller, S. A.,
'Cincinnati Quart. Journ. Sci.,' vol. i, p. 343.

In the year 1874 Miller described under the name of **Megalograptus** a peculiar "large cylindrical form, not a graptolite, bearing fronds with spinose processes, and covered with cellular openings."

1874.
Etheridge, R., junr.,
" Observations on a
few Graptolites from
the Lower Silurian
Rocks of Victoria,"
' Ann. Mag. Nat. Hist.,'
ser. 4, vol. xiv.

An important addition to our knowledge of the Graptolitic fauna of Australia was made by Etheridge, junior, in 1874. Several species are recorded, figured, and described: (1) " *Tetragraptus bryonoides* " (*Tetra. serra* and *Didymo. caduceus* (*gibberulus*)), (2) *Tetrag. quadribrachiatus*, (3) *Phyllog. typus*, (4) *Loganograptus Logani* (*Goniograptus*), (5) *Climacog. sp.*, (6) *Diplog. mucronatus* (*Glossograptus* and *Lasiograptus*), (7) *D. pristis*, (8) *Didymog.? fruticosus*, (9) *D. nitidus*, (10) *D. Pantoni?* (= *D. v. fractus*), (11) *G. latus*, which he regards as a fragment of a *Dichograptus*, (12) *Graptolithus sp.*

1874.
M'Coy, Fred.,
' Prodromus of the
Palæontology of
Victoria,' dec. 1.

The same year M'Coy figured several fine examples of Graptolites from Victoria in the first part of his ' Prodromus of the Palæontology of Victoria.' The only new form named by him is a variety of *G. Logani, i. e.* (1) var. **australis**. The other species are very similar to those given in Etheridge's paper. (2) *Phyllograptus folium*, His. var. *typus*, Hall, (3) *Diplog. mucronatus* (= *Glossograptus*), (4) *D. pristis*, (5) *D. rectangularis*, (6) *Diplog.* (*Climacograptus*) *bicornis*, (7) *Graptolithus* (*Didymog.*) *fruticosus*, (8) *Gr.* (*Didymog.*) *quadribrachiatus*, (9) *Gr.* (*Didymog.*) *bryonoides*, (10) *Gr.* (*Didymog.*) *octobrachiatus*.

1874.
Lossen,
" Ueber Graptolithen
aus dem Harz," ' Zeit.
d. deutsch. geol.
Gesell.,' bd. xxvi.

Lossen, during the same year, recorded Graptolites from seven new localities in the Harz, corresponding in age to the Upper Thuringian-Fichtelgebirge graptolite horizon.

1875.
M'Coy, Fred.,
' Prodromus of the
Palæontology of
Victoria,' dec. 2.

Several additional forms of Australian Graptolites are given in the second part of M'Coy's " Prodromus," which appeared in 1875. One new species is named: (1) *Retiolites* **australis,** from the Wenlock. The other forms described and figured are: (2) *Didymog. extensus*, (3) (*Didymog.*) *caduceus* (under this name are figured specimens of *Didymog. gibberulus* and *Tetrag. Bigsbyi*), (4) *Diplog. palmeus*, (5) *Cladog. ramosus*, (6) *Cladog. furcatus*. All these occur in the Llandeilo. *Gr.* (*Didymog.*) *gracilis*, which he suggests might be made the type of a new sub-genus, is recorded from the Bala. He refuses to accept the genera *Tetragraptus*, *Dichograptus*, etc., considering that the number of stems conjoined is a " character certainly not of generic value."

1875.
Swanston,
" Graptolites, with
special reference to
those found in co.
Down," ' Proc. Belfast
Naturalists' Field
Club,' ser. 2.

In a paper read before the Belfast Naturalists' Field Club in 1875, Swanston recorded some fifteen species of Graptolites from the Silurian strata at Coalpit Bay, co. Down.

1875.
Nicholson, H. A.,
"On a new Genus and
some new Species of
Graptolites from the
Skiddaw Slates," 'Ann.
Mag. Nat. Hist.,'
ser. 4, vol. xvi.

A new genus of Graptolites, from the Skiddaw slates, was described by Nicholson in 1875, *viz.* **Azygograptus,** with its type species *A.* **Lapworthi.** This genus, which has but one branch, with the sicula in a similar position to that of *Didymograptus*, Nicholson considers to be intermediate in its characters between *Monograptus Nilssoni* and the *Nemagraptus* of Emmons. A new species of *Thamnograptus*, *T.* **Doveri,** is named. The specific name *Didymog.* **gibberulus** is given to one of the forms originally included under Salter's *D. caduceus.*

1875.
Nicholson and Lapworth,
"On the Central Group
of the Silurian Series
of the North of Eng-
land," 'Brit. Ass.
Report.'

In a joint paper read before the British Association at Bristol, Nicholson and Lapworth divided the Coniston Mudstones into two distinct groups, the Skelgill and the Knock beds, correlating the former with the Scottish Birkhill beds on account of the similarity of their Graptolites, and the latter with the Gala and Hawick beds. The Coniston Flags are regarded as the equivalents of the Denbighshire Flags and the Riccarton Beds.

1875.
Mallada,
"Sinopsis de las
Especies Fósiles que se
han encontrado en
España," 'Boletin de
la Com. Geol.'

In 1875 Mallada gave brief descriptions and figures of some species of Graptolites met with in Spain. These include : (1) *Monograpsus Nilssoni,* (2) *M. latus,* (3) *M. Halli,* (4) *M. Becki,* (5) *M. priodon,* (6) *M. convolutus,* (7) *Diplograpsus palmeus,* and (8) *D. pristis.* The figures are not original, but copies of those of previous authors.

1875.
White, C. A.

Several new species of Graptolites are described by White in the 'Report of the 100th Meridian.' These are : (1) *Phyllograptus* **Loringi** (very closely allied to *P. typus*) from Utah, from beds belonging to what he calls the Canadian Period. From beds of the Trenton Period he describes (2) *Graptolithus* (*Climacograptus*) **ramulus** (*Dicranograptus*), (3) *Diplograpsus* **hypniformis** (*D. foliaceus*), (4) *D. pristis?* (5) *D. quadrimucronatus?* These occur near Belmont, Nevada.

1875.
Richter, R.,
"Aus dem Thüring-
ischen Schiefergebirge,"
'Zeit. d. deutsch. geol.
Gesell.,' bd. xxvii,
heft 2.

The same year Richter recognised two horizons of Graptolite shales (a Lower and an Upper) in Thüringia, but noted that nearly all the forms met with in the lower occur also in the upper. Two new species are figured and described. (1) *Dicranograptus* **posthumus,** (2) *Monog.* **microdon,** and also (3) *Monog. ludensis* (= *M. Flemingii*), (4) *M. fugax,* (5) *Cyrtograptus* sp. (*C. Murchisoni?*). *M. Nilssoni* is described, but not figured.

The results of Hopkinson's and Hicks' discovery of Graptolites in the St. David's district, which had been noted from time to time, were collected and published in 1875.

1875.

*Hopkinson and
Lapworth,*
" The Graptolites of
the Arenig and
Llandeilo Rocks of St.
David's,"'Quart.Journ.
Geol. Soc.,' vol. xxxi.

The classification of the Rhabdophora, proposed by Lapworth in 1873, is adopted, but in addition a new sub-order, entitled **Cladophora,** is proposed by Hopkinson to embrace all the remaining Graptolithina.   This Sub-order is divided into two sections: (1) Thamnoidea, with its family, Thamnograptidæ and its genera *Thamnograptus* and *Butho-graptus,* and (2) Dendroidea, with its families Ptilograptidæ and Callograptidæ, the latter family including the genera *Dendrograptus*, *Callograptus*, and *Dictyograptus*, with its new Sub-genus **Desmograptus.**

An attempt is made to render the nomenclature of the Graptolites more uniform, by employing the termination " graptus " in the names of all the genera, as, for example, *Rastrograptus, Gladiograptus*, etc.

Terminology.—A section of the paper is devoted to the terminology employed in describing the Graptolites. The " sicula " is carefully distinguished from the " radicle "; but the latter term is still employed in Hall's original significance, for the " proximal prolongation of the virgula " whatever its form.   The various forms of appendages—*viz.* 1, lateral or peridermal; 2, ventral or thecal; 3, proximal and mesial; 4, apertural; 5, proximal; or 6, radicular—are defined and distinguished.

Description of Species.—The descriptions of the species are concise and accurate, but the figures are poor, and their identification is in some cases a matter of difficulty. The following genera, species, and varieties are given as new: (1) *Didymograptus* **sparsus,** (2) *D.* **Nicholsoni,** (3) *D.* **euodus,** (4) *D. indentus* var. **nanus,** (5) *D.* **furcillatus,** (6) *Tetragraptus* **Halli** (= *T. serra*), (7) *T.* **Hicksii** (*Azygograptus*), **Clematograptus** n.g., (8) *C.* **implicatus,** (9) *Climacograptus* **cœlatus,** (10) *C.* **confertus,** (11) *Phyllograptus* **stella** (= *Didymog. gibberulus*), (12) *Trigono-graptus* **truncatus,** (13) *Ptilograptus* **cristula,** (14) *P.* **Hicksii,** (15) *P.* **acutus,** (16) *Dendrograptus* **persculptus,** (17) *D.* **arbuscula,** (18) *D.* **Ramsayi,** (19) *D.* **serpens,** (20) *D.* **flexuosus** var. **recurvus,** (21) *Callograptus* **radiatus,** (22) *C.* **radicans,** (23) *Dictyograptus* **Homfrayi, Desmograptus** n.g., (24) *D.* **cancellatus.**

Other species described and figured are: (25) *Didymograptus affinis*, (26) *D. extensus*, (27) *D. indentus*, (28) *D. patulus*, (29) *D. Murchisoni*, (30) *D. pennatulus*, (31) *Tetragraptus quadribrachiatus*, (32) *T. serra*, (33) *Nemagraptus capillaris*, (34) *Dicellograptus moffatensis*, (35) *Diplograptus dentatus*, (36) *D. foliaceus*, (37) *D. tricornis*, (38) *Glossograptus ciliatus*, (39) *Trigonograptus ensiformis*, (40) *Dendro-graptus diffusus*, (41) *D. divergens*, (42) *D. flexuosus*, (43) *Callograptus elegans*, (44) *C. Salteri*, (45) *Dictyograptus irregularis.*

Range and Distribution.—The range and distribution of these species are discussed. The division of the Arenig as proposed by Hicks into Lower, Middle, and Upper is adopted, and the Arenig and Llandeilo beds are compared as respects their Graptolites with their American and British equivalents.

1875.
Tromelin et Lebesconte,
" Essai d'un Catalogue
Raisonné des Fossiles
Siluriens," etc., ' Assoc.
Franc. pour l'avance-
ment des Sciences.'

recorded.

From the " Schistes ardoisiers " Tromelin and Lebesconte record *Didymograpsus Murchisoni*, *Gr. Hisingeri*, and *Gr. Sedgwicki.* In the lowest part of the third fauna, namely the zone of ampelites and phtanites of Anjou and Lower Loire, they recognise *Gr. colonus*, *Gr. Becki*, *Gr. Nilssoni*, *Gr. spiralis*, and *Diplo. folium*, while from the upper zone or " ampelite nodules " *Gr. Bohemicus*, *Gr. Becki*, and *Gr. priodon* are

1876.
Linnarsson,
" On the Vertical
Range of Graptolites
in Sweden," ' Geol.
Mag.,' dec. 2,
vol. iii.

In a valuable paper " On the Vertical Range of the Graptolites in Sweden," Linnarsson showed that Graptolites occur on at least six distinct horizons in the Cambrian and Silurian rocks of Norway and Sweden each horizon being marked by different forms, and he satisfactorily parallels them with known horizons in Britain : (1) the Olenus Shales with the Upper Dolgelly of North Wales; (2) the Dictyonema Shales with the Dictyonema-bearing beds in Britain and the Baltic provinces; (3) the Lower Graptolitic Shales with the Skiddaw and Quebec groups; (4) the Middle Graptolitic Shales with the Moffat series; (5) the Upper Graptolitic Shales with the Coniston and Gala beds; (6) the Higher Silurian with the Wenlock and Ludlow.

1876.
Nicholson,
" Notes on the Corre-
lation of the Grapto-
litic Deposits of Sweden
with those of Britain,"
' Geol. Mag.,' dec. 2,
vol. iii.

C. annulatus.

A general review of Linnarsson's paper, by Nicholson, accompanied it. In addition to discussing the various British strata corresponding to Linnarsson's five Swedish zones, Nicholson gives diagnoses and figures of four new genera from the Skiddaw Slates, previously included by him under the single genus *Dichograptus.* These are (*a*) **Trichograptus**, type *T. fragilis*; (*b*) **Temnograptus**, *T. multiplex*; (*c*) **Schizograptus**, *S. reticulatus*; (*d*) **Ctenograptus**,

1876.
M'Coy,
" On a new Victorian
Graptolite," ' Ann.
Mag. Nat. Hist.,' ser.
4, vol. xviii.

A new species of Graptolite was described and figured by M'Coy from the Bendigo beds of Victoria, under the name *Didymograpsus* **Thureaui**, but he suggests that, if the genus *Didymograpsus* is to be restricted to the two-branched forms, this species should be made the type of a new genus— **Goniograptus**.

1876.
Törnquist,
" Nyblottad profil med
Phyllograptus skiffer i
Dalarne," ' Geol. Fören.
i Stock. Förh.,' no. 36,
bd. 3, no. 8.

A short stratigraphical paper appeared this year by Törnquist, on the Phyllograptus shales of Dalarne and their fossils.

1876.
Lapworth,
" On Scottish Mono-
graptidæ," ' Geol.
Mag.,' dec. 2, vol. iii.

In 1876 Lapworth described in detail the various species of Scottish Monograptidæ, revising the synonymy and classification, and figuring the species, many of which were new.

Synonymy.—Four genera are included in the family of the Monograptidæ, *Monograptus, Cyrtograptus,* and *Rastrites,* and a new genus **Dimorphograptus.** The author discusses at some length the respective merits of the generic names *Graptolithus, Monoprion,* and *Monograptus,* for those unilateral Graptoloidea in which the thecæ are in contact with one another, and decides in favour of *Monograptus.* He retains the genus *Rastrites,* and notes the presence of the virgula in this genus.

Development.—The development of the Monograptidæ is here stated to be similar to that of the bilateral Graptoloidea, but "the polypary originates from a point near the broad end of the sicula," and " grows backward along the distal portion of the sicula itself, to which it adheres." He considers it " probable that all the Graptoloidea ought to be regarded as colonies of siculæ, held together by a common body."

Description of Species.—The following species are described and figured, and the localities and ranges of each are given.

Genus *Rastrites.*—(1) *R. peregrinus,* (2) *R. capillaris,* (3) *R. maximus,* (4) *R.* **distans,** (5) *R. peregrinus* var. **hybridus.**

Genus *Monograptus :*

Group I.—Type *M. Nilssoni.* (6) *M. Nilssoni,* (7) *M. intermedius,* and (8) var. **involutus,** (9) *M.* **gregarius,** (10) *M. attenuatus,* (11) *M. Salteri,* (12) *M.* **argutus,** (13) *M. tenuis,* (14) *M.* **Sandersoni,** (15) *M.* **concinnus.**

Group II.—Type *M. Hisingeri.* (16) *M. Hisingeri* and (17) var. **jaculum,** (18) *M.* **cyphus,** (19) *M.* **leptotheca,** (20) *M. vomerinus.*

Group III.—Type *M. Halli.* (21) *M. Halli,* (22) *M.* **Riccartonensis,** (23) *M.* **galaensis,** (24) *M. priodon,* (25) *M. Flemingii,* (26) *M. colonus,* (27) var. *dubius.*

Group IV.—Type *M. Sedgwickii.* (28) *M. Sedgwickii,* (29) *M. convolutus,* (30) var. (*a*) **communis,** (31) (*b*) *fimbriatus,* (32) (*c*) *proteus,* (33) (*d*) *spiralis,* (34) *M. triangulatus,* (35) *M. turriculatus.*

Group V.—Type *M. lobiferus.* (36) *M. lobiferus,* (37) *M. Becki,* (38) *M. Clingani,* (39) *M.* **runcinatus,** (40) *M. Barrandei,* (41) *M. exiguus,* (42) *M.* **crispus.**

Genus *Cyrtograptus.*—(43) *C.* **Carruthersi,** (44) *C.* **Grayi.**

Genus *Dimorphograptus.*—(45) *D.* **elongatus** and (46) *D.* **Swanstoni.**

The discovery of the genus *Dimorphograptus,* intermediate between *Monograptus* and *Diplograptus,* led Lapworth to abandon his former theory that in *Diplograptus,* etc., the sicula gave origin to two buds, and to suggest a simpler one. " According to this new theory the sicula in *all* the Graptoloidea throws off a single bud only, and this theoretically invariably originates a single cœnosarcal tube." In *Diplograptus*, *Pleurograptus*, etc., this is divided immediately after origin; but in *Tetragraptus* division takes place twice. The horizontal bar between the proximal ends of the two pairs of primary branches in *Tetragraptus*, etc., is Hall's true "funicle," and is the only non-polypiferous portion of the polypary.

Lapworth considers that the structure of *Dimorphograptus* "lends support to the theory that the Monograptidæ are the direct descendants of the Diplograptidæ, and not of any of the compound monoprionidian genera."

Range and Distribution.—The paper concludes with a general account of the range and distribution of the Monograptidæ in Britain and abroad. The author recognises three successive specific groups, viz., those of the (1) Birkhill Shales (= Coniston Mudstones = Lower Llandovery), (2) Gala and Girvan, (3) Upper Silurian of Riccarton and the Pentland Hills (= Coniston Flagstones and Wenlock and Ludlow Beds). Each of these possesses a well-marked and distinctive assemblage of Monograptidæ, and the vertical distribution of species in Scotland is in complete agreement with that in England, Ireland, and Europe.

He emphasises the fact that not a single species of any genus of the Monograptidæ occurs in beds lower than the Bala limestone and therefore that the family is exclusively a Middle and Upper Silurian one.

1876.
Lapworth,
'Catalogue of Western Scottish Fossils.'
"*Graptolites.*"

A complete illustrated list of the Scottish Graptolites recognised in 1876 was given by Lapworth in the ' Catalogue of Western Scottish Fossils ' prepared for the Meeting of the British Association at Glasgow. Among them were several new species which were figured but not described. It will suffice here to mention the figured forms then new to science : (1) *Diplograptus pristis* var. **truncatus**, (2) *D. foliaceus* var. **calcaratus**, (3) *D.* **perexcavatus**, (4) *D. quadrimucronatus* var. **spinigerus**, (5) *D.* **aculeatus**, (6) *Climacograptus* **tubuliferus**, (7) *C. bicornis* var. **tridentatus** and (8) var. **peltifer**, (9) *C.* **Scharenbergi**, (10) *C.* **cœlatus**, (11) *Cœnograptus* **nitidulus**, (12) *C.* **pertenuis**, (13) *C.* **explanatus**, (14) *Amphigraptus* **radiatus**, (15) *Didymograptus* **superstes**, (16) *Dicranograptus* **ziczac** and (17) var. **minimus**, (18) *Dicellograptus* **pumilus**, (19) *D.* **caduceus**, (20) *D. Forchammeri* var. **articulatus** and (21) var. **flexuosus**, (22) *Corynoides* **curtus**, (23) *Thamnograptus* **scoticus**, and (24) *Dictyograptus* **moffatensis**.

1874–6.
Dairon,
" Notes on the Silurian Rocks of Dumfriesshire and their Fossil Remains," ' Trans. Geol. Soc. Glasgow.'

Two papers on the Silurian rocks of Dumfriesshire and their fossil remains were read by Dairon before the Geological Society of Glasgow in 1874—1876. In these a general account of the Graptolites is given, their affinities, structure, development, etc. A number of species are given and figured, and three new ones are figured but not described : (1) *Thamnograptus* **crucifer**, (2) *Retiolites* **branchiatus**, and (3) *Dicellog.* **guilloche.**

1876.
Swanston & Lapworth,
" On the Silurian
Rocks of Co. Down,
with Appendix,"
' Proc. Belf. Nat. Field
Club.'

The results of the researches of Swanston among the Graptolite-bearing Silurian rocks of co. Down, together with a list of the Graptolites that they afford, were published in 1876—1877. Many new localities for Graptolites in Ireland are given, and the existence of the representatives of the Glenkiln shales, Lower and Upper Hartfell, Lower and Upper Birkhill (with the exception of the *Rastrites maximus* zone), and the Gala beds is proved.

To Swanston's paper there is an Appendix by Lapworth in which he figures and describes, mainly from his own collections, a large number of the species common to Ireland and Scotland. The majority of these were forms already named, while a few new forms are described in addition. (Those marked S.F. below had already been figured by Lapworth in his ' Catalogue of Western Scottish Fossils,' and are here described for the first time.) The list includes: (1) *Rastrites peregrinus*, (2) var. *hybridus*, (3) *Monograptus triangulatus*, (4) *M. spiralis*, (5) var. *fimbriatus*, (6) var. *communis*, (7) var. *proteus?*, (8) *M. Sedgwicki*, (9) *M. turriculatus*, (10) *M. crispus*, (11) *M. exiguus*, (12) *M. Barrandei*, (13) *M. runcinatus*, (14) *M. lobiferus*, (15) var. **pandus**, (16) *M. priodon*, (17) *M. riccartonensis*, (18) *M. galaensis*, (19) *M.* **M'Coyi**, (20) *M. Hisingeri* var. *jaculum*, (21) *M. cyphus*, (22) *M. leptotheca*, (23) *M. concinnus*, (24) *M. gregarius*, (25) *M. argutus*, (26) *M. Sandersoni*, (27) *M. tenuis*, (28) *M. attenuatus*.

Dimorphograptus.—(29) *D. Swanstoni*, (30) *D. elongatus*.

Cephalograptus.—(31) *C. cometa.*

Diplograptus (in the sub-genus *Glyptograptus* (*G. tamariscus*) Lapworth believes that there is no vertical septum, the cœnosarcal tube being apparently undivided).—(32) *D. acuminatus*, (33) *D. modestus* (S. F.), (34) *D. sinuatus*, (35) *D. tricornis*, (36) *D. angustifolius*, (37) *D. tamariscus*, (38) *D. dentatus*, (39) *D. Hughesi*, (40) *D. insectiformis*, (41) *D. folium*, (42) *D. truncatus*, (43) *D. foliaceus*, (44) *D. vesiculosus*, (45) *D. quadrimucronatus*, (46) *D. Whitfieldi*, (47) *D. (Lasiograptus?) mucronatus*, (48) *D. (Hallograptus) bimucronatus.*

Glossograptus.—(49) *G. Hincksii.*

Lasiograptus.—(50) *L.* **margaritatus**, (51) *L. Harknessi.*

Clathrograptus.—(52) *C. cuneiformis* (S. F.).

Retiolites.—(53) *R.* **fibratus** (54) *R. perlatus?* (55) *var.* **Daironi** (56) *var.* *obesus.*

Climacograptus.—*C. scalaris*, var. *tectus*, (56) *var.* **normalis** (57) var. *rectangularis*, (58) *var.* **caudatus**, (59) *var. tubuliferus* (S.F.), (60) *C. Scharenbergi* (S.F.), (61) *C. bicornis*, (62) var. *tridentatus* (S.F.), (63) *var. peltifer* (S.F.), (64) *C. cœlatus*, (65) *C. Wilsoni* (S.F.), (66) *C. perexcavatus* (S.F.), (67) *C. innotatus.*

Dicranograptus.—(68) *D. ramosus*, (69) *D. Nicholsoni*, (70) *D. formosus*, (71) *D. clingani*, (72) *D. ziczac* (S.F.), (73) var. *minimus* (S.F.).

Dicellograptus.—(74) *D. elegans,* (75) *D. Forchammeri,* (76) *D. moffatensis,* (77) var. *divaricatus,* (78) *D. caduceus* (S.F.).

Didymograptus.—(79) *D. superstes* (S.F.)

Leptograptus.—(80) *L. flaccidus.*

Cœnograptus.—(81) *C. gracilis,* (82) *C. surcularis,* (83) *C. pertenuis* (S.F.).

Thamnograptus.—(84) *T. typus ?*

Dictyonema.—(85) *D. moffatensis* (S.F.).

1878.
Linnarsson,
" Om Graptolitskiffern
vid Kongslena i Vester-
götland," ' Geol. Fören.
Förh.,' no. 41, bd. 3.

In 1877 Linnarsson recorded the presence of a large number of species of Graptolites in the " *Lobiferus* shales" at Kongslena, in Vestergötland.   He only gives the names of those already described, but mentions that there are others, probably new.

He discusses at considerable length the probable equivalents of these beds in Great Britain, Bohemia, etc.

1878.
Haupt,
" Die Fauna des Grap-
tolithengesteines,"
' Neues Lausitzisches
Mag.' bd. 54.

Some addition to our knowledge of the fauna of the " Graptolithengestein " of North Germany was made by Haupt in 1878.   He records 14 Graptolites, and the majority are described and figured.   One genus and species is new. He describes, but gives no figures of, *Monograpsus priodon, M. bohemicus, M. colonus* and *M. sagittarius.*   He describes and figures (1) *M. distans ?* (= *M. scanicus* Tullb.), (2) *M. Salteri ?* Gein., (3) *M. Nilssoni,* (4) *M. proteus ?* (5) M. sp. 1, and (6) sp. 2, (unidentifiable), (7) *M. turriculatus,* (8) *Rastrites* sp., (9) *Dendrograpsus* sp. (10) **Quadruplograpsus rhomboidalis** is the name given by him to a form which he supposes to have four rows of cells.   (The identity of this form is dubious, and it would seem that the four-rowed character is only apparent, not real.)

From other isolated blocks in the Drift he figures the new forms (11) *Retiolites* **gracilis,** (12) *Retiolites* sp., (both Diplograptid, probably of the amplexicaul type) and (13) *Diplograpsus* sp. (which resembles *Cryptograptus*).

1878.
M'Coy,
' Prodromus of the
Palæontology of
Victoria,' dec. 5.

The fifth part of M'Coy's ' Prodromus ' appeared in 1878. Good specimens of *Didymograptus* (*Goniograptus*) *Thureaui* are figured, and one of *Didymograptus* (*Tetragraptus*) *Headi.*

1878.
Lapworth,
" The Moffat Series,"
' Quart. Journ.
Geol. Soc.,'
vol. xxxiv.

In this year Lapworth published a paper on the " Moffat Series of South Scotland."   In this memoir the species of Graptolites are for the first time employed throughout in the grouping and correlation of the stratigraphical subdivisions; and the conclusion is drawn that this Moffat Series, which is only some 300 feet in collective thickness in the typical area, embraces three formations—the Glenkiln, Hartfell, and Birkhill—answering respectively to the Upper Llandeilo, Bala, and Lower Llandovery formations of Southern Britain and is separable into at least eleven Graptolitic zones.

1878.
Gümbel,
" Einige Bemerkungen über Graptolithen," ' Neues Jahrb.,' 1878.

An attempt to study the minute structure of Graptolites by means of dissolving away the calcareous matrix, and thus isolating the Graptolite, was made in 1878 by Gümbel. The experiments were made with specimens of *M. priodon*, and accurate drawings of some of the cells, both in relief and in section, are given in his paper. Gümbel notices particularly the thickening of the cell-wall in three places—(1) at its proximal end, (2) at the point of junction with the cell-wall next above, and (3) at the edge of the aperture.

As a result of his tests he believes that the skeleton is formed of a " structureless, membranous substance, consisting of several thin layers," like the chitinous skeleton of the Sertularia.

He records the existence of two distinct Graptolite horizons in the Fichtelgebirge. He notes the discovery of *Cyrtograptus ? Murchisoni* and *Pleurograptus* cfr. *linearis*, from the Upper Graptolite shales.

1878.
Richter,
" Notize über die Graptoliten d. H. Gümbels," ' Neues Jahrb.'

Gümbel's paper was briefly referred to and criticised by Richter in another number of the ' Neues Jahrbuch ' for the same year, and his conclusions are in the main accepted.

1878.
Kayser,
" Die ältesten Devon. Ablagerungen des Harzes," ' Abhandl. zur geol. Specialkarte v. Preussen u. d. Thüringischen Staaten, bd. 2, heft 4.

In this year also, Kayser described and figured all the Graptolite forms hitherto recorded from the Harz Mountains from the beds lying at the top of the so-called " Untere Wieder Schiefer," and below the Haupt Quartzite. These Graptolite-bearing beds he still considered to be of Lower Devonian age. The forms figured are only fragmentary, but they appear to be correctly identified on the whole; they include : (1) *M. Nilssoni*, (2) *M. convolutus*, (3) *M. Halli*, (4), *M. colonus*, (5) *M. dubius*, (6) *M. sagittarius*, and (7) *M. jaculum?* He considers that *M. dubius* is identical with Roemer's species *M. Jüngsti, M. polydonta, M. obliquo-truncatus,* and *M. subdentatus* (pars).

1878–1879.
Spencer,
" Graptolites of the Niagara formation," ' Canadian Naturalist.'

A large number of genera and species of the Dendroidea are described by Spencer, from the Niagara formation. No figures are given.

The new genera are **Calyptograpsus, Rhizograpsus, Acanthograpsus**. The new species are *Calyptograpsus* **cyathiformis,** *C.* **subretiformis,** *Rhizograpsus* **bulbosus,** *Dictyonema* **tenella,** *Acan-*

thograpsus **Granti**, *Ptilograpsus* **foliaceus**, *Thamnograpsus* **bartonensis**, and *Callograpsus* **niagarensis.**

1879.
Linnarsson,
"Jakttagelser öfver de graptolitförande skiffrarne i Skåne,"
'Geol. Fören, Förh.,'
no. 50, bd. 4.

In 1879 Linnarsson described the various graptolite horizons of Sweden, and compared them with those of Great Britain and elsewhere. He recognises three main divisions : (1) the Lower, (2) the Middle, and (3) Upper Graptolite shales, corresponding in the main to (1) Arenig, (2) Llandeilo-Hartfell, (3) Llandovery to Lower Ludlow.

1879.
Linnarsson.
"Om Gotlands Graptoliter," 'Kongl. Vetensk-Akad. Förh.,'
no. 5.

In the same year Linnarsson described three species of Graptolites from the Visby and Middle Gotland Groups of the Silurian formation in Gotland. The Graptolites, though rare, are well preserved in limestone, and Linnarsson was able to give detailed descriptions of their structure, illustrated by excellent drawings.

Structure.—In his description of *Monograpsus priodon* he calls attention to the variation in form of the proximal portion of the polypary, as figured by different authors ; and he inclines to the view that the straight shape is the more typical.

1879.
Zittel.
'Handbuch der Palæontologie'
(1876–80), band I.

An admirable summary and digest of the chief results of research and opinion with respect to the Graptolites to the close of the year 1878 was given by Zittel in the second part (issued in 1879) of his 'Handbuch der Palæontologie.' He assigns the Cladophora (Hopk.) to the Campanulariæ, and classes the Rhabdophora (Allmann) as a distinct sub-order of the Hydroida, but under the title of the Graptolithidæ. The structure of the polypary in the Graptolithidæ is described and illustrated ; and the mode of existence and zoological position, etc., of the Graptolites in general discussed. The classification adopted agrees essentially with that of Lapworth (1873). The geological ranges of the various genera are noted, and it is shown that six main Graptolitic horizons or zones are already recognisable in Europe and North America. The text is illustrated by several good figures.

His observations on the structure of *Retiolites Geinitzianus* are somewhat indefinite, owing to the fact that he had only one specimen for examination. The chitinous network he believes to be quite superficial, and the stronger strands which mark the boundaries of the thecæ probably mere threads, not lamellæ as is the case with other Graptolites, so that the interior of the polypary is not divided up into separate thecæ. No virgula was observed by him in this Gotland specimen, and he is inclined to doubt the existence of two virgulæ as described by Barrande. A species of *Dictyonema* is also recorded by him, but is not described or figured.

1879.
Törnquist,
" Några Iakttagelser
öfver Dalarnes Grapto-
litskiffar," 'Geol. Fören.
Förh.,' bd. iv, no. 4.

A paper by Törnquist on the Graptolite shales of Dalarne is mainly stratigraphical, but two new species are described and figured, *viz. Didymograptus minutus* and *Phyllograptus* **densus,** while a new variety of *Diplograptus palmeus* var. **superstes,** and one of *M. spiralis* var. **subconicus** are recorded.

In Dalarne he recognises the following graptolite horizons: 1, Phyllograptus skiffer with *Phyllograptus, Tetragraptus,* etc.; 2, Trinucleus skiffer, corresponding to the Hartfell; 3, Lobiferus skiffer, with five zones: (*a*) *M. leptotheca,* (*b*) *Diplo. cometa,* (*c*) *M. Sedgwickii,* (*d*) *M. turriculatus,* (*e*) *M. priodon* and *D. palmeus* var. *superstes;* and 4, Retiolites skiffer with *Ret. Geinitzianus* and *M. spiralis* var. *subconicus.*

1879—1881.
Quenstedt,
' Petrefactenkunde
Deutschlands,' bd. vi,
Graptolithi.

In his text-book of fossils Quenstedt gives a general account of the Graptolites, and figures of many forms, but the majority of these are copies from earlier authors. He classes the Graptolites among the corals, between the " Rind " corals and the Bryozoa, and retains the old nomenclature.

Description of Species.—(1) *Dictyonema flabelliforme* he describes under the old name of *Gorgonia.* His figures of (2) *Graptolithus serratus* include forms of the group *M. colonus,* etc. His (3) *Gr. ludensis* includes *M. priodon* and *M. testis,* his (4) *Gr. colonus* = *M. Roemeri, M. dubius, Didymo. pennatulus* and *D. Murchisoni,* his (5) *Gr. scalaris* embraces *M. Nilssoni, M. bohemicus,* etc.

The Didymograpti noticed by him include *Tetragraptus,* and all the allied branched forms, together with *Dendrograptus,* while *Diplograptus* and *Climacograptus,* etc., are classed as *Digrapti.*

1880.
Lapworth,
" On new British
Graptolites," ' Ann.
Mag. Nat. Hist.,' ser. 5,
vol. v.

In 1880 Lapworth described and figured a number of new British species of Graptolites, revised a few forms already recognised but of which little was known, and he suggested some new generic and sub-generic names. The species figured include the following : (1) *Monograptus leintwardinensis,* Hopk. MS., (2) *M. Salweyi,* Hopk. MS., (3) *M. Roemeri,* (4) *M. colonus,* (5) *M. galaensis* var. **basilicus,** (6) *M.* **crenularis,** (7) *M.* **crassus,** (8) *M. riccartonensis,* (9) *M. Flemingii,* (10) *M. Hisingeri* var. **nudus,** *M. Salteri,* (11) *Cyrtograptus* **Linnarssoni,** (12) *Azygograptus* **cœlebs,** (13) *Dicellograptus* **complanatus,** (14) *Dicello.* **intortus,** (15) *Dicello.* **patulosus,** (16) *Dicello. divaricatus* var. **rigidus.** The new genus **Bryograptus** is founded, and two new species described belonging to this genus, viz. (17) *Bryo.* **Kjerulfi** and (18) *Bryo.* **Callavei.**

Under *Diplograptus* he describes Nicholson's (19) *D. physophora,* (20) *D.* **socialis,** nov. d., (21) *D.* **(Glyptograptus) euglyphus,** (22) *D. perexcavatus, D. rugosus, Emm*? (23) *Climacog.* **confertus.**

He proposes the new sub-generic name **Idiograptus** for forms typified by (24) *D.* **aculeatus.**

The peculiar structure presented by the curious forms (25) *D. tricornis, D. marcidus*, and *D. Etheridgii* is described and explained, the new generic title **Cryptograptus** is suggested for them, and a new variety (26) var. **Schäferi** is figured.

A new species of *Lasiograptus*, (27) *L.* **retusus** is figured and described.

1880.
Marr.
" On the Pre-Devonian
Rocks of Bohemia,"
' Quart. Journ. Geol.
Soc ,' vol. xxxvi.

In a paper giving the results of his personal researches among the Pre-Devonian Rocks of Bohemia, Marr shows that the Band E. *e.* 1 of Barrande includes three distinct Graptolitic zones (1, *Diplograptus* zone; 2, *priodon* zone; 3, *colonus* zone), and that the same three zones occur in the so-called "colonies" in the same order, thus affording " grounds for the supposition that these are only portions of the Band E. *e.* 1 faulted down among the grits and shales at the summit of the Cambrian (Ordovician) Series."

1879 – 1880.
Lapworth,
" On the Geological
Distribution
of the Rhabdophora,"
' Ann. and Mag. of Nat.
Hist.,' ser. 5, vols.
iii, iv, v, and vi.

During the years 1879—1880 appeared Lapworth's paper "On the Geological Distribution of the Rhabdophora," which was published in parts in the 'Annals and Magazine of Natural History.'

This memoir is devoted to an exhaustive examination and discussion of the available facts bearing upon the distribution and vertical range of all the known Graptolite species in Britain and abroad, with a view of correcting the prevalent neglect of these fossils by geologists in general and of showing their importance as constituting probably the most reliable chronological indices available in working out the detailed stratigraphy of the Lower Palæozoic formations.

PART I.—*Historical.*—In the introductory part Lapworth points out and illustrates the geological and palæontological difficulties which had caused these fossils to fall into disrepute, enters into a critical discussion of previous opinions, and summarises the latest views upon the subject.

PART II.—*Data.*—The second part of the work is devoted to the fixation of the actual localities and horizons in Britain and abroad from which known species of Graptolites had been obtained, so far as could be gathered from previous publications and personal researches, and the special association of forms is given in each case.

PART III.—*Results.*—In the third part the results deducible from the foregoing data are discussed and tabulated both from the stratigraphical and the palæontographical points of view. In the geological section the fauna of each Graptolite-bearing formation is fixed, and in the palæontological section each Graptolite family is dealt with and the ranges of the component genera and species deduced and shown in illustrative Tables.

PART IV.—*Conclusions.*—In the fourth part Lapworth points out that stratigraphically and morphologically the Graptolites arrange themselves into four groups :

(1) Monograpta (including the family of the Monograptidæ only).

(2) Diplograpta (including the families of the Diplograptidæ, Lasiograptidæ, and Retiolitidæ).

(3) Didymograpta (including the Dichograptidæ and Phyllograptidæ).

(4) Dicellograpta (including the Dicranograptidæ and Leptograptidæ).

He gives a table showing the Vertical Distribution of the component genera, and he considers the following propositions as established :

(1) " The Rhabdophora, or true Graptolites, are exclusively Lower Palæozoic fossils, coming into visible existence in the Upper Cambrian, and disappearing from sight in the Upper Silurian."

(2) " They attain their maximum, both in genera and species, about the middle of this range, i. e., in the Llandeilo formation ; and there is a gradual decrease in forms in proportion as we pass upwards or downwards from this horizon."

(3) " The three grand groups of the Didymograpta, Dicellograpta, and Monograpta are so restricted in their vertical range that each distinguishes a certain portion of the ascending succession of formations. The Didymograpta are essentially Lower Ordovician fossils, the Dicellograpta Upper Ordovician, while the Monograpta are confined exclusively to the Silurian proper."

(4) " With but two exceptions, each of the families of the Rhabdophora ranges through a fraction only of the entire succession of the Lower Palæozoic rocks, nowhere exceeding in vertical extent that of an entire system. The Dichograptidæ are Upper Cambrian and Lower Ordovician fossils ; the Phyllograptidæ are exclusively Arenig ; the Leptograptidæ and Dicranograptidæ are essentially Upper Ordovician ; while the Lasiograptidæ are as rigidly confined to the Ordovician itself as the Monograptidæ are to the succeeding Silurian."

(5) " Among the genera this limitation in time is carried out even more minutely. *Loganograptus*, *Tetragraptus*, *Dichograptus*, *Retiograptus*, and several others are exclusively Arenig genera. *Pleurograptus*, *Amphigraptus*, *Cœnograptus*, etc., are peculiar to the Bala. *Rastrites* distinguishes the Valentian, and *Cyrtograptus* the Salopian."

(6) " Descending to the species of the Rhabdophora, we find that they are so restricted in vertical distribution that few have a more extended range than that which is covered by a single formation in the vertical series ; while the vast majority are peculiar to a single sub-formation, or mark certain special horizons outside of which they are unknown. The forms which have the greatest longevity present us with the greatest number of recognisable varieties, while the species of shorter range rarely show any notable departure from the primitive type."

(7) " The ascertained restriction of the divisions, families, and genera of the

Rhabdophora in time necessarily gives to the collective Graptolitic fauna of each of the subsystems or major formations of the Lower Palæozoic rocks a special and distinctive aspect that renders it capable of immediate identification all over the world. The Arenig division is recognisable at a glance by its crowds of Phyllograptidæ and Dichograptidæ; the Bala, by the absence of these families and the presence of multitudes of Dicellograpta and Diplograpta; the Valentian, by the absence of the former and the presence of the latter in association with Monograptidæ; and the higher Silurians by the absence of the Diplograptidæ and the presence of Monograptidæ alone."

(8) " The further restriction in time and vertical extension of the species and varieties of the Rhabdophora places in our hands the material available for a more minute subdivision of the formations of the Lower Palæozoic rocks than has hitherto been attempted. These subdivisions, or Graptolite horizons, answer roughly to the Ammonite zones of the Jurassic rocks of Europe, and will, in all probability, prove of equal value in the correlation of widely separated deposits." At present the following zones are recognisable, many of them of extraordinary geographical range :

Upper Cambrian.

1. Zone of *Bryograptus Callavei.*

Lower Ordovician.

2. Zone of *Tetragraptus bryonoides.*
3. Zone of *Didymograptus bifidus.*
4. Zone of *Didymograptus Murchisoni.*

Upper Ordovician.

5. Zone of *Cænograptus gracilis.*
6. Zone of *Dicranograptus Clingani.*
7. Zone of *Pleurograptus linearis.*
8. Zone of *Dicellograptus complanatus.*
9. Zone of *Dicellograptus anceps.*

Silurian System.

10. Zone of *Diplograptus acuminatus.*
11. Zone of *Diplograptus vesiculosus.*
12. Zone of *Monograptus gregarius.*
13. Zone of *Monograptus spinigerus.*
14. Zone of *Rastrites maximus.*
15. Zone of *Monograptus exiguus.*
16. Zone of *Cyrtograptus Grayæ.*

(Wenlock-Ludlow Series.)

17. Zone of *Cyrtograptus Murchisoni*.
18. Zone of *Cyrtograptus Linnarssoni*.
19. Zone of *Monograptus testis*.
20. Zone of *Monograptus Nilssoni*.

(A table is given showing the Geographical Range of these zones.)

(9) "The several zones common to two or more regions occupy invariably the same relative position with respect to each other and the same vertical place in the ascending series of formations. Hence we have no choice but to regard them as homotaxially or synchronologically identical."

(10) "In the face of these results, the host of proofs formerly supposed to be afforded by the abnormalities of the vertical distribution of the Graptolithina, in favour of the doctrines of migration and colonies, vanish. We have at present no evidence to show that any Graptolite group, or even a single species or variety, made its appearance at an earlier date in one region than in another; and, as a consequence, the place of its origin and the direction of its extension in space are at present equally incapable of recognition. The Graptolite appears to be as restricted in its vertical range, and as widely extended in its horizontal distribution, as any known form of life hitherto recognised as existent in Palæozoic times. It is one of the most suitable of fossils for the purposes of the working geologist and systematist; its short vertical range affording elements for the subdivision of the accepted Lower Palæozoic formations into their component zones; its wide horizontal distribution allowing of the exact parallelism of synchronous deposits in areas now geographically separated; and its universal dissemination rendering it easy of collection and study."

1880.
Tullberg,
" Några *Didymograp-tus*-arter vid Kiviks-Esperöd," ' Geol. Fören. Förh.,' bd. 5.

Three papers by Tullberg appeared in the year 1880. In the first he describes and figures five new species of *Didymograptus* from the Lower Graptolite shales at Kiviks-Esperöd : *D.* **balticus**, *D.* **vacillans**, *D.* **pusillus**, *D.* **filiformis**, and *D.* **suecicus**.

1880.
Tullberg.
" Tvenne nya grapto-litslägten," ' Geol. Fören. Förh.,' bd. 5.

In a second paper he describes two new genera and species of Graptolites: (1) **Lonchograptus ovatus** from the *Didymog. geminus* zone and **Janograptus laxatus** from the *D. mucronatus* zone at Fågelsang.

1880.
Tullberg,
" Lagerföljden vid Röstånga," ' Geol. Fören. Förh.,' bd. 5.

Tullberg's third paper is entirely stratigraphical, and is preliminary to his great work—the " Skånes Graptoliter." He recognises several graptolite zones in the Silurian *Cyrtograptus*, *Retiolites* and *Lobiferus*-skiffer, and also in the beds belonging to the Ordovician. He correlates these with the zones previously worked out in Britain by Lapworth.

1880.
Törnquist,
" Studier öfver *Retio-
lites,*" ' Geol. Fören.
Förh.,' bd. 5.

A most important paper by Törnquist bearing on the structure of *Retiolites* appeared in 1880. By means of sections, transverse and longitudinal, he was able to elucidate many points which Linnarsson, Barrande, Hall and others had left unsettled. His materials consisted not only of specimens of the true *Retiolites*, but also of another form, which was separated later by Tullberg as a distinct genus—*Stomatograptus.*

Structure.—The main result arrived at by Törnquist is that the polypary in these genera consists of two different elements : (1) " the outer polyparium or periderm, consisting of the reticulate skin, with its parietal and apertural threads, strands, and two virgulas; (2) the inner polyparium, with smooth and thinner walls."

The inner polyparium is very similar to that of a diprionidian Graptolite, and possesses distinct interthecal walls of a trough-like shape. The thecæ arise from an extremely narrow common canal, and open outward and upward.

This inner polyparium was presumably attached to the outer in such a way that " the free wide thecal apertures of the former are fastened to the parietal threads of the latter."

The network-like periderm, which may have been covered by a thin skin, is not in close contact with the inner polyparium, and the space between the two Törnquist calls the " parietal canal."

CHAPTER IV.

1881 to 1907.

1881.
Törnquist,
" Om några graptolit-
arter från Dalarne,"
' Geol. Fören. Förh.,'
bd. 5.

Törnquist, in the year 1881, described and figured five new species of *Monograptus* from the *Retiolites* shales of Dalarne. These are *M.* **cultellus,** *M.* **nodifer,** *M.* **crenulatus,** *M.* **continens,** and *M.* **sartorius.** In addition to these he figures and describes specimens of *Diplog. folium, D. pristis,* and *Monog. priodon.*

1881.
Linnarsson,
" Graptolitskiffrar med
*Monograptus turricu-
latus* vid Klubbudden
nära Motala," ' Geol.
Fören. Förh.,' bd. 5.

The same year Linnarsson described a number of Grapto-lites from Klubbudd in beds which he considers answer in stratigraphical position to the Lower Gala of Scotland. Four of his species are given as new, viz. : *Monog.* **rhyncophorus,** *M.* **dextrorsus,** *M.* **tortilis,** and *M.* **resurgens.** The other species described and figured by Linnarsson are : *M. jaculum, M. priodon, M. cf. crassus, M. cf. lobifer, M. runcinatus, M. turri-culatus, Rastrites Linnæi, Diplograptus palmeus,* and *Retiolites perlatus ?*

1881.
Holm.
" Tvenne nya slägten af familjen Dichograptidæ," 'Öfvers. K. Vetenskaps Akad. Förhandl.,' no. 9.

Two new genera were founded by Holm in 1881 for certain forms of Dichograptidæ with four primary stipes: **Holographtus.** His genus (1) (type *H.* **expansus**) includes those species in which the four primary stipes branch irregularly and on both sides. The type species occurs in the *Phyllograptus* shales of Vestergötland.

In his genus (2) **Trochograptus** (type *T.* **diffusus**) the four primary stipes give off branches on one side only at fairly regular intervals, and these may branch again in their turn.

Holm shows that the so-called " funicle " bears cells in *T. diffusus*, and is not non-celluliferous, as originally supposed.

1881.
Holm,
" *Pterograptus*, ett nytt graptolitslägte," 'Öfvers. K. Vetenskaps Akad. Förhandl.,' no. 4.

In 1881 Holm described a third genus, namely, **Pterograptus**, with its species *P.* **elegans.**

He again discusses the question of the celluliferous or non-celluliferous character of the " funicle " in Graptolites. He thinks it not improbable that the "funicle" will prove to be always celluliferous, even in the case of *Cœnograptus gracilis.*

1881.
Barrois,
" Sur le Terrain Silurien superieur de la presqu'île de Crozon," 'Ann. Soc. Géol. du Nord,' vol. vii.

Barrois notices the occurrence of *Monog. colonus, M. Sedgwickii, M. priodon,* and *M. Hisingeri* in the Ampelite Shales of Crozon. A previous record of the existence of Graptolites at this horizon had been made by M. Guillier.

1881.
Keeping,
" The Geology of Central Wales," 'Quart. Journ. Geol. Soc.,' vol. xxxvii.

In 1881 Keeping made known the existence of various graptolitiferous horizons in the Silurian of Central Wales, and utilised the Graptolites for the purpose of working out the geological succession in that region.

In addition to the Rhabdophora collected by him, Keeping found a number of Cladophora, most of them new to science.

1881.
Lapworth,
Ibid. Appendix " On some New Species of Cladophora," 'Quart. Journ. Geol. Soc.,' vol. xxxvii.

In an appendix to this paper, Lapworth described and figured seven new species of Cladophora discovered by Keeping. These are: *Dictyonema* **venustum**, *D.* **delicatulum**, *D.* **corrugatellum**, *Calyptograptus?* **plumosus**, *C. ?* **digitatus**, *Acanthograptus* **ramosus**, and *Odontocaulis* **Keepingii**.

A new genus **Odontocaulis** is also described and figured.

1882.
Brögger,
'Die Silurischen Etagen 2 und 3.'

Brögger, in a classic work on " Die Silurischen Etagen 2 und 3," described and figured two new species of *Bryograptus* from Norway: *Bryograptus* **ramosus** and *B.* **retroflexus.** He also re-figured Kjerulf's *Grapt. tenuis* (*Bryog. Kjerulfi,* Lapw.).

Brögger also figured Norwegian specimens of *Dictyograptus flabelliformis*

(Eichwald), and discussed the structure of this genus in considerable detail. He shows that the polypary originates from a sicula, is basket-shaped, that the branches bear cells along their whole length on their inner side, and that *Dictyonema* " is therefore much more closely allied to the true Rhabdophora than is usually supposed."

One of the forms described by Kjerulf as *Dictyonema norvegicum* is regarded as a mutation of *D. flabelliformis*. The others are identical with it, while another variety, var. **conferta,** had been recognised by Linnarsson (MS). Many other species of Graptolites found in Norway are recorded, and their synonymies given, but no descriptions or figures.

He corrects previous views as to the age of the *Bryograptus* beds, and shows that they occur in the upper part of the *Dictyonema* beds, in the passage beds between Etagen 2 and 3.

1882.
Tullberg,
" On the Graptolites described by Hisinger,"
' Bihang K. Vetenskaps Akad. Handl.,' bd. 6, no. 13.

A paper of decided importance was published by Tullberg in this year. Considerable differences of opinion had arisen as to the exact nature of the species figured and described in Hisinger's ' Lethea Suecica,' which was issued as far back as 1837–40. It had become impossible to identify many of his forms until they had been re-figured and re-described in the light of modern knowledge. This much-needed revision was undertaken by Tullberg in this paper.

The paper commences with an excellent historical account of the work of the older Swedish palæontologists and of the part that each played in developing our knowledge of the nature of Graptolites. Eight species of Graptolites are described and figured, and in each case Hisinger's original type specimens are re-drawn and re-described. The species include *Climacograptus scalaris, Diplograptus pristis, Diplog. teretiusculus, Monograptus sagittarius, M. convolutus, Cephalograptus folium, Didymog. geminus* and *Dictyonema flabelliforme.*

1882.
Kurck,
" Några nya Graptolit-arter från Skåne,"
' Geol. Fören. Förh.,' bd. 6.

Four new species of Swedish Graptolites were described and figured in 1882 by Kurck from the *M. cyphus* beds at Bollerup, Scania. These were *Monograptus* **revolutus,** *Diplog.* **longissimus,** *Cephalog.* **ovatoelongatus** and *Climacog.* **undulatus,** and two associated forms are compared with *Dimorphog. Swanstoni* and *M. cyphus* respectively. The figures are good and the structure well represented.

1882.
Herrmann,
" Vorläufige Mittheilung über eine neue Graptolitenart," ' Nyt Mag. for Naturvid.,' bd. xxvii.

Some very interesting forms of Dichograptids from Christiania were described by Herrmann in 1882, which illustrate well the powers of variation of one species as regards the number of its branches. In a new species described by him as *Loganograptus* **Kjerulfi,** he notes every gradation from forms with five stipes to those with twelve and perhaps sixteen. All these are characterised by a large disc, the development of which he traces.

Two specimens of a form doubtfully referred to *Dichograptus Milesi* are figured, and fragments of *Dendrograptus* (?), *Pleurograptus* (?), and *Cœnograptus* (?).

1882.
Hopkinson,
"On some Points in the Morphology of the Rhabdophora or true Graptolites," 'Ann. Mag. Nat. Hist.,' ser. 5, vol. ix.

A paper by Hopkinson on the "Morphology of the Rhabdophora," previously read before the British Association in 1881, was published in full in the early part of 1882. In this paper he figures and describes a specimen of *Tetragraptus serra,* indicating the presence of an internal ridge and constriction at the base of each hydrotheca.

He concludes (1) "that in certain Graptolites the calycles seem to be completely cut off from their supporting perisarc, this appearance being due to a constriction, or the presence of a partially dividing ridge," and (2) that "in these same forms there are at least constrictions in the perisarc, dividing it into sections, from each of which a calycle is produced."

He holds that these phenomena show that the calycles of the Graptolites are true hydrothecæ, and that the Graptolites are the "Palæozoic representatives of the recent Hydrophora."

Three stratigraphical papers in which Graptolites are referred to also appeared during the year 1882.

1882.
Marr,
"On the Cambrian and Silurian Rocks of Scandinavia," 'Quart. Journ. Geol. Soc.,' vol. xxxviii.

The first was by Marr, who gave a generalised account of his visit to the classic localities of the Cambrian and Silurian rocks of Scandinavia, summarised the discoveries of the Scandinavian geologists, and correlated anew the fossil-bearing formations with the British and Bohemian deposits. He notes the Graptolite zones previously correlated by earlier observers, and gives a full list of the Graptolites from the *Retiolites* beds of Bornholm.

1882.
Schmidt,
"On the Silurian (and Cambrian) Strata of the Baltic Provinces of Russia," 'Quart. Journ. Geol. Soc.,' vol. xxxviii.

The second was by Schmidt, who gave a description of the "Silurian (and Cambrian) Strata of the Baltic Provinces of Russia," and their successive fossiliferous zones as developed by himself during a lifetime of research. He compares them with those of Scandinavia and the British Isles and fixes the exact horizons of *Dictyonema* and *Phyllograptus* in the sequence in the Baltic Provinces.

1882.
Lapworth,
"On the Girvan Succession," 'Quart. Journ. Geol. Soc.,' vol. xxxviii.

The third was by Lapworth, on "The Girvan Succession in South Scotland," in which the value and reliability of the Graptolites as "zone fossils" were again demonstrated, by the discovery in the Girvan district of the same series of Graptolitic zones in the same order as in the previously described Moffat area, notwithstanding the great differences between the two regions as respects the thickness, lithology, and palæontological features of the formations present.

1882.
Tullberg,
"Skånes Graptoliter,"
'Sver. Geol. Unders.,'
ser. C, no. 50.

In this year also Tullberg brought out the first part of his great work on the "Skånes Graptoliter." In this part of his paper Tullberg divided the whole of the Cambrian and Silurian Beds of Scania into successive zones, the majority being characterised by distinctive Graptolites.

The zoning of the strata containing Graptolites is even more detailed than that by Lapworth in his Moffat paper, and the result of Tullberg's work not only added to the proof of the value and utility of Graptolites as zone indices, and as a means of correlating synchronous deposits in countries geographically distant, but pointed the way to a closely detailed correlation of the British and Scandinavian deposits.

1883.
Tullberg,
"Skånes Graptoliter,"
'Sver. Geol. Unders.,'
ser. C, no. 55.

In the second part of this work, published the following year, Tullberg dealt with the classification of the Graptolites in general, and the various species found in the *Cyrtograptus* and *Cardiola* shales in particular.

Classification.—A modification of Lapworth's classification is suggested. Nine families are recognised. Six of these—Dictyograptidæ, Dichograptidæ, Nemagraptidæ, Monoprionidæ, Mono-diprionidæ, and Diprionidæ —are grouped together under the Class **Monophyontes**; one—Heteroprionidæ (including the single genus *Dimorphograptus*)—under the Class **Mono-Amphiphyontes**; and two—the Glossograptidæ and Retiolitidæ—under the Class **Amphiphyontes.**

Under the family Dichograptidæ Tullberg includes not only the true Dichograptids, but also *Pleurograptus*, *Cladograptus*, and *Phyllograptus*. The family Retiolitidæ is left unchanged, and contains *Trigonograptus* and *Clathrograptus*, in addition to *Retiolites* and a new genus **Stomatograptus.**

The genus *Monograptus* the author divides into six groups, according to the general form of the polypary and the shape of the thecæ, and for these he proposes names—*i. e.* (1) Leptopodes (*M. Nilssoni,* etc.), (2) Orthopodes (*M. Hisingeri*), (3) Helicopodes (*M. convolutus,* etc.), (4) Opisopodes (*M. lobifer,* etc.), (5) Kamtopodes (*M. testis,* etc.), (6) Prosopodes (*M. colonus,* etc.). These groups, it will be seen, correspond fairly closely with those suggested by Lapworth. ('Scottish Monograptidæ,' 1876.)

Description of Species.—A large part of this paper is devoted to the description of the Scanian species, many of which are new. These are admirably figured in two plates, and include :

Leptopodes.—Monograptus Nilssoni.

Orthopodes.—*M. Hisingeri, M. vomerinus,* *M.* **personatus,** *M.* **Linnarssoni,** *M.* **spinulosus,** *M.* **speciosus.**

Opisopodes.—M. priodon, M. Flemingii, M. riccartonensis, M. nodifer, M. sartorius, M. **scanicus,** *M.* **capillaceus,** *M.* **flexuosus,** *M.* **retroflexus.**

Kamptopodes.—M. testis, M. bohemicus.

Prosopodes.—M. dubius, M. colonus, M. cultellus, M. **uncinatus.**

Cyrtograptus.—C. Grayi, C. spiralis, C. Murchisoni, var. **crassiusculus,** *C. Carruthersi,* and the new species : *C.* **dubius,** *C.* **Lapworthi,** *C.* **pulchellus,** *C.* **flaccidus,** *C.* **moniliformis,** *C.* **rigidus,** *C.* **Lundgreni.**

Retiolites Geinitzianus and the new genus **Stomatograptus** with its species *S.* **Törnquisti** are also described and figured. This last genus, originally described by Törnquist as a *Retiolites,* is characterised by the presence of large pores on the sides of the polypary by means of which the "lateral canals are in connection with the exterior."

Structure.—A detailed account of the structure of *Retiolites* is given, which in some respects marks an advance on the results already arrived at by Törnquist, but no figures are given illustrating the points of structure described.

1883.
Törnquist.
" Bergbygnaden inom Siljansområdet i Dalarne," ' Sver. Geol. Unders.,' ser. C, no. 57.

A stratigraphical paper was also published by Törnquist during the year, which was the forerunner of a series of papers by him entitled " Siljansområdets Graptoliter." In this memoir the various zones of the *Rastrites* and *Retiolites* shales are given and correlated with those abroad.

1883–84.
Postlethwaite,
" Graptolites of the Skiddaw Slates,"
' Trans. Cumb. Assoc. for Advancement of Literature and Science,' vol. viii.

In 1883–84 Postlethwaite published a paper in which he dealt with the structure, classification, reproduction, and state of preservation of the Graptolites found in the Skiddaw Slates, and gave a complete list of the forty species already obtained from these beds.

He figures a few of the most typical species, and among them a new form of *Tetragraptus,* which was subsequently described by Elles (in 1898) as *Tetrag. Postlethwaitii.*

1884.
Spencer,
' Niagara Fossils,' pt. i ;
" Graptolitidæ of the Upper Silurian System," ' Bull. Mus. of the Univ. of Missouri.'

Dr. Spencer, who had previously described (in 1878) three new genera and nine new species of Dendroid Graptolites from the Niagara formation, but had given no figures of these new forms, figured and re-described them in a paper published in 1884 ; and in addition one new genus and twenty-two new species, making a total of thirty species and four genera. A brief account is given by him of the distribution, zoological affinities, structure, reproduction, and classification of the Graptolites. The descriptions are brief and the figures poor.

Description of Species.—The species described are *Phyllograptus* ? **dubius,** *Retiolites venosus, Dendrograptus* **ramosus,** *D.* **simplex,** *D.* **Dawsoni,** *D.* **frondosus,** *D.* **prægracilis,** *D.* **spinosus,** *D.* (s.g. **Chanograptus**) **novellus,** *Callograptus Niagarensis, C.* **Granti,** *C.* **multicaulis,** *C.* **minutus** ; *Dictyonema retiforme, D. gracile, D.*

expansum, *D. Websteri, D. tenellum, D. splendens, D. pergracile*; *Calyptograptus cyathiformis, C. subretiformis, C.* **micronematoides,** *C.* ? **radiatus,** *Rhizograptus bulbosus, Acanthograptus Granti, A.* **pulcher**; *Inocaulis plumulosus, I. bellus, I. divaricatus, I.* **problematicus,** *I.* **diffusus,** *I.* **cervicornis,** *I.* **phycoides,** *I.* **ramulosus**; *Thamnograptus bartonensis, T.* **multiformis** (not a *Thamnograptus*), *Ptilograptus foliaceus*; **Cyclograptus n.g.,** *C.* **rotadentacus.**

The new species are mainly founded on the features presented by their general outlines.

1885.

Herrmann,

" Die Graptolithenfamilie Dichograptidæ, Lapw.," 'Nyt Mag. for Naturvid.,' bd. xxix.

In the year 1885 Herrmann published a monograph on the family of the Dichograptidæ. This work is valuable as giving a fairly complete summary of our knowledge of the Graptolites in general, and of the Dichograptidæ in particular, at the time it was written.

Classification. — The family Dichograptidæ includes, according to Herrmann, sixteen genera, which he groups as follows: 1, *Didymograptus*; 1a, *Trichograptus*; 1b, *Bryograptus*; 1c, *Pterograptus*; 1d, *Pleurograptus*; 2, *Janograptus*; 3, *Tetragraptus*; 3a, *Schizograptus*; 3b, *Trochograptus*; 3c, *Ctenograptus*; 3d, Type *Gr. Richardsoni*; 3e, *Holograptus*; 3f, *Goniograptus*; 4, *Dichograptus*; 4a, *Clematograptus*; 5, *Clonograptus*.

He gives all the previously described species of each genus, and transfers some species from genera in which they had been placed by previous authors to other genera. The only new species described by Herrmann is *Pterograptus* ? **dilaceratus,** from the *Phyllograptus* shales of Norway.

Organisation.—In a chapter dealing with the organisation and economy of the Graptolites, the sicula is described, and its function as an organ of attachment denied. The question of the " angle of divergence " of the stipes is discussed in much detail, and Herrmann suggests, for the first time, the employment of the ventral instead of the dorsal angle as the conventional " angle of divergence," a plan which has since been generally adopted by Graptolithologists. He considers that the Graptolites probably lived with the sicula below and the branches growing upwards. He supports Holm's opinion that in some cases, at any rate, the " funicle " bears thecæ; and gives additional examples. He deals with the function and development of the central disc at considerable length; attaches considerable importance to Hopkinson's discovery of an apparent dividing septum between the theca and common canal, and agrees with him that it places the " hydrothecal " nature of the cells beyond doubt.

1885.

Roemer,

" Lethaea erratica," 'Palæont. Abh. v. Dames und Kayser,' bd. ii, heft 5.

Roemer describes and figures several species of Graptolites from the greenish-grey Graptolithen-Gestein of North Germany. These are: *M. ludensis, M. testis, M. scanicus, Retiolites Geinitzianus,* and two unidentified species of *Monograptus.*

In a short note referring to " Graptolites of the Quebec Group," collected by Mr. Richardson for the Montreal Museum, Dawson describes a new species of *Dictyonema* (*D.* **delicatulum**) from the Point Levis at Fort, No 2, and records the occurrence of *D. sociale* in the black shales of Matane.

In the following year Lapworth published a " Preliminary Report on some Graptolites " forwarded to him by the Canadian Survey from the Lower Palæozoic Rocks on the south side of the St. Lawrence—from Cape Rosier to Tartigo River, from the north shore of the Island of Orleans, one mile above Cape Rouge, and from the Cove Fields, Quebec.

He gives detailed lists of Graptolites from a large number of Canadian localities, arranges them in stratigraphical order, and correlates the containing beds with the graptolite bearing formations of America, Britain, and Europe.

He recognises the existence in Canada of two main faunas, each divided into two sub-faunas :

(A) Quebec or Calciferous—Chazy Fauna :

Sub-fauna 1.—Cape Rosier (and Barrasois River) Zone of Calciferous age= Tremadoc rocks and *Dictyonema* beds.

Sub-fauna 2.—St. Anne River Zone=Arenig.

(B) Trentonian, Marsouin River (or Norman's Kill) fauna :

Sub-fauna 1.—*Cœnograptus* zone of Griffin Cove and Marsouin River= Middle Llandeilo and Glenkiln.

Sub-fauna 2.—Cove Fields and Orleans Island Zone=Highest Llandeilo or Lowest Caradoc.

Törnquist, in a paper on the " Older Palæozoic Rocks of Ostthüringia and Voigtland," classified and described their Graptolite zones, and correlated them with those of Scandinavia and Britain. He recognises three main horizons : (*a*) Lower *Rastrites* shales, (*b*) Upper *Rastrites* shales, and (*c*) *Retiolites* shales. In an appendix to this paper he describes and figures three new species : (1) *Rastrites* **phleoides,** (2) *Cyrtograptus* **radians,** and (3) *Retiolites* **macilentus.**

In 1888 a very important paper from the stratigraphical aspect of graptolithic literature was published by Marr and Nicholson. They divide the whole of the Stockdale Shales of the Lake District into their component zones, the majority of these being distinguished by special Graptolites. They demonstrate the presence in the Stockdale Shales of the same Graptolite zones as those in the Birkhill and Gala beds of South Scotland in the same order of sequence, and parallel them with those of many other districts, British and Foreign.

1889.
Marr,
" Notes on the Lower Palæozoic Rocks of the Fichtelgebirge," ' Geol. Mag.,' dec. 3, vol. vi.

In the following year Marr gave a revised list of the Frankenwald and Thüringerwald Graptolites preserved in the Dresden Museum. This list is of special interest, as many of these forms are the originals of the figures on the plates of Geinitz's work, ' Die Graptolithen,' 1852. Marr also made known the presence of Graptolites in the representatives of the Wenlock formation in Thüringia and the neighbouring parts of Bavaria.

1889.
De Rouville,
" Note sur un nouvel Horizon de Graptolites dans le Silurien de Cabriéres," ' Bull. Soc. Géol. de France,' t. xviii.

De Rouville made known the presence of Graptolites in a rich Arenig fauna discovered near Cabriéres. No specific names of Graptolites are given.

1889.
Lapworth,
" On the Ballantræ Rocks of S. Scotland and their place in the Upland Sequence," ' Geol. Mag.,' dec. 3, vol. vi.

The same year Lapworth, after pointing out the systematic importance of the existence of a typical Arenig fauna in the Ballantræ Rocks of South Scotland, the discovery of which had been made known by him in 1886 (Jukes-Browne, ' Historical Geology,' 1st Edition), gave a generalised account of his conclusions respecting the structure and sequence throughout the Southern Uplands, and paralleled in tabulated form their graptolite bearing formations with their equivalents in England and Wales.

1889.
Jaekel,
" Ueber das Alter des sogennanten Graptolithengesteins mit besonderer Berücksichtigung der in demselben enthaltenen Graptolithen," ' Zeitschr. deutsch. geol. Gesell.,' bd. xli.

A paper on the fauna of the Graptolithengestein of the German Drift erratics, with special relation to their Graptolites, was published by Jaekel in 1889 (compare the subsequent memoir by Jahn).

Several Graptolite species are described and figured, and it is proposed to divide the genus *Monograptus* into two new groups or genera—**Pristiograptus** and **Pomatograptus**, a similar division being regarded as probably equally applicable to the two- and many-rowed forms of Graptolites.

Pristiograptus is characterised by a straight or convexly curved polypary, cylindrical cells in contact throughout, having their apertures occupying the whole of the upper end of the cell, and their apertural processes, if present, occurring as spines on the lower edge of the aperture.

In *Pomatograptus* the polypary is straight or concavely curved, the cells free at their outer end, while the aperture is small and protected by a roof-like process arising from the upper part of the cell.

Under the name *Pristiograptus* the author describes *P. bohemicus*, *P. Roemeri* ?, *P. Nilssoni*, *P. colonus*, *P. testis*, and a new species *P.* **frequens.** Under the name *Pomatograptus*, he describes *P. priodon*, *P. Becki*, *P. Barrandei*, and a new form, *P.* **micropoma.**

The structure of *Retiolites* is also described.

Mode of Life.—In discussing the mode of life of the Graptolites, Jaekel returns to the idea of their rooted attachment to the floor of the sea, and also revives the opinion of Hall that all single Graptolites "usually described under the name *Monograptus*" are torn-off pieces of larger colonies. He maintains that the narrow end is never complete, and never shows well-developed cells.

Affinities.—Jaekel agrees with Neumayr that the Graptolites should be placed in a special class, which might be compared with the Corals.

1890.
Holm,
"Gotlands Graptoliter,"
' Bihang K. Svenska
Vet.-Akad. Handl.,' bd.
xvi, af. 4, no. 7.

During this year, a brilliant paper on the minute structure of the Graptolites was published by the Swedish palæontologist, Holm.

Previous researchers had attempted to isolate the Graptolite from the rock in which it was imbedded, but the process was brought to great perfection by Holm, and the value of the results thus obtained can hardly be over-estimated. In this paper ("Gotlands Graptoliter") Holm gives the results of his investigations of the structure of *Dictyonema*, *Monograptus dubius*, *Retiolites*, and *Stomatograptus*.

Dictyonema **cervicorne.** The structure of a branch of *Dictyonema* is worked out in this new species. The thecæ are turned inwards. From one side of the thecæ, alternately on the left- and right-hand side, there grows out a "bird-nest"-like theca, which Holm terms a "bitheca" (? gonangium). Between each pair of thecæ there is one dissepiment. Holm points out that the forms referred to the genus *Dictyonema* are both siculate and non-siculate, and holds that it may be found necessary in the future to sub-divide this genus.

M. priodon and *M. sub-conicus* are recorded from Gothland, and *M. Flemingii* (which he regards as only a mutation of *M. priodon*) and *M. dubius* are figured. The structure of the latter species is very accurately represented.

Retiolites Geinitzianus. The isolation methods adopted by Holm also enabled him to give a far more complete description, and a more detailed and perfect figure of the structure of *Retiolites* than had hitherto been possible. The nature of the outer net-like periderm is more particularly elucidated, as well as the relations between the straight and zigzag "virgulas," and "threads" of the skeleton generally.

Stomatograptus Törnquisti. Many further points of structure, obscure in

Retiolites, are cleared up by Holm's development of specimens of *Stomatograptus* (*S. Törnquisti*). This genus he considers to be more distinct from *Retiolites* than was previously supposed, for in *Stomatograptus*, in addition to the "pores," the form of the apertures is different. He proves that the periderm in this genus, as in other Graptolites, is composed of three layers; "the outer and inner are smooth and without spaces, while the middle one is formed of a network of chitinous threads."

Retiolites nassa. Another interesting form, *Retiolites* **nassa** (later made the type of a new genus—*Gothograptus*—by Frech.), is shortly described. This form bears the same relation to the Retiolitidæ that *Climacograptus* does to the Diplograptidæ.

1890.
Törnquist,
"Siljansområdets Graptoliter," i, 'Lunds Univ. Årsskrift,' bd. xxvi.

In 1890 Törnquist brought out the first part of his "Siljansområdets Graptoliter." In this paper several species of Graptolites (some new) are carefully described and figured. The author recites concisely his own views on the structure of *Retiolites*, and discusses Tullberg's opinions on the subject.

Description of Species.—*Retiolites Geinitzianus*, *R. cfr. perlatus*, and *R. obesus* are described and figured, and he thinks that, when better material is obtained, the last species will form the type of a new genus. He refers the species described by Suess ('Böhmische Graptolithen,' 1851) as *R. grandis*, to the genus *Stomatograptus*. In addition, he figures *Lasiograptus margaritatus*, *Dichog. octobrachiatus*, *Clonograptus* **robustus** (allied to *Gr. ramulus*, Hall), *Tetragraptus serra*, *Tetrag. curvatus*, *Didymograptus minutus*, *D.* **gracilis**, and *D.* **decens**. In his description of *D. minutus*, he notes the long thread extending from the apex of the sicula, and thinks that this might correspond more nearly to the virgula in the genera *Diplograptus* and *Monograptus*. He suggests that the genus *Didymograptus* should be made the type of a new family—the Didymograptidæ—distinct from that of the Dichograptidæ on account of apparent differences in the development of the branches.

The other species described in the paper are *Dicellograptus anceps*, *Phyllograptus densus*, *Climacograptus scalaris*, *C.* **internexus**, and a form which the author thinks further research will prove to be a distinct species, and for which he suggests a name, *C.* **phrygionius**; *Diplograptus pristis*, *D. truncatus*, *D. palmeus*, a new species, *D.* **bellulus**, *Cephalograptus folium* and *C. cometa*.

In most of the specimens the sicula and its relations to the proximal thecæ are carefully figured.

1890.
Malaise,
"Sur les Graptolites de Belgique," 'Bull. Acad. Roy. de Belgique,' ser. 3, vol. xx.

Malaise in 1890 summarised his work among the graptolite bearing Palæozoic rocks in Belgium, and correlated the zones with those of Britain. He recognises in Belgium four successive chronological groups of Graptolites, viz. those of the Arenig, Bala, Llandovery, and Wenlock-Ludlow, but the presence of two others, *i. e.* the *Dictyonema sociale* zone and that of the Llandeilo, he regards as yet problematical.

1890.
Dodge,
" Some Silurian Grap
tolites from Northern
Maine," 'Amer. Journ.
Sci.,' vol. xl.

Dodge described a few species of Graptolites found by himself in Maine in beds of Norman's Kill age. They include *Helicograptus gracilis*, *Dicellograptus* sp., *Diplograptus* sp., *Cryptograptus marcidus*, and *Glossograptus spinulosus*.

1890.
Gurley,
" Geological Age of the
Graptolite Shales of
Arkansas and some new
Species of Graptolites,"
'Ann. Rep. Geol. Surv.
Arkansas '

The geological age of the Graptolite shales of Arkansas is discussed by Gurley in a paper published in 1890. He recognises two horizons—the Trenton and the Calciferous. The latter, he considers, corresponds to the Point Levis beds, the former to the upper part of the Norman's Kill fauna. The stratum marked out by the latter distinctive fauna is termed by him the " *Dicellograptus* zone," and is divided into (*a*) Lower *Dicellograptus* sub-zone with *Cœnograptus gracilis*, and (*b*) Upper *Dicellograptus* sub-zone without *Cœnog. gracilis*. The Arkansas fauna thus corresponds to that of the Upper Zone. He describes four new forms : *Dicranograptus* **arkansasensis**, *Dicranog.* **Nicholsoni** var. **parvangulus**, *Diplog.* **trifidus**, and *Dictyonema* **obovatum**. Only two of the species are figured. A complete list of the Graptolites obtained from Arkansas is also given.

1890.
Getz,
" Graptolitförende ski-
ferzoner i det Trondh-
jemske," ' Nyt Mag.
for Naturvid.,' bd. xxxi.

In 1890 Getz described a few specimens obtained (mainly by Herrmann) from two distinct Graptolite horizons in the Trondhjem district, corresponding the one to Upper Glenklin or Lower Hartfell, and the other to the Birkhill. The Graptolites are poorly preserved owing to the cleavage of the beds, and it is impossible to identify more than *Dicranograptus ramosus* and *Climacog. bicornis*, but the author records in addition a species of *Didymograptus*, *Diplograptus*, and *Dicellograptus* from the lower horizon. From the Upper Birkhill zone he figures *M. cfr. convolutus*, *Monograptus Halli* and *Rastrites* sp.

1890.
Nicholson, H. O.,
" Note on the Occur-
rence of *Trigonograptus
ensiformis*, etc.," ' Geol.
Mag.,' dec. 3, vol. vii.

The same year H. O. Nicholson recorded the existence of *Trigonograptus ensiformis* in the Ellergill Beds of the Lake District, and also described a new variety of *Didymograptus* under the name *D. v-fractus* var. **volucer.**

1890.
Geinitz,
' Die Graptolithen des
k. Mineralog. Mus. in
Dresden.'

Geinitz, in a paper published in this year, revised the fine collection of Graptolites in the Dresden Museum, and gave descriptions and figures of some of the species identified. He retains all the genera and species of Graptolites originally described by him, with the exception of *Nereograptus*. Very brief (if any) descriptions of the forms illustrated are given, and the figures are mainly copies from the illustrations to his ' Graptolithen,' 1852, or from other authors; but the synonomy of each species is given fully. The species referred to are : *Monograptus sagittarius, M. Hisingeri, M. nuntius, M. Nilssoni, M. Salteri, M. tenuis, M. bohemicus, M. latus, M. virgulatus, M. Barrandei, M. colonus, M. frequens, M. testis, M. priodon, M. millipeda, M. Becki, M. Halli, M. Flemingi, M. clintonensis, M. Sedgwickii, M. convolutus, M. turriculatus, M. proteus, M. peregrinus, M. Linnæi, M. gemmatus; Cyrtograptus Murchisoni, C. radians; Pterograptus dilaceratus* (which he regards as identical with *Cœnog. gracilis*) ; *Didymograptus Murchisoni* and *D. Forchammeri; Tetragraptus serra* and *T. fruticosus; Diplograptus ovatus, D. palmeus, D. folium, D. cometa, D. foliaceus, D. teretiusculus, D. pristis, D. secalinus, D. mucronatus, D. Swanstoni; Phyllograptus cfr. angustifolius; Triplograptus Nereitarum; Retiolites Geinitzianus.* Geinitz's views appear to have undergone but little change since 1862, and he refuses to accept such genera as *Rastrites, Dicellograptus, Dimorphograptus, Climacograptus,* etc.

1891.
Hall, T. S.,
" On a new Species of *Dictyonema*," 'Proc. Roy. Soc. Vict.,' n.s., vol. iv.

In 1891 T. S. Hall described a new species of *Dictyonema* —*D.* **grande** from the Lancefield beds of Australia.

1891.
Björlykke,
" Graptolitförende Skifere i v. Gausdal," ' Norges Geol. Undersögelse Aarbog,' no. 1.

In 1891 Björlykke gave brief descriptions and sketch-figures of some species of *Didymograptus*, etc., found by himself in Western Gausdal. These include *D. geminus, D. cfr. nitidus* or *balticus, D. cfr. euodus, D. hirundo* or *patulus, D. pusillus, Diplog. teretiusculus, Climacograptus* sp., *Pterograptus elegans,* and *Tetragraptus bryonoides.*

1891.
Matthew,
" On a new Horizon in the St. John's Group," ' Canad. Record of Science.'

Matthew during this year recorded the discovery of *Dictyonema flabelliforme* in the St. John's Group. He discusses the importance of this fossil as a zone index in Europe.

He considers that *Dictyonema* was a free, floating organism, and " began life as a *Bryograptus.*"

1891.
Pritchard,
" On a new Species of the Graptolitidæ-*Temnograptus magnificus*," ' Proc. Roy. Soc. Vict.'

During the same year Pritchard recorded what he considered to be a new species of *Temnograptus* from Australian deposits, under the name *T.* **magnificus.**

1891.
Gurley,
" On some recent Graptolitic Literature," ' Amer. Geol.,' vol. viii.

A general review of Jaekel's, Geinitz's, Törnquist's, and Holm's papers issued during the previous two years, was published by Gurley in the ' American Geologist ' for 1891.

1891.
Moberg,
" Om ett par Synono-
mier," ' Geol. Fören.
Förh.,' bd. xiii.

Two questions of synonymy were discussed by Moberg in 1891. The first deals with *Dictyonema* and *Dictyograptus*, and Moberg urges the adoption of the latter name for the same reason as that given by Hopkinson and Lapworth, namely, that *Dictyonema* is an " old-established name for a genus of plants." The second deals with *Didymog. caduceus* or *D. gibberulus.* As Salter's name, *D. caduceus*, has been used for two species, and it is impossible to say to which of the two the original name corresponded, Moberg agrees with Nicholson that it can no longer stand, and he proposes the general adoption of Nicholson's specific name, *D. gibberulus.*

1892.
Törnquist,
' Siljansområdets
Graptoliter,' pt. ii.

Numerous papers dealing with Graptolites appeared in 1892, the majority of them being by Swedish writers.

Another part of Törnquist's great work on the " Siljansområdets Graptoliter " was published during this year. It deals wholly with the Monograptidæ. The author discusses the three genera usually included under this family—viz. *Cyrtograptus*, *Rastrites*, and *Monograptus*, and classifies the various species belonging to the last-named genus, rejecting Jaekel's division of *Monograptus* into *Pomatograptus* and *Pristiograptus.*

Törnquist's classification is based on the principle adopted by Lapworth (1876)—namely, on the form of the polypary and the form of the thecæ, but it is far more detailed. The proposed divisions are as follows :

(A) Thecæ tube-like, of prismatic form, with upper walls in contact throughout.

    (1) Rhabdosome narrow, flexible ; type *M. Nilssoni.*

    (2) Rhabdosome wide, rigid ; type *M. Hisingeri.*

(B) Upper wall of theca partly free.

    (1) Rhabdosome straight or curved.

        (*a*) Theca with only a narrow border near the aperture free ; apertures hook-shaped, pressed into the polypary (*M. crenulatus*, etc.).

        (*b*) Upper wall of theca prolonged into a lip-shape, type *M. priodon.*

        (*c*) Outer part of theca bent round and grown to the lower walls (*M. lobiferus*).

        (*d*) Free part of theca bent round in a loop shape (*M. sartorius*).

        (*e*) Free part of theca bent double into an **S**-shape (*M. nodifer*).

        (*f*) Upper free wall of theca passing over without a boundary into the lower wall of the theca above (*M. runcinatus*).

        (*g*) Upper wall of theca prolonged into a long spine (M. *Sedgwickii*).

    (2) Polypary spirally curved.

        (*a*) Theca growing out centripetally (*M. discus*, etc.).

The main part of the paper is occupied with careful descriptions illustrated by excellent figures, of numerous species of *Monograptus*, etc., some of the species being new. The following are described: *Rastrites peregrinus*, *R. hybridus*, *Monograptus gregarius*, M. **limatulus**, *M. leptotheca*, *M. crenulatus*, *M. continens*, *M. priodon* (Törnquist discusses and criticises the view of Jaekel as to the structure of the theca in this species), M. **cygneus**, *M. cultellus*, *M. lobiferus*, *M. cfr. Becki*, M. **singularis**, *M. sartorius*, M. **ansulosus**, *M. cfr. dextrorsus*, *M. exiguus*, *M. nodifer*, *M. runcinatus*, *M. Sedgwicki*, *M. convolutus*, *M. spiralis*, var. *subconicus*, *M. turriculatus*, *M. discus*, *M. proteus*, M. **flagellaris.** In all the illustrations great attention is paid to the correct figuring of the proximal end.

1892.
Törnquist,
" Ett inlägg i en synonymifråga," 'Geol. Fören. Förh.,' bd. xiv.

The proposal of Moberg to adopt Hopkinson's and Lapworth's suggested name of *Dictyograptus* in place of *Dictyonema* was discussed and criticised by Törnquist in 1892. He considers that the fact that "*Dictyonema*" is an old-established name for a genus of plants is not a sufficient reason for its rejection.

1892 ?.
James,
" Manual of Palæon-tology of the Cincinnati Group, pt. 2, Cœlenterata."

In 1892 (?) James published the second part of his Manual on the Palæontology of the Cincinnati Group dealing with the Cœlenterata.

Classification.—James considers that in the sub-class Hydroida there is only one order—the Thecaphora—and that this includes the two genera *Dendrograptus* and *Dictyonema.* Under the genus *Dendrograptus* (which according to James includes *Buthotrephis*, Hall, pars, and *Psilophyton*, Lesqx.) he describes *D. gracillimum*, Lesqx., *D. tenuiramosus*. The genus also includes *D. arbusculum*, Ulrich (= *D. irregulare*, Hall).

According to the author the sub-class Graptoloidea comprises seven genera : (1) *Graptolithus* ; (2) *Diplograptus*, including the two species *D. spinulosus* and *D. Whitfieldii* ; (3) *Climacograptus* with its species *C. typicalis* and *C. bicornis* ; (4) *Dicranograptus*, type *D. ramosus*; (5) *Megalograptus* (*M. Welchi* being entirely distinct from any other known Graptolite) ; (6) *Inocaulis* ; and (7) *Dawsonia* (*Lockeia*), *D. siliquaria.*

1892.
Jahn,
" Vorläufiger Bericht über die Dendroiden des böhmischen Silur," 'K. Akad. Wiss. Wien, Mathem. Naturw. Classe,' bd. ci, heft 7.

In 1892 Jahn published a preliminary report of his investigations among the Dendroidea of the Silurian Rocks of Bohemia. He discusses the history of our knowledge of the genus *Dictyonema* up to the year 1875, but ignores the more recent work. He considers that many species have been included under this genus which really belong to other genera, and he suggests that *Desmograptus*, regarded by Hopkinson and Lapworth as a sub-genus of *Dictyonema*, should be regarded as a distinct genus. He also proposes a new genus— **Damesograptus.** Jahn names eight new species of Dendroidea, mainly from Stage E. e. 2 in Bohemia, but gives no descriptions or figures. These are *Callograptus* **tenuissimus**, *Callog.* **bohemicus**, *Callog.* **palmeus**, *Desmograptus*

giganteus, *Desmog.* **diffusus,** *Desmog.* **bohemicus,** *Desmog.* **frondescens,** *Dictyonema*
Barrandei. Owing to the absence either of descriptions or figures none of these
forms can be identified.

1892.
Barrois,
" Mémoire sur la Dis-
tribution des Grapto-
lites en France," ' Ann.
Soc. Géol. du Nord,'
t. xx.

A most important paper by Barrois appeared in 1892
dealing with the Distribution of Graptolites in France. In
this paper Barrois collected together all the scattered information obtained by himself and others for many past years
as to the occurrence of Graptolites in France. In addition
to recording all such Graptolite species as had been discovered
and identified, several species are fully described by him,
but no figures are given.

In Languedoc three horizons are recognisable. From the lowest (Arenig)
the following species are described: *Didymog. balticus, D. v-fractus, D. pennatulus,*
D. nitidus, D. bifidus, D. indentus, D. **Escoti,** *Tetrag. serra* and *T. quadribra-*
chiatus; from the middle one (Llandeilo) only *D. euodus*; from the highest
(Wenlock Shales) *M. priodon, M. bohemicus, M. colonus, M. Roemeri, M. Nilssoni.*

From the Pyrenees the Graptolites are mainly of Birkhill-Tarannon age, and
the following species are described: *Monog. Becki, M. priodon, M. attenuatus,*
M. crispus, M. proteus, M. Barrandei, M. spiralis, M. communis, M. nodifer, M.
sartorius, M. runcinatus, M. Salteri, M. crassus, M. Halli (?), *M. Roemeri, M. riccar-*
tonensis, M. vomerinus, M. basilicus, M. Nilssoni, M. concinnus, M. discus, M.
Lapworthi, *Cyrtog. Murchisoni, Cyrtog. Grayi, Retiolites Geinitzianus, R. perlatus.*

From the Ardennes, Graptolites of Cambrian, Arenig, Caradoc, Lower Llandovery, Tarannon, Wenlock and L. Ludlow age are recorded. From Normandy
only Wenlock fossils are known.

In Brittany, *Didymog. Murchisoni, D. euodus, D. nanus, D. furcillatus* are
described from Angers, while *Diplog. foliaceus* and *D. angustifolius* represent the
Glenkiln horizon. The Llandovery strata of Anjou contain *Monog. spiralis, M.*
crenularis, M. lobiferus, M. **sublobiferus,** *M. Sedgwickii, M. cyphus, M. Clingani,*
Climacog. normalis, Cephalog. folium, Diplog. Hughesi, Rastrites peregrinus, and
R. Linnæi.

The Ampelite schists of Western France are shown by their fossils to be of
Tarannon-Wenlock age, and the Ampelite limestones to occupy a somewhat higher
horizon.

1892.
Gürich,
" Ueber die Zellenöff-
nung von *Monograptus*
priodon," ' Sitzungs-
bericht d. Schles.
Gesell., Naturw. Sect.,
Breslau.'

Gürich, in a paper published during this year, discussed
Jaekel's proposed division of the genus *Monograptus* into
Pomatograptus and *Pristiograptus*, and suggested that for the
two groups typified by these names the titles of (*a*) *Mono-*
grapti **reversi,** and (*b*) *Monograpti* **erecti,** should be substituted, as the " question of their systematic relationships was
as yet not ripe for decision." He gives a " schematic

representation of the conjectural connection of the *Monograpti reversi* and *M. erecti* with a hypothetical disc."

1892.
Moberg,
" Om skiffern med
Clonograptus tenellus,
dess Fauna och Geolo-
giska Alder,"
' Geol. Fören. Förh.,'
bd. xiv.

In 1892 Moberg described the fauna of the *Clonograptus* Beds of Sweden, and added two new species. He discusses briefly the various genera included in the family of the Dichograptidæ, and points out that nearly all these genera are founded on the somewhat unsatisfactory character of differences in the mode of branching, and are as a rule typified by only one species. He refers his new species doubtfully to *Bryograptus*.

Clonograptus tenellus is fully described, and a variety of it—var. **hians**—is proposed. The new species are : *Bryograptus ?* **hunnebergensis**, and *Bryo. ?* **sarmentosus**. The author discusses at length the resemblances and differences between *Clonograptus* and *Bryograptus*, and points out that it is probable that *Didymograptus* was developed from the latter genus.

Moberg also enters fully into the question of the age of the *Clonograptus tenellus* beds.

1892.
Moberg,
" Om några nya Grap-
toliter från Skånes
Undre Graptolits-
kiffer," ' Geol. Fören.
Förh.,' bd. xiv.

In a second paper published during the same year Moberg described two new genera of Dichograptidæ from the Lower Graptolite shales of Scania. One of these is **Mæandrograptus**, with its species *M.* **Schmalenseei**, a form intermediate between *Didymograptus* and *Dicellograptus*, the " sicula giving rise directly (near its broader end and on the same side) to many thecæ."

The structure of the proximal end of *Didymog. gibberulus* is also worked out by Moberg with great care, and on account of its peculiarities he suggests that the species should be made the type of a new genus **Isograptus.**

In addition to the two new genera, he describes and figures a new species of *Azygograptus* (*A.* **suecicus**).

1892.
Moberg,
Referat von G. F.
Matthew " On a new
Horizon in the St.
John's Group," ' Geol.
Fören. Förh.,' bd. xiv.

A third paper published by Moberg during this year consists of a review of Matthew's paper " On a New Horizon in the St. John's Group." He gives a summary of Matthew's results and compares them with the occurrence of the *Dictyonema* and *Bryograptus* beds in Sweden.

1893.
Moberg,
" En *Monograptus*
försedd med discus,"
' Geol. Fören. Förh.,'
bd. xv.

In 1893 Moberg described a very curious species of *Monograptus*—*M.* **pala**—provided with a disc at the proximal end. The presence of a disc had not been previously observed in *Monograptus*.

During this same year, also, several important papers were published dealing with the details of the structure of *Diplograptus* and *Monograptus*.

1893.
Törnquist,
" Observations on the
Structure of some
Diprionidæ," ' Lunds
Univ. Årsskrift,'
bd. xxix.
The first paper was one by Törnquist, " On the Structure of some Diprionidæ." By means of transverse and longitudinal sections of specimens preserved in iron pyrites he worked out many details of structures previously obscure. Like Moberg he distinguishes carefully between the two aspects of the polypary. For these he adopts the terms " obverse " (in which the sicula is completely visible) and " reverse " (in which the sicula is partly concealed from view). The row of thecæ which contains the first theca is called by him the " primordial " series, and the other the " second " series.

As regards the development of the first thecæ he shows that the sicula gives rise (on the left side in the obverse aspect) to a small lobe which develops into a conical space similar to that of the sicula, but " shorter, and communicating with the common cavity of the rhabdosoma." This he distinguishes as the " connecting canal." " Surrounding not only both sides of the sicula, but also its reverse wall," there is an undivided space which he calls the " biserial chamber." The structure of the median septum as respects its complete or incomplete nature, as shown in certain species, is also worked out. In all cases the author apparently implies that each theca is developed from the theca immediately below belonging to the same series.

These details are worked out in the case of *Climacog. scalaris*, *C. internexus*, *Diplog. bellulus*, *D. palmeus*, and *Cephalog. cometa*.

1893.
Wiman,
" Ueber Diplograpti-
dæ," ' Bull. of the Geol.
Instit. of Upsala,' no. 2,
vol. i.
The structure of *Diplograptus* was still further elucidated by Wiman during the same year. He obtained his results by carrying out the methods previously employed by Holm in the case of *Dictyonema* and *Retiolites*, namely, of completely isolating the specimen from the matrix.

The most important discoveries made known by Wiman in this paper are with reference to the sicula and the virgula.

The sicula is shown to consist of two parts : (*a*) a proximal cup-like portion similar in structure to an ordinary theca and possessing growth lines, and (*b*) " a distal portion consisting of coarse, longitudinal branching or anastomosing thickenings or threads." The apertural spines (two in number) of the sicula are also described, and are shown to be flat, not cylindrical. The bilateral symmetry of the sicula thus produced reminds one of a " Bryozoan rather than a modern Hydroid polypary."

The relations of the virgula, as then accepted, are fully discussed, and it is shown to consist of two distinct and separate parts, the distal part (or virgula proper as now understood), having its origin in the union of the longitudinal lines of the distal portion of the sicula.

As regards the development of the proximal thecæ, Wiman agrees in the main with Törnquist's observations, but differs in the interpretation of the facts. He regards the "connecting canal" as merely a bud which develops into the proximal part of the first theca. In the special form of *Diplograptus* figured by Wiman in illustration, the second theca develops directly from the first theca, and each subsequent theca "develops from the next more proximally situated theca of the opposite side." There is, therefore, no special structure such as a common canal to give origin to the theca in this species, nor is there any double longitudinal septum.

1893.
Wiman,
"Ueber *Monograptus,* Geinitz," 'Bull. of the Geol. Instit. of Upsala,' no. 2, vol. i.

This important paper was followed a few months later by another by the same author on the structure of *Monograptus.* The structural details described in this paper are worked out in specimens of *M. dubius.*

The sicula in this *Monograptus* is shown to be essentially similar in structure to that in *Diplograptus;* the mode of development of the virgula is also similar.

The growth lines at the point of origin of the first theca show a very marked discontinuity; the passage between sicula and theca is wider than in *Diplograptus,* and the first theca arises from the left side in the obverse aspect. This first theca does not grow downward at first and then curve upward as in *Diplograptus,* but grows upward at once. Each subsequent theca develops from the one immediately below, and not from a common canal. Wiman observes the same zig-zag-like union of the growth lines on the thecæ as noted in the case of *Diplograptus.*

1893.
Lake and Groom,
"On the Llandovery and Associated Rocks of Corwen," 'Quart. Journ. Geol. Soc.,' vol. xlix.

In 1893 Lake and Groom noted the occurrence of the upper part of the *Monog. gregarius* zone (Lower Birkhill) in the Corwen district of North Wales.

1893.
Sollas,
"On the Minute Structure of the Skeleton of *Monog. priodon,*" 'Geol. Mag.,' dec. 3, vol. x.

A short note on the structure of the skeleton of *Monograptus priodon* was published by Sollas in 1893. By means of transverse sections he was able to recognise in British specimens the three layers (comp. Richter, 1871): "an outer and inner, which are very thin, separated by a space now filled with calcite, and a thicker middle layer." He notices also and explains the thickening of the wall of the theca near its inner opening into the common canal and also along the free edges of the theca, but points out that this is caused "partly by an enlargement of the space between the layers, and partly by a thickening of the middle layer."

"The virgula," he holds, "would appear to possess no independent existence; it seems to be merely a thickening of the middle layer."

1893.
Hall, T. S.,
" Note on the Distribution of the Graptolitidæ in the Rocks of Castlemaine," 'Australian Assoc. for the Advance. of Science.'

T. S. Hall in this year contributed a short note on the " Distribution of the Graptolitidæ in the Rocks of Castlemaine."

The zones which he recognises from below upward are as follows : (1) *Tetragraptus fruticosus* with *Tetrag. quadribrachiatus* and *Phyllog. typus*, etc. ; (2) *Didymograptus bifidus* ; (3) *Didymog. caduceus* and *Phyllog. typus* ; (4) *Didymog. caduceus* without *Phyllog. typus* ; (5) *Loganograptus Logani* occurs at a high horizon, and probably occupies a fifth zone.

1893.
Barrois,
" Sur le *Rouvilligraptus Richardsoni* de Cabriéres," 'Ann. Soc. Géol. du Nord,' t. 21.

The same year Barrois described and figured a specimen of *Graptolithus Richardsoni*, from Cabriéres, and suggested the generic name of **Rouvilligraptus** for species of this type.

1894.
Perner,
' Études sur les Graptolites de Bohême,' pt. i.

In 1894 Perner published the first part of his extensive work on the " Graptolites de Bohême," a memoir in which he revised and carried on the Graptolite work so brilliantly begun by Barrande in 1851. This first part is devoted mainly to the study and description of the minute structure of the Graptolite skeleton.

An early section of the paper is devoted in part to a defence of Jaekel's work and his division of the genus *Monograptus* into *Pomatograptus* and *Pristiograptus*, and in part to a study of the thecal wall and its ornaments.

The greater portion of the memoir is, however, devoted to an account of the microscopic structure of the skeleton in *Monograptus*. Perner recognises four distinct layers (in distinction to the three described by previous observers) : (1) the epidermic layer, (2) the black layer, (3) the layer with coigns, and (4) the layer with " colonnettes," all of which are described in detail.

Perner shows that the virgula or " solid axis " does not lie in the dorsal sinus, as was believed by Barrande, but occurs in the third layer, under the black layer. It was very thin and elastic, but hardly ever loses its continuity in the fossil.

After giving an account of the structure of *Retiolites* as worked out by Holm and other Swedish authors, Perner concludes with a description of the microscopical structure of the skeleton.

1894.
Pritchard,
" Notes on some Lancefield Graptolites," 'Proc. Roy. Soc. Vict.,' n. s., vol. vii.

In 1894 Pritchard gave a description of a few forms of Graptolites from the Lancefield Beds of Victoria. He also records *Clonograptus flexilis* and *Tetragraptus quadribrachiatus* from the Lancefield locality.

1894.

Marr,

" Notes on the Skiddaw
Slates," ' Geol. Mag.,'
dec. 4, vol. l.

An important paper, dealing with the range and distribution of Graptolites in Britain, was published in 1894 by Marr, who gave a complete list of the Graptolites hitherto found in the Skiddaw Slates. He divides the Skiddaw Slates for the first time into the following horizons :

(2)
- (d) Millburn Beds—Uppermost Arenig or Llandeilo (Lower).
- (c) Ellergill Beds, characterised by *Didymog. fasciculatus, Azygog. cœlebs, Trigonograptus,* etc.
- (b) *Tetragraptus* Beds { Upper, with *Didymog. nanus.* { Lower.
- (a) *Dichograptus* Beds.

(1) *Bryograptus* Beds = Tremadoc Slates.

This paper and its zonal divisions formed the geological foundation for most of the subsequent graptolitic literature concerning the fossils of the Skiddaw Slates.

In this paper the author figures from the *Bryograptus* Beds specimens of *Bryograptus ramosus* and *Bryo. Callavei ?.*

1894.

Hall, T. S.,

" Note on the Distribution of the Graptolitidæ in the Rocks of Castlemaine," ' Australian Assoc. for the Advance. of Science,'

Hall, in 1894, published a more detailed account of the Graptolite zones of Castlemaine.

He argues in favour of the distinctness of the genera *Goniograptus* and *Loganograptus,* and points out that the auriferous bands of the colony begin above the base of the *T. fruticosus* zone, and range as high as the beds with *Phyllograptus.*

1894.

Lundgren,

Geol. Fören. Förh.,'
bd. xvi, heft 4.

Three letters dealing with the controversy on the terminology—*Dictyonema* or *Dictyograptus* — appeared in 1894.

In the first of these Lundgren considers that on the strict ground of priority Salter's name, *Graptopora,* should be substituted for *Dictyonema,* but thinks that it is unnecessary to adhere so rigorously to the laws of priority, and that in this case it would be mere pedantry to change a name so well established for a genus characteristic of so widespread an horizon.

1894.

Moberg,

" *Dictyonema* contra *Dictyograptus,*" ' Geol. Fören. Förh.,' bd. xvi.

In the second, Moberg enters into the question in considerable detail, and replies to the arguments brought forward, adopting " Hopkinson's name of *Dictyograptus,* on the ground that I find it to be the oldest name in the list of synonymies which is free from objections."

1894.

Törnquist,

" *Dictyonema* contra *Dictyograptus,*" ' Geol. Fören. Förh.,' bd. xvi.

The third and final communication on the subject was a letter written by Törnquist, who confines himself to the argument that the use of *Dictyonema* as the generic name of both a plant and an animal is quite in accordance with the rules laid down by the German Zoological Society.

1894.
Törnquist,
" Några anmärkingar
om Graptoliternas Ter-
minologi," ' Geol.
Fören. Förh.,' bd. xvi.

In a paper published during the same year Törnquist discussed some of the points in which he differs from Wiman as to the structure of *Diplograptus*. He controverts Wiman's statement that a " double longitudinal septum does not exist," as in most species of *Diplograptus* it certainly is present, either in a complete or incomplete form, " though in no known case does it extend to the extreme proximal part of the polypary."

He also discusses in detail Wiman's statement that there is no " common canal," and that the " connecting canal " is only part of the first theca. According to Wiman the whole rhabdosome consists of thecæ only (excluding the virgula and sicula), and Törnquist discusses whether this complete change in the nomenclature previously adopted is justified.

Törnquist also urges that the question depends entirely on what significance is attached to the word " theca," and he objects, in the third place, to Wiman's statement that because the sicula gives rise to only one bud, therefore *Diplograptus* is " monoprionidian."

1895.
Holm,
" Om *Didymograptus,*
Tetragraptus, och
Phyllograptus," 'Geol.
Fören. Förh.,' bd. xvii.

Two papers of first-rate importance, dealing with the structure of the Graptolites, were published in 1895.

The first was by Holm, " On the Structure of *Didymograptus, Tetragraptus,* and *Phyllograptus.*" The important results arrived at by him were due largely to the extreme skill with which he was able to isolate specimens from the matrix.

The first conclusion arrived at is, that " in the main, a complete conformity exists in the first stages of development of the polypary, both in the genera *Didymograptus, Tetragraptus,* and *Phyllograptus,* and in the family of the Diplograptidæ." In all these the sicula gives rise to one bud only on one side (left), and " from this bud is developed partly the second theca, partly the ' connecting canal,' which connects both halves of the polypary, and which in the first place gives origin to the third theca, and partly also the common canal, which connects the second theca with the succeeding ones." In the Monograptidæ, which Holm regards as degenerate from the Diplograptidæ, the " connecting canal " is entirely absent.

He agrees with Wiman that the sicula consists of two parts, but he differs in regarding the pointed end (initial part) as the original and oldest portion from which the apertural part was developed. This apertural part, having the same function as a theca, " might therefore be considered justly as the first theca of the polypary." He uses throughout the terms first, second, and third theca, but holds that the old name " sicula " is, nevertheless, a " convenient and significant one," and should be retained. Referring to the common canal, he writes: " The common canal, by which all the cells of the polypary are connected with one another—whether the different individuals have been developed by budding from the cœnosarc, or whether they are developed from, and connected with the one immediately preceding—must be considered to begin already in the sicula, even if it has been convenient, for the sake of description, to distinguish one part as the 'connecting canal.'"

The "*connecting canal*," as defined by Holm, is the canal which "arises almost simultaneously with the left theca and the common canal for the left half of the polypary," and which "crosses the dorsal side of the sicula and gives rise to the third theca and the common canal for the right half of the polypary." (This canal is not the same as that for which Törnquist used the term "connecting canal," and Törnquist has later proposed the name "crossing canal" for Holm's structure.)

The *Virgula.*—His observations lead him to the conclusion that "a virgula corresponding to that in *Diplograptus* and *Monograptus* cannot occur in the Dichograptidæ," at any rate "embedded in the dorsal side of the branches." Even in those cases in which the "sicula is embedded in the polypary, a virgula need not of necessity be present," and Holm fails to find any trace of one in *Phyllograptus.*

He points out that the "cylindrical chitinous thread which originates as a result of growth within the apertural end of the sicula" in *Diplograptus*, etc., as described by Wiman, "stands evidently in no relation whatever to the real virgula, but may be regarded as an apertural spine." And he draws especial attention to the fact that "the presence of a virgula has hitherto been considered as the main character of Graptolites" (Rhabdophora), "although such was never described or expressly mentioned except in the groups Diplograptidæ, Monograptidæ, and Retiolitidæ."

The structure of *Didymograptus minutus*, *D. gracilis*, and *D. gibberulus* is described in detail.

The genus *Tetragraptus*, of the development of which little was known, is shown by Holm to pass through the same early stage as a *Didymograptus* ("*Didymograptus* stage"). The four stipes arise by a "direct splitting of the common canal by a vertical wall" on each side of the connecting canal. This structure is worked out in specimens of *T. Bigsbyi.*

The development of the genus *Phyllograptus* is proved to be practically identical with that of *Tetragraptus*, but the branches, instead of having "four independent periderm walls, form a single, cruciform, four-winged, longitudinal septum." The sicula is embedded in the polypary, but no virgula has been detected.

This memoir is illustrated by excellent figures.

The second important paper, published in 1895, was by Wiman, and was perhaps even more far-reaching in its results.

The author commences with an account of the methods adopted by him for preparing the specimens examined.

1895.
Wiman,
" Ueber die Grapto-
liten," ' Bull. Geol.
Instit. Upsala,'
art. no. 6.

He next discusses certain controversial points with reference to the work of Törnquist and Holm. He accepts the facts obtained by Törnquist, and considers that their differences of opinion concern questions of terminology only.

By the term " theca " he means " a part of the test of a bilaterally symmetrical animal," and he repeats that " the individuals corresponding to the thecæ were developed from other similar individuals, and not from a substance contained in any common canal." " This," he writes, " is not only clear from the course of the growth-lines," but receives additional support from the recent discoveries made by himself as to the structure of the Dendroidea, in which there is no common canal.

He acknowledges the correctness of Törnquist's view that there is possibly always a longitudinal septum in *Diplograptus.*

He agrees generally with Holm as to the structure of *Didymograptus,* but considers it advisable to retain the word " sicula," though he regards it as probably the first theca, and he accepts Holm's view that the apical part of the sicula is the initial part and the youngest.

Wiman then gives a classification of the Graptolitidæ in general. This agrees in essentials with that proposed by Lapworth (1873), but is modified in some particulars.

The family of the Monograptidæ is first described, and the typical structure of a *Monograptus* is exemplified in *Monog. dubius, M. lobifer,* and *M. discus.*

He places the genus *Azygograptus* in the family of the Monograptidæ on account of its having only one row of cells, but he considers it to be a *Didymograptus*-like form in which one branch is missing and that it probably belongs to the Dichograptidæ, with which it is also contemporaneous.

Dimorphograptus may be considered as a transition form between the Diplograptidæ and the Monograptidæ.

The Dichograptidæ he divides into two sub-groups according as they resemble *Didymograptus* or *Tetragraptus.*

He points out that in *Didymograptus* the opening between the sicula and the first theca may not only occur on the initial or the apertural part, but may occupy very different positions on the bilaterally symmetrical sicula.

In the Graptoloidea he believes there is no " essential difference " between monopodial and dichotomous branching, though in the Dendroidea it would have more significance.

Wiman attaches considerable importance to Hopkinson's discovery of partition walls in *Tetrag. serra* between the common canal and the thecæ, and he indicates the analogy between this structure and that in the Dendroidea. He suggests that the " Graptoloidea are only the most superficial periderm of the Dendroidea ": " the proximal projections of the thecæ in the Dendroidea, which fill the common canal with many delicate tubes," were "still thinner in the Graptoloidea and less likely to be preserved, so that they have almost entirely disappeared, and have only exceptionally left behind traces of their existence."

The structure of the Diplograptidæ is next discussed, and *Diplograptus* **uplandicus**, n.sp., and *Climacograptus Kuckserianus*, Holm, are taken as examples. An interesting new form—*Climacog.* **retioloides**—is also described.

The family of the Phyllograptidæ is retained by Wiman with full knowledge of the demonstrations of Holm that its structure is essentially similar to that of *Tetragraptus*.

The histology of the Graptolite periderm is next discussed, and Wiman concludes that in *Monograptus* there are only three layers, a middle thick one and a thin one on each side of it.

Retioloidea.—In the group of the Retioloidea Wiman gives a full description of the structure of *Retiolites nassa*.

Dendroidea.—The latter part of Wiman's paper is of especial interest, containing many new facts of far-reaching importance connected with the group of the Dendroidea.

Wiman shows by means of sections that "in all Dendroidea there can be distinguished three different kinds of individuals: *nourishing individuals* (which he also calls thecæ, since they doubtless correspond to the thecæ in the Graptoloidea), *budding individuals*, and *sexual individuals* or *gonangia.*"

The Dendroid structure is described in great detail in the case of several species and genera of the Dendroidea. In all cases the budding individual never opens to the exterior, but itself gives rise to three new individuals, and these, as they grow, gradually fill up the whole of the cavity of the mother budding theca.

The species described include *Dictyonema* **rarum**, *D.* **peltatum**, *D.* **tuberosum**, and *D. flabelliforme*. *Dendrograptus* (?) **œlandicus**, *D.* (?) **balticus**, *D.* (?) **bottnicus**, and *Ptilograptus* **suecicus**.

The method of branching in the Dendroidea is carefully worked out in *Dendrog.* (?) *œlandicus*.

In *Ptilograptus suecicus* the structure is somewhat different. "The branches carry twigs" which spring out alternately to right and left, and "these consist of four individuals, opening one after the other."

The mode of growth of these various forms of Dendroidea is very different, some having a sicula, others having a disc from which a stem proceeds. In *Dictyonema peltatum* "a large number of branches spread centrifugally within a disc," and then rise up, "branch, anastomose, and join again by means of the ordinary connecting fibres." "The proximal ends of the branches" in this species do not "possess the intricate structure that characterises the distal parts," and resemble those in a *Monograptus*.

Wiman shows clearly from the foregoing that the older generic diagnoses of the

Dendroidea, drawn up as they were when practically nothing was known of their structure, are now of little value, but he admits that a new classification would as yet be impracticable and inadvisable.

As respects the systematic position of the Graptolites he considers that it is impossible to say more than that the " Graptolites are bilaterally symmetrical Invertebrates."

Phylogeny.—Wiman discusses the relationship between the Graptoloidea and the Dendroidea, and considers that " they are two parallel stocks of equal value in which the division of labour is performed in somewhat different ways." " In the Graptoloidea the different functions (while all the individuals of the first order remain the same) are shared among different organs." " In the Dendroidea, on the other hand, the different functions are shared by three different forms of individuals of the first order."

The theory that the Graptoloidea are descended from the Dendroidea seems to him very improbable; while the reverse idea, namely that the Dendroidea are descended from the Graptoloidea, meets with greater favour, as it is usual for " division of labour in a colony to bring about a difference of individuals."

The mode of life of the Graptolites is next dealt with, and Wiman concludes that the only possible view to take is that " the Graptolites, in some way or other, stood upright " and lived in the " deeper littoral regions."

The paper concludes with an Appendix giving an abstract of Ruedemann's discoveries of colonies of *Diplograptus* attached by their virgulas, and some of the points referred to there are discussed and criticised. Exception in particular is taken to Ruedemann's idea that these colonies were provided with a swimming bladder.

1895.
Ruedemann,
" Synopsis of the Mode of Growth and Development of the Graptolitic genus *Diplograptus,*"
' Amer. Journ. of Sci.,' ser. 3, vol xlix, no. 294.

Considerable light had already been thrown on the mode of life and development of the Diplograptidæ by Ruedemann's discovery of some remarkable specimens of forms referred by him to *Diplograptus pristis* and *Diplog. pristiniformis* (afterwards named *Ruedemanni*). The first notice of this was given in an abstract published by him in the American Journal of Science.

Ruedemann summarises his conclusions as follows:

(1) These two species grow in " compound colonial stocks which appear in the fossil state as stellate groups."

(2) " The virgulæ are joined to a central connecting stem, the 'funicle' of Hall, which is mostly extended to a vesicle of quadrangular shape." The funicle is " enclosed in a central disc " which is a " thick, chitinous capsule " also quadrangular in shape.

(3) " The central disc is surrounded by a verticil of oval capsules," in number four to eight or more. Some of these oval appendages are seen to contain siculæ

" which radiate from an axial club-shaped protuberance within the vesicle, to which they are joined by the filiform prolongation of their pointed ends." Ruedemann compares these vesicles with the gonangia of recent Hydrozoa.

(4) Overlapping the gonangia and even the proximal ends of the stipes, there is an organ which he compares with the air-bladder or pneumatocyst of the Discoideæ and which he regards as having acted as a float.

(5) The siculæ " at the time of developing the first two hydrothecæ, possess a quadrangular plate, joined by a small node in the centre to the end of the filiform proximal process "; while at a slightly later stage of development four oval impressions can be seen around the central node. This quadrangular plate (or probably vesicle) develops into the pneumatocyst, the central node into the funicle and central disc, and the small oval impressions probably indicate the gonangia.

(6) From the position of the siculæ at the remote end of the stipes the " so-called proximal sicula-bearing end of the single stipes appears in the compound colonial stock as the distal one." " The stipe grows backward towards the centre and the sicula is carried to the distal end."

(7) With regard to the affinities of the Graptolites, Ruedemann points out that by the " possession of a pneumatocyst and the arrangement of the reproductive organs at the bases of the stipes, the colonial stocks of *Diplograptus* had a general similarity to those of certain Siphonophora, while the chitinous structure of the hydrothecæ and gonangia can be only referred to the Sertularians."

1895.
Matthew, G. F.,
" Two new Cambrian Graptolites with Notes on other Species of Graptolitidæ of that Age," 'New York Acad. Sci. Trans.,' August 29th.

Matthew described in 1895 some new species of *Clonograptus*, *Bryograptus*, etc., from the lower part of Division 3 of the St. John Group. *Clonograptus* **proximatus**, sp. nov., resembles *Clonog. tenellus* in many respects; unlike the latter it occurs in association with *Dictyonema flabelliforme* and not above it. Matthew distinguishes *Clonograptus* from *Bryograptus* by its being " devoid of the sicula, or with the sicula obscure, absorbed, or merged in the funicle." Four species of *Bryograptus* are described and figured : *Bryog.* **patens**, *B.* **spinosus**, *B.* **lentus**, *B.* **retroflexus?** A fragment of *Callograptus* is figured and two specimens of *Dictyonema flabelliforme*, showing " short rootlets developed from the proximal end of the sicula." As regards the occurrence of the last-named form in America, Matthew states that the species was not a " solitary Graptolite," as in some parts of Europe, but was associated sparingly with *Bryograptus* and *Clonograptus*.

As to the phylogenetic relationships of the Graptolites he writes, " the succession of the Dichograptidæ in the Cambrian and Lower Ordovician is a good exemplification of increased condensation of structure due to selection ; for the many-branched forms of the former are gradually replaced by the *Tetragrapti* and these by the *Didymograpti* of the Upper Arenig." He repeats his former view that the *Bryograpti* were the ancestors of *Dictyonema*.

1895.

Lake,

" The Denbighshire
Series of South
Denbighshire," 'Quart.
Journ. Geol. Soc.,'
vol. li.

During the same year Lake noted the existence of both a Wenlock and Lower Ludlow Graptolitic fauna in the Denbighshire series of S. Denbighshire.

1895.

Perner,

"Études sur les Grapto-
lites de Boheme," pt.
ii ; "Monographie des
Graptolites de l'Étage
D."

The second part of Perner's work on the Bohemian Graptolites, including the species found in Étage D, was published in 1895. Several new species are described and figured, but owing to the poor state of preservation and fragmentary condition in which these are found in Bohemia, any certain identification of them is a matter of difficulty. The following are described and figured :

Dichograptus (?) **leptotheca**, n. sp.

Tetragraptus caduceus.

Didymograptus.—(Group A)—*D. bifidus, D. Murchisoni, D. denticulatus, D.* **oligotheca,** *D. indentus* var. *nanus, D.* **spinulosus,** *D.* **clavulus,** *D.* **Barrandei,** *D.* **Lapworthi,** *D. bifidus* var. **incertus,** *D.* **vacillanoides.** (Group B)—*D. v-fractus.* (Group C)—*D. pennatulus : D.* **linguatus,** *D.* **lonchotheca,** *D. pennatulus* var. **hamatus,** *D.* **retroflexus.**

Dicellograptus anceps.

Cryptograptus tricornis.

Climacograptus tectus.

Diplograptus pristis, D. euglyphus var. **angustus,** *D.* **lobatus,** *D.* **lingulitheca,** *D.* **terres,** *D.* **insculptus,** *D. rugosus* var. **Fritschi,** *D. truncatus, D. foliaceus* var. *vulgatus.*

Dendrograptus constrictus.

A table is given showing the range of each species.

A useful list of works published on the Graptolites in general and an historical account of the Graptolites in Bohemia, are prefixed to this second part of Perner's work.

1895.

Nicholson and Marr,

" Phylogeny of the
Graptolites," 'Geol.
Mag.,' dec. 4, vol. ii.

An important paper bearing on " The Phylogeny of the Graptolites " was published by Nicholson and Marr in 1895. The authors conclude (1) that " the character of the hydrothecæ is the most important point to retain in view in separating different families of the Graptoloidea "; and (2) that the next most important point as "indicating genetic relationship," is the angle of divergence of the stipes; while the number of stipes, on which the present classification of the Graptolites largely depends, is relatively insignificant.

In consonance with these conclusions, the authors take the eight known species of *Tetragraptus,* and group round each of them those species of *Dichograptus,*

Bryograptus and *Didymograptus* which agree with them most closely in the character of the thecæ and the "angle of divergence." Group 1 contains *Bryograptus Callavei*, *Tetrag. Hicksii*, and *Didymog. affinis*; Group 3 *Bryog. ramosus*, *Tetrag. fruticosus* and *Didymog. Murchisoni*; Group 6 *Tetrag. Bigsbyi*, *Didymog. gibberulus*, and an unknown *Dichograptus*, and so on. In those cases where, in any corresponding place in a given group, there is no species with the required characters to fill the gap, the authors confidently assert that further research will probably reveal its existence. The authors hold that the members of each of these groups are phylogenetically related, and that it is very difficult to understand how the "extraordinary resemblances between the various species of *Tetragraptus* and *Didymograptus* have arisen, if, as usually supposed, all the species of these genera have descended from a common ancestral form for each genus, in the one case four-branched, in the other case two-branched;" "on the other hand, it is comparatively easy to explain the more or less simultaneous existence of forms possessing the same number of stipes, but otherwise only distantly related, if we imagine them to be the result of the variation of a number of different ancestral types along similar lines." They suggest that the genus *Monograptus* also may contain "descendants of more than one family."

The authors point out that if their conclusions are correct, the present nomenclature would have to be eventually altered. Meanwhile they propose to retain such names as *Monograptus*, *Didymograptus*, etc., as "generic" names, but the "species placed under these different groups do not belong to definite genera" (in the strict biological sense of the word): they constitute cases of what Buckman terms the "hetero-genetic homœomorphy" of forms which are only distantly allied to one another.

They adduce briefly reasons for this "special case of mimicry, and endorse Clement Reid's suggestion that the variations in form may be connected with the supply of food"; the necessity of providing food brought about a reduction in the number of stipes, and also a change in the direction of these stipes. Those series of hydrothecæ which were farthest apart would have a better chance of obtaining food, and thus the "angle of divergence" increased from a very small angle until it reached its maximum of 360° in *Phyllograptus*, *Diplograptus*, etc. Variations in the form of the hydrothecæ may also be explained on the same ground.

In a note to this paper, Nicholson and Marr suggest the new specific name of *Tetragraptus* **inosculans** for those forms which resemble *Tetrag. Bigsbyi*, but in which the stipes are in contact or even more or less fused.

1895.
Hall,
" Notes on *Didymograptus caduceus*, with remarks on its synonymy," 'Proc. Roy. Soc. Victoria,' vol. viii.

T. S. Hall discussed at considerable length the question of the synonymy of *Didymograptus caduceus* in a paper published in 1895. He concludes that Salter's name *D. caduceus* has priority over *D. gibberulus*, and therefore the latter should fall out of use.

1896.

Ruedemann,

" Development and
Mode of Growth of
Diplograptus," 'N. Y.
State Geol. Annual
Report' for 1894.

Ruedemann's preliminary notice on the development of *Diplograptus* was followed about a year later by a more complete paper fully illustrated. His previous views are here repeated and amplified, but in a few cases they are somewhat modified as the result of a further investigation of additional material.

With regard to the general form of the complete frond of *D. pristis* and *D. Ruedemanni,* which consists of many stipes arranged so as to radiate outwards from a central point—the funicle, these stipes are of three different lengths, and are connected together to an approximate central form by their virgulas, "or more properly hydrocauli." Ruedemann distinguishes between the virgula proper and the hydrocaulus, "which forms the connecting stem and is a canal containing the virgula of the rhabdosome included in its distal part."

As to the function of the central disc "which encloses the funicle," it may have served to support the bases of the stipes, it was "certainly a protection to the funicle," and it may have served as a float.

The *basal cyst* consists of "two segments resting in the middle on both sides of a subquadrate base, and the test is comparatively thin. The gonangia and rhabdosomes which proceed from the central disc and funicle, occur below the basal cyst. Ruedemann at first regarded this organ as a "float" or swimming bladder, and believed that the Graptolites floated, on account of (1) the extreme length and thickness of the hydrocaulus in some specimens, which "makes it difficult to imagine how such an extremely thin stem could have supported the long and broad rhabdosome in any other than a suspended position"; (2) the absence of any evidence of the sessile nature of the colonies; and (3) their wide distribution, which would be accounted for by their floating habit. This view of their floating habit, however, Ruedemann relinquished in this second paper, on account of the discovery of a large slab in which more than a hundred colonies of *D. Ruedemanni* are spread out regularly. He considers that the "improbability of such an array of nicely ordered, apparently undisturbed stellate groups having been drifted together is obvious." The hydrocauli and rhabdosomes possess only a very slight flexibility, and therefore it was only where there were no currents in the sea that one could hope to find entire colonies.

He abandons the floating theory previously held by him, and suggests that the basal cyst was "buried in the detritus" on the floor of the ocean, and served to procure stability for the colony.

He compares the gonangia or vesicles containing the siculæ with certain organs in the Sertularians, and considers that they resemble in all the more important features the Sertularid gonangium, which contains a cylindrical column, the "blastostyle," and he thinks that the possession of these organs and also their structure are arguments for the hydrozoan nature of the Graptolites.

Ruedemann next discusses briefly the various supposed reproductive organs described by previous authors, and suggests that the "bi-thecæ" observed by Holm in *Dictyonema* should rather be compared with the nematophores of the Plumularians than with gonangia.

He deals with the development of the sicula at some length, and considers that while there is "conclusive" evidence that numerous siculæ left the gonangia, it is also clear that others did not sever their connection with the parent colony, but grew out into new rhabdosomes.

The development of *Diplog. pristis* is worked out by him in detail, and his results may be summarised as follows:

(1) The detached sicula is attached by means of a small round node to a basal appendage.

(2) The hydrocaulus gradually lengthens and more and more thecæ are formed.

(3) "The node becomes the central disc and funicle. The sicula produces at first one theca, then a second, third, etc."

(4) The growth of the gonangia (four small capsules) begins with the budding of the first thecæ.

(5) The gonangia mature and open, the siculæ, however, remaining connected with the parent colony, the basal cyst, funicle, etc., are all present.

(6) The siculæ grow out to rhabdosomes.

(7) A second generation of gonangia begin to grow, and the process is continued.

The "number and length of the rhabdosomes increase with the age of the whole colony."

Affinities.—As respects the affinities of the Graptolites, he merely states that they should be placed in a distinct class—the Rhabdophora.

He concludes his paper with a reply to some of the objections raised by Wiman, especially with regard to the terms employed for the various structures. He maintains that the "central discs" of *Dichograptus* and *Diplograptus* are "genetically identical," but he relinquishes the employment of the term "funicle" for the connecting stem of *Diplograptus*. He argues also in favour of the "gonangia"-like nature of the capsules described by him.

1896.
Gürich,
" Bemerkungen zur
Gattung *Monograptus*,"
' Zeitsch. d. deutsch.
geol. Gesell.,' vol. xlviii.

In the same year, 1896, Gürich published a paper, "Remarks on the genus *Monograptus*," in which he discussed the structure, the shape of the thecal aperture, and also the biology of the Monograptidæ in general.

As regards the histology of the Graptolite skeleton, he recognises the four structural layers described by Perner, but considers that the appearances are capable of a different explanation. He adds many new facts regarding these, and considers that the layer " with coigns " is " not the organic structure of a special layer of the rhabdosome wall, but a peculiar calcite deposit formed during the process of fossilisation, the formation of which was only possible when the skeleton of the rhabdosome wall—namely the black layer—was surrounded by an organic integument."

The " palisade " layer (layer with colonnettes) he also regards as another form of the calcite deposit surrounding the black layer. Gürich concludes from his investigations that the chitinous skeleton of the rhabdosome in the living condition was surrounded by a skin, but that it is impossible to say of how many layers this skin consisted, or what were their particular histological peculiarities. The existence of such an outer skin is proved by the presence of growth-lines, and the chitinous skeleton, instead of being external, is mesodermal. These conclusions of Gürich are in accordance with Wiman's views ('Ueber die Graptolithen,' 1895).

The form of the aperture in *Monograptus priodon* is dealt with, and Gurich disputes Jaekel's idea of a laterally expanded projection, pointing out that the theca in this form is merely a " tube whose open oval end is bent back towards the sicula." He gives a figure showing a schematic reconstruction of the cells of this species.

The paper concludes with remarks on the mode of life of the Monograptidæ. He considers that it is very improbable that they were attached to the sea floor, and argues that their geological distribution, their method of preservation, etc., speak in favour of their being Plankton.

Ruedemann's discovery of colonies of *Diplograptus* justifies us, he considers, in concluding that the Monograptidæ possessed a swimming bladder, and also that a large number arose from one and the same stock. In this case the relationship between the rhabdosome, sicula and disc becomes of primary importance, the form of the aperture is secondary, and is the result of such a relationship. He re-figures and discusses his schematic representation of the differences between the *Monograptus erecti* and *M. reversi* groups.

1896.
Gurley,
" North American Graptolites," ' Journ. Geol.,' vol. iv, no. 1.

In 1896 Gurley published a paper entitled " North American Graptolites," in which he gives a complete list of American forms, discusses the synonymy of many of the genera and species, and describes a large number of new species, without, however, figuring them.

Description of Species.—The following forms are referred to or described: *Phyllograptus?* **cambrensis,** *Bryograptus?* **multiramosus,** *Dichograpsus* **remotus,** *D. abnormis,* *Tetragrapsus* **acanthanotus,** *Didymograpsus* **bipunctatus,** *D.* **perflexus,** *D. geminus,* *D. hirundo,* *D.* **convexus** and *D.* **sagitticaulis.**

The generic name *Stephanograptus* should, he considers, take precedence of *Helicograptus* and *Cœnograptus* : two new species are described : *S.* **crassiusculus** and *S.* **exilis.** *Azygograptus* is represented doubtfully by one species *A. ? Walcotti.*

Other forms noted are *Leptograptus?* **macrotheca,** *Dicellograptus intortus* var. **poly-thecatus,** *D.* **Gurleyi,** *D. elegans;* *Dicranograptus furcatus, D. Nicholsoni,* var. **arkansensis,** var. *whitianus,* var. **parvangulus,** var. **diapason;** *Climacograptus antiquus, C. caudatus,* var. **laticaulis,** *C.* **oligotheca,** *C. cœlatus, C.* **phyllophorus;** *Diplograpsus stenosus; Glossograptus* **arthracanthus;** *Lomatoceras* (he thinks that this name has clear priority over *Monoprion* or *Monograptus,* and so far as he can ascertain has never been used for the name of an insect); *Gladiolites* (instead of *Retiolites*) *venosus; Reteograptus Geinitzianus; Dictyonema cf. neenah, D.* **perexile,** (=*D. delicatulum,* Dawson, preoccupied), *D.* **actinotum,** *D.* **Blairi;** *Desmograptus* **macrodictyum,** *D.* **devonicus;** *Dendrograptus* **unilateralis,** *D.* **arundinaceus,** *D. cf. serpens.*

Gurley describes three species of *Caryocaris* which "from its resemblance to *Dawsonia* may be a Graptolite"; the species are *C. Wrightii, C.* **oblongus,** and *C.* **curvilatus.** *Dawsonia* is represented by two new species: *D.* **monodon** and *D.* **tridens.** A new genus—**Phycograptus**—is proposed and two species of this genus are described: *P.* **brachymera** and *P.* **lævis.** *Thamnograptus Barrandii* is also referred to, and it is suggested that the "thecæ appear to have been excavated out of the substance of the branch."

1896. *Gurley,* " North American Graptolites," ' Journ. Geol.,' vol. iv, no. 3.	The second part of this paper, which was published three months later, deals mainly with the " Vertical Range of the Graptolites in America," and detailed tables are given, showing their distribution and range.

In addition, a new species is described, viz. *Diplograpsus* **Ruedemanni,** being one of the forms mentioned by Ruedemann as *D. pristiniformis* in his paper on the "Mode of Life of the Graptolites."

1896. *Elles and Wood,* " On the Llandovery and Associated Rocks of Conway," ' Quart. Journ. Geol. Soc.,' vol. lii.	In 1896 Elles and Wood recorded the existence of an Upper Birkhill graptolitic fauna at Conway, North Wales, including the zone of *Rastrites maximus.* They also found representatives of the faunas characteristic of the overlying Tarannon and Wenlock Shales.

1896. *Hall, T. S.,* " On the Occurrence of Graptolites in North-Eastern Victoria," ' Proc. R. S. Victoria,' vol. ix (new series).	In the year 1896 T. S. Hall recorded the existence of Ordovician Graptolites from two or three localities in North-Eastern Victoria. He considers that judging from the species of *Dicellograptus, Dicranograptus, Diplograptus* and *Climograptus* identified by him the beds appear to belong to the " higher part of the Ordovician."

In 1896 Wiman published the results of his researches on the structure of the Dendroidea by a paper on a new species of *Dictyonema*—**D. cavernosum.**

1896.
Wiman,
" Ueber *Dictyonema*
cavernosum," 'Bull.
Geol. Inst. Upsala,'
vol. iii, art. no. 1.

He gives special attention to the structure of the proximal end and shows how the first thecæ originate from the disk of attachment (" haftscheibe ").

By means of cross sections he finds that two individuals of different sizes appear to arise from the disc of attachment, and he gives various explanations as to their origin. He inclines to the view that the larger individual was the older and was originally free-swimming, and that from it the smaller budding individual was developed. Another explanation which he considers probable is, that both thecæ were produced from an older and non-chitinous individual which was originally free-swimming.

1897.
Törnquist,
" On the Diplograptidæ
and the Heteroprionidæ
of the Scanian *Rastrites*
Beds," 'Acta Reg. Soc.
Physiog. Lund.,' vol.
viii.

In 1897 Törnquist published the first part of his Monograph on " The Graptolites of the *Rastrites* Beds." In this he deals with the Diplograptidæ and the Heteroprionidæ, and describes and figures several species of *Diplograptus* and *Climacograptus*, some of which are new to science.

Description of Species.—The descriptions and illustrations are excellent, and are specially concerned with the elucidation of the detailed structure of the proximal end, which had previously remained almost unnoticed. The species described include the well-known forms : *Climacograptus scalaris, C. rectangularis, C. undulatus, Diplograptus palmeus, D. folium, D. acuminatus, D. cometa, D. tamariscus, D. bellulus, D. longissimus,* and in addition two new species : *Climacog.* **medius** and *Diplog.* **cyperoides**, and a new varietal form, *Dimorphog. Swanstoni* var. **Kurcki.**

Törnquist considers that " at present it is advisable to retain the genus *Diplograptus* undivided," and he therefore does not adopt the sub-generic names of *Petalograptus* and *Cephalograptus.*

Terminology.—Törnquist discusses various questions of terminology, and endeavours to bring the nomenclature employed by Wiman, Holm and himself into uniformity. He considers that the terms " obverse " and " reverse " aspects are liable to less ambiguity than those of " sicula " and " anti-sicula " side, and he also prefers the names " primary " and " secondary " to the " left " and " right " for distinguishing between the two series of thecæ. He proposes the term " prolific side " for that side of the sicula which " communicates with the proximal cavity of the rhabdosome." The opposite side he calls the " dorsal " side.

He also suggests the new name " virgella " for the " so-called proximal prolongation of the virgula."

He discusses in some detail the question of the exact application of the words " thecæ " and " common canal," and thinks that the term " theca " is a convenient one for that part of the common chamber " which is capable of being broken off." He therefore considers that it is " advisable to retain the word theca in its original sense," and if a new word be necessary, to give one to " that portion of the periderm which corresponds to an individual zooid once living within " (that is to say the theca and its contributory part of the common canal combined).

Although Törnquist agrees with Holm that the sicula is the covering of the first zooid, he considers it very advisable to distinguish " in practice, between sicula and thecæ," and therefore does not apply the term " first theca " to the sicula.

Range and Distribution.—Törnquist prefaces his paper with an account of the seven Graptolite Zones in the *Rastrites* Beds of Scania. These are very similar to those given by Tullberg, with the addition that the lowest zone, *i. e.* that of *Diplograptus acuminatus*, is here recognised in Scania for the first time.

1897.
Perner,
" Études sur les Graptolites de Boheme," Prague, pt. iii, sect. *a.*

In 1897 Perner published the first section of the third part of his Monograph on the " Graptolites of Bohemia." This part contains a description of the species of Graptolites found in the lower layers of the band E. i., which corresponds to the Llandovery-Tarannon beds of England.

Description of Species.—In the genus *Diplograptus* Perner describes and figures the well-recognised species of *Diplograptus palmeus, D. bellulus, D. (Glyptograptus) vesiculosus, D. tamariscus, D. sinuatus, D. ovatus,* and *D. modestus.*

In the genus *Cephalograptus* he includes *C. cometa* and *C. folium.*

The genus *Climacograptus* comprises *C. phrygionius, C. scalaris,* and the new species *C.* **bohemicus.**

Rastrites is represented by *R. Linnæi* (= *R. fugax,* Barr.), *R. peregrinus,* including two new varieties, var. **longispinus** and var. **approximatus,** and a new species, *R.* **Richteri.**

The group of Leptopodes in the genus *Monograptus* includes *M. argutus, M. attenuatus, M. cyphus, M. limatulus,* and a new species, *M.* **tubiferus.**

In the group of the Orthopodes Perner describes *M. leptotheca, M. Hisingeri, M. crenulatus, M. Sedgwicki, M. Halli,* and a new variety of *M. jaculum, i.e.* var. **variabilis.**

The group of Helicopodes contains *M. planus* (= *M. resurgens,* Linnars.), *M. convolutus, M. proteus, M. triangulatus, M. turriculatus, M. communis* and a new species, *M.* **mirus,** Barr., sp. ms.

In the group of the Opisopodes, Perner discusses at some length the exact identity of *Monograptus Becki,* and shows that Barrande had confused three distinct forms, all from different zones, under this single specific name. He refigures the true *M. Becki* and also describes *M. lobiferus,* two new varieties (var. **Lapworthi,** and var. **undulatus**), *M. runcinatus, M. crispus, M. dextrorsus, M. distans, M. Clingani, M. (Rastrites) gemmatus,* and the following new species: *M.* **retusus,**

M. **Marri,** *M.* **Holmi,** *M. densus,* *M.* **Nicholsoni,** *M. Clingani* var. **tenera,** and var. **Hopkinsoni.**

The group of Kamptopodes contains *M. nuntius.*

The genus *Retiolites* is represented by *R. perlatus* and *R. obesus.*

1897.
Lapworth,
" Die Lebensweise der Graptolithen," in *Walther's* " Ueber die Lebensweise fossiler Meeresthiere," ' Zeitsch. d. deutsch. geol. Gesell.,' vol. xlix, Heft. 2.

In the year 1897, Walther, of Jena, published in the pages of the ' Zeitschrift der deutschen geologischen Gesellschaft' a memoir on the " Mode of Life of Fossil Sea Animals." This memoir includes (pp 238–258) an article by Lapworth on " The Mode of Life of the Graptolites." In this, Lapworth dealt with this subject in the same comprehensive manner as he had already dealt with the classification of the Graptolites in 1873, and their distribution in 1879–80.

He adduces the facts known with reference to the relative distribution of the Graptolites in the various types of sediment within the British Isles, and shows that these facts go to prove that :

(1) The presence of Graptolites in any of our British rock-layers stands in some way related to the presence of carbonaceous matter in the sediments in which the Graptolites occur.

(2) Although Graptolites are found in all our Proterozoic sediments, yet they are normally and typically restricted to regions where much carbonaceous matter was deposited.

(3) The relative abundance of Graptolites in any single layer or rock-group is in some way connected with the calm of the sea-floor where the carbonaceous deposits were laid down (for the material in which the Graptolites lie embedded is usually so impalpable in grain that the gentlest current would have removed it) ; and that the most typical and richest British Graptolite-bearing beds are those which accumulated at the slowest rate.

It is next shown that the Graptolites themselves did not supply the carbonaceous matter in the sediments, nor did they live where they are now found. The carbon-producing organisms must also have been strangers to the locality, and it is inferred that these must have been floating sea-weeds.

The distribution of the black sediments and their thinness both point in the same direction ; they are deposits formed mainly from the relics of floating sea-weeds, arranged in quiet waters parallel to the shore, having been drifted by currents and sinking when waterlogged to the bottom. The presence of Graptolites associated with these is in harmony also with the abundance of Hydroid organisms living on the fronds of the Sargassum sea-weed of the present day, which have been drifted from the shore, and become accumulated in special regions of the ocean or swept by currents into almost all latitudes.

These conclusions being conceded, we have what appears to be the clue to the mode of life and the general line of evolution of the Graptolites, including both virgulate and non-virgulate forms. The Cladophora or non-virgulate forms, like the modern Sertularians and their allies, must have been fixed to rocks in the shallow parts of the sea-shore, and therefore stationary, or to floating objects of a comparatively large size. The Rhabdophora or virgula-bearing Graptolites, on the other hand, were attached to floating sea-weeds, and were therefore drifted far and wide over the waters of the sea at the mercy of winds and currents. The non-virgulate forms grew vertically upwards, and like their modern representatives, were more or less tree-like. The virgulate forms hung vertically downwards, being pendent to the under side of the sea-weed by a thread-like fibre, which in its earliest stages constituted the " nema " proceeding from the apex of the sicula, and which, in the later forms of the Rhabdophora, growing with the general growth of the rhabdosome, constituted the " solid axis or virgula." In other words the Rhabdophora form a special section of the Graptolites, modified for a pseudo-planktonic mode of existence. The modification commenced in later Cambrian times, within the limits of the genus *Dictyonema*. Some forms of this genus are provided with a short stem and a disc of attachment, and some examples, even of the same species, may have grown vertically, while others may have assumed a pendent position. Abundant examples, however, are met with in which the stem is lengthened out into a long, thread-like hydrocaulus or nema. In these forms the pendent mode of attachment is the only one possible. In harmony with this we find that once this change from dendroid to pendent is initiated, the Graptolites become world-wide in their distribution and remarkable for their abundance.

In the successive stages of the evolution of the Rhabdophora in time, the number of branches is gradually reduced, and they become turned more and more backwards and upwards towards the light. A first stage is typified by the oldest family (the Dichograptidæ), in which the nema is lengthened, and within the limits of which the branches bearing thecæ, originally turned downwards owing to their pendent position, turn in the successive genera backwards and upwards towards the line of the nema. The angle of divergence of the branches gradually increases thus from 0° to 360°, and in the Phyllograptidæ the branches, which by this time have been reduced to four in number, attach themselves to each other dorsally and grow backwards up the line of the nema, and the thecæ have practically recovered their upward direction.

In a succeeding stage (the Diplograptidæ) the branches are reduced to two in number, and the nema, which apparently lengthens with the growth of the organism, has become a typical virgula.

In the final stage (the Monograptidæ) the branches are typically reduced to one, and the evolutionary series is closed.

It is pointed out that difficulties exist, especially as regards the Dicello-grapta (Leptograptidæ and Dicranograptidæ), but if it be accepted, even as a broad generalisation, that the typical nema- and virgula-bearing Rhabdophora were pendent forms attached to floating sea-weed, this generalisation harmonises the previously known facts as respects their special mode of occurrence, their universal dissemination, their superabundance in carbonaceous deposits, their restricted geological range, and their broad lines of evolution in time.

1897.

Frech,
"Lethæa Geognostica,"
in continuation of
Roemer's 'Lethæa
Palæozoica,' vol. i.

In 1897 Frech published an extended monograph on the Graptolites in general, in his continuation of Roemer's great work 'Lethæa Palæozoica,' which was left unfinished at his death.

In addition to epitomising and illustrating the discoveries and conclusions of previous observers, Frech made many new theoretical suggestions, especially as regards the classification of the Graptolites.

Organisation of the Graptolites.

Frech commences with discussing the organisation of the Graptolites, dealing first with :

A. *Organisation of the Fully-grown Animal.*—Broadly speaking, he adopts the views of Ruedemann with relation to the so-called pneumatophores, gonangia, etc., and extends them to all the Diplograptidæ. In the Dendroidea, however, he regards the gonothecæ of Wiman and Holm as corresponding to the nematophores of the Hydrozoa.

As regards the structure of the test, Frech does not adopt Perner's view of the existence of a fourth layer.

Frech lays great stress on the free-swimming or floating character of the Rhabdophora, and explains many of the peculiar structures found in Graptolites by the assumption that they were connected with their swimming mode of life. He recognises four different modifications of swimming organs :

1. The bladder in *Diplograptus physophora* is a "rudder-like propelling organ." The same is the case with the vesicle in *Monograptus pala*, of which he gives a theoretical drawing of colonies attached to a float. 2. The so-called disc at the base of *Climacograptus bicornis* he regards as having served in some way for the movement of the animal. 3. A third modification is found in *Cephalograptus*, where the "whole surface of the hydrothecæ has widened and taken on a rudder-like form." He gives a theoretical drawing of *Petalograptus folium* attached to a float. 4. A fourth modification occurs in *Dicellograptus divaricatus*, in which a membrane exists between the branches.

All these aided in giving the Plankton colony-animals an undulating up-and-down movement rather than a forward one. As it is doubtful whether all the *Monograpti* possessed floats, the "float," therefore, must not be regarded as an organ of systematic importance. In those forms that have a float, "the axis is the rudder-stem, and the float itself the rudder-fins;" the fixed Dendroidea have no such organ.

Frech does not regard the spine-like appendages to the apertures of the cells, as found in the Glossograptidæ, as of systematic importance, but as protective organs, "perhaps also sensory."

Frech divides the Graptolites into two main groups, differing from each other in the development of the axis, the rudder floats, and common canal, and also in their embryonic stages.

ORDER 1: *Axonophora.*—This includes *Diplograptus, Climacograptus, Dicranograptus, Dicellograptus* and *Monograptus.* The sicula is distal in position, and the later polyps insert themselves between the apex of the sicula and the central bladder. The apertures are directed inwards (proximally), a virgula is present, a common canal absent. The mode of life is planktonic, with a passive or active movement.

ORDER 2: *Axonolipa.*—This includes the Dichograptidæ and the Dendroidea. The sicula is proximal, and the younger cells grow distally, their apertures being directed outwards. A common canal for the cœnosarc exists in the Dichograptidæ, but not in the Dendroidea. "A virgula has not been observed in any of the main types of this order, in spite of numerous microscopic sections."

The Retiolitidi, according to Frech, correspond in the structure and arrangement of the hydrothecæ, and in the presence of an axis, to *Diplograptus;* and *Retiolites* is a "younger derived form" of that genus. As regards the Dendrograptidi, Frech accepts Wiman's opinion that they had no axis, and he considers that there are many points of contact between the Dendrograptidi and the Dichograptidi.

B. *Embryonal Development of Graptolites.* — Frech gives a summary of Ruedemann's and of Wiman's work, and accepts their conclusions. As regards Ruedemann's work, he seems to think that in addition to the primary hydro-rhabdosomes, there should be "secondary hydro-rhabdosomes," arising direct from the proximal part of the virgula, or from the central plate, and having no siculæ, thus producing a colony like that seen in *Retiograptus,* and he tries to account for the paucity of these non-siculate secondary hydro-rhabdosomes.

Frech emphasises strongly his opinion that "an analogy exists between the development of the Axonolipa and the Tabulate Corals," while "the embryonal polyp of *Phyllograptus* has the greatest similarity to the primary calyx of *Pleurodictyum.*" He considers that the terms "Hydrozoa" and "Anthozoa," which are founded on living forms, are in no way applicable to their Palæozoic ancestors. He places the Graptolithidæ as the third member of the following series :

(A) 1, Archarocyathinia. (B) 2, Acalephæ; 3, Graptolithidæ; 4, Tabulata; 5, Stromatoporoidea. (C) 6, Pterocorallia. All except the first and last "take

t

the place of the modern Hydrozoa, and are perhaps phylogenetically related to them."

C. *The Position of the Graptolites in the Zoological System.*—The Axonophora and Axonolipa are very distinct, the only point of similarity between them being the form of the sicula. He discusses at great length the relationship of the Graptolites to the Sertularians, but thinks that all the resemblances are superficial. Between the Dendrograptidi and living Plumularians, however, there is much direct relationship in the organisation of the grown animal, and the only main distinction between them is the want of a common canal in the former, and the shape of the embryonic polyps.

D. *Classification.*—The classification adopted by Frech differs in many respects from that proposed by British and Swedish workers, and the number of genera and species, are, in the majority of cases, materially decreased. His classification is as follows :

ORDER I.—AXONOLIPA.

1. Dendrograptidi.—(Hydrothecæ dimorphous, larger nutritive and smaller protective polyps, branching irregular.)

       (a) *Dictyonema.*—Nineteen species are recorded, and *D. flabelliforme* and *D. tuberosum* are re-described.

       (b) *Callograptus.*—*C. Salteri* and *C. elegans* are described.

       (c) *Dendrograptus.*—Nine species are recognised and several figured.

       (d) *Ptilograptus.*—Four species are recognised, and *P. acutus* is described and figured.

(Frech considers that *Thamnograpsus, Inocaulis* and *Corynoides* are " incompletely known, and their systematic position uncertain.")

2. Dichograptidi.—(This includes the Dichograptidæ, Leptograptidæ, and the Phyllograptidæ.) Free-swimming hydrothecæ, one kind only, branching dichotomous.

    A. *Sub-family* Didymograptini (two main branches) :

       (a) *Bryograptus.*—*B. Kjerulfi* (= *B. Callavei*), *B. retroflexus, B. ramosus, B. Hunnebergensis* and *B. sarmentosus.* (He gives a table showing the phylogenetic relationship of the genus.)

       (b) *Cœnograptus* s. str.. *C. gracilis* and *C. fragilis*, and including *Trichograptus.*

       Pterograptus, sub-gen.—*P. elegans.*

       Pleurograptus, sub-gen.—*P. linearis, Amphigraptus* (*A. divergens*).

       (c) *Didymograptus.*—In the group of *D. Murchisoni* he includes *D. Murchisoni* var. *gemina, D. dentatus* (*indentus*), *D. v-fractus, D. nitidus :* in the group of *D. flaccidus* (*Leptograptus ex parte*), *D. extensus, D. minutus, D. flaccidus ;* and in the group (or subgenus) of *D. gibberulus* that species only.

B. *Sub-family* Tetragraptini (four main branches).

    (a) *Dichograptus.*—*D. octobrachiatus* and *D. Logani.*

        Temnograptus, sub-gen.—*T. Milesi, T. reticulatus, T. annulatus, T. diffusus, T. expansus, T. Richardsoni*, and a new species, *T.* **Barroisi** (=*Rouvilligraptus Richardsoni*, pars).

        Clonograptus, sub-gen.—*C. tenellus, C. flexilis, C. rigidus, C. multifasciatus, C. Thureaui.*

    (b) *Tetragraptus.*—In the group of *T. Bigsbyi*,—*T. Bigsbyi, T. bryonoides, T. denticulatus, T. fruticosus*, and *T. octonarius.* In the group of *T. Headi*,—*T. Headi, T. alatus, T. quadribrachiatus.*

c. *Sub-family* Phyllograptini (four main branches which grow together dorsally).

    Phyllograptus.—*P. typus, P. ilicifolius, P. Anna, P. angustifolius, P. Loringi.*

ORDER II : AXONOPHORA.

3. Climacograptidi.—(Hydrothecæ at right angles, outer edge straight, indented by the thecal apertures.) This includes, in addition to *Climacograptus*, the Dicranograptidæ, and the Glossograptidæ.

    (a) *Retiograptus.*—*R. eucharis, R. tentaculatus, R. ? Geinitzianus.*

    (b) *Climacograptus.*—The group of *C. bicornis* includes *C. bicornis*, the new species *C.* **Nicholsoni,** and *C. antennarius.* The group of *C. scalaris, C. Scharenbergi, C. typicus, C. estonus mut. Kuckersianus, C. estonus, C. scalaris, C. caudatus, C. internexus, C. Wilsoni* and *C. retioloides.*

    (c) *Dicranograptus.*—*D. ramosus, D. Clingani, D. Nicholsoni.*

        Dicellograptus, sub-gen.—*D. anceps, D. complanatus, D. divaricatus, D. Morrisi, D. moffatensis, D. intortus, D. Forchammeri, D. patulosus, D. sextans, D. elegans.*

    (d) **Monoclimacis,** n. g.—Rhabdosome with only one row of cells, hydrothecæ like *Climacograptus.* *M. vomerinus, M. personata, M. crenularis, M. continens, M. spinulosa.*

    The genus *Trigonograptus* probably belongs to this group.

4. Diplograptidi.—(Rhabdosome with two rows of cells ; hydrothecæ oblique ; outer edge toothed.)

    (a) *Diplograptus.*—

    Group I includes *D. pristis, D. foliaceus, D. teretiusculus, D.* **sertularoides** n.s., *D. physophora, D. bellulus, D. gracilis*, and *D. palmeus.*

    Group II (=*Glossograptus*) includes *D. Whitfieldii, D. uplandicus* and *D. cf. aculeatus.*

    Glyptograptus, sub-gen.—*D. amplexicaulis, D. tamariscus.*

    Orthograptus, sub-gen.—*O. quadrimucronatus.*

    Petalograptus, sub-gen.—*P. folium* mut. *ovato-elongata, P. folium*, and *P. ovatus.*

Cephalograptus, sub-gen.—*C. cometa.*

(b) *Dimorphograptus.*—*D. elongatus, D. Swanstoni.*

5. Monograptidi.—(Rhabdosome with one row of cells, rarely branched. Hydrothecæ of many kinds.)

(a) Rhabdosome simple :

i. *Monograptus.*—(Hydrothecæ of many forms and bent round in a distal direction, attached to the axis.)

In the group of *M. priodon* he includes *M. priodon* mut. *Clintonensis* var. *Flemingi, M. galænsis, M. riccartonensis, M. cultellus.*

In the group of *M. Becki,*—*M. Becki, M. cygneus, M. scanicus, M. Barrandei, M. attenuatus.*

In the group of *M. runcinatus,*—*M. runcinatus, M. dextrorsus, M. nodifer.*

In the group of *M. turriculatus,*—*M. turriculatus, M. proteus, M. spiralis, M. triangulatus.* In addition he describes *M. resurgens* and *M. Clingani.*

ii. *Pristiograptus.*—(Hydrothecæ, as in *Diplograptus,* neither elongated nor bent round.)

In the group of *P. frequens* he includes *P. frequens, P. dubius, P. colonus, P. Roemeri, P. jaculum, P. Hisingeri, P. leptotheca, P. uncinatus, P. leintwardinensis, P. pala.*

In the group of *P. gregarius,*—*P. gregarius, P. cyphus.*

In the group of *P. testis,*—*P. testis,* and *P. discus.*

iii. **Linograptus** n. g.—(Like *Pristiograptus,* but the sicula and hydrothecæ both open distally.) *L. Nilssoni, L. concinnus, L. Sandersoni, L. tenuis, L. bohemicus.*

iv. *Rastrites.*—(Hydrothecæ straight, not connected with the axis.)

R. Linnæi, R. maximus, R. fugax var. *distans, R. peregrinus, R. hybridus, R. gemmatus, R. capillaris, R. Barrandei.*

(b) Rhabdosome branched :

Cyrtograptus.—*C. Grayiæ, C. Murchisoni, C. rigidus, C. pulchellus, C. Linnarssoni, C. Carruthersi.*

6. Retiolitidi.—(Rhabdosome with two rows of cells, perisarc consisting of a network of chitinous threads.)

(a) (Hydrothecæ oblique) :

i. *Retiolites.*—*R. Geinitzianus, R. venosus, R. australis.*

ii. *Stomatograptus,* sub-gen.—*S. grandis.*

iii. *Lasiograptus.*—*L. bimucronatus, L. costatus, L. margaritatus.*

(b) **Gothograptus,** n. g.—(Hydrothecæ vertical.) *G. nassa.*

The range and distribution of the Graptolites and dealt with in some detail, and a table is given of their geological distribution.

1897.
Elles,
" The Sub-genera
Petalograptus and
Cephalograptus,"
' Quart. Journ. Geol.
Soc.,' vol. liii.

In 1897 Elles published a paper on the " Sub-genera *Petalograptus* and *Cephalograptus,"* in which she adduced evidences that these sub-genera are quite distinct, and readily distinguished the one from the other. She worked out carefully the structure of the various species belonging to the sub-genera, especially that of the proximal end, showing that it differs in important particulars in the two groups.

From her study of these species she concludes that the *Petalograpti* have been derived from *Orthograptus foliaceus* through *O. truncatus,* and the *Cephalograpti* direct from the *Petalograpti, Cephalograptus petalum* being the intermediate form.

Description of Species.—The following species are re-described in detail and re-figured, *Petalog. folium, P. palmeus,* var. *latus,* var. *tenuis,* and var. *ovato-elongatus, P. ovatus, Cephalog. cometa,* and two new species, *P.* **minor** and *C.* **petalum.**

1897.
Wiman,
" Ueber den Bau einiger
Gotländischen Grapto-
liten," 'Bull. Geol.
Inst. Upsala,' vol. iii,
art. no. 10.

In this year also Wiman gave a further account of his extended researches on the structure of the Graptolites describing and figuring a large series of cross-sections which he had made of some Graptolites from Wisby in Gothland preserved in silicified limestone.

He figures a good specimen of *Dictyonema cavernosum* provided with stolons, one of *Dictyonema (?) tuberosum* and one of *Climacograptus.* Other forms dealt with are isolated specimens of Dendroidea, which, however, he does not attempt to refer to definite species.

1898.
Meek, A.,
" On Graptolites,"
' Proc. Univ. Durham
Phil. Soc.,' vol. i, pt. 2.

In a paper read before the Durham Philosophical Society A. Meek summarised the researches of Holm and Ruedemann respecting the structure and mode of growth of the Graptolites.

He lays stress on the supposed absence of a virgula in the Dichograptidæ, and suggests that as they " do not seem to have possessed a means of fixing themselves," " it must be supposed that they had the power of movement and temporary attachment with whatever the living contents of the thecæ provided." Forms like *Phyllograptus* he thinks were " purely crawling forms—say by means of tentacles or pseudopodia."

He tentatively suggests that the presence or absence of a virgula might form the basis of a new classification.

1898.
Hall, T. S.,
' Victorian Graptolites,'
part ii, " Graptolites of
the Lancefield Beds,"
' Proc. Roy. Soc. Vic-
toria,' n.s., vol. xi, pt. 2.

In 1898 T. S. Hall made a further contribution to the graptolitic fauna of the Lancefield Beds, Victoria, which confirmed his original views that they underlie the *Tetragraptus fruticosus* zone.

Description of Species. — He describes and figures a number of new species of Graptolites, including *Bryograptus*

Victoriæ, *Bryog.* **Clarki,** *Leptog.* **antiquus,** *Didymog.* **Pritchardi** (a form occasionally possessing three branches), *D.* **Taylori,** *Tetrag.* **decipiens,** *Dictyonema* **pulchellum,** and he re-describes and figures *Clonog. flexilis, C. magnificus, C. rigidus* var. *tenellus, Phyllograptus* species and *Dictyonema Macgillivrayi.*

According to Hall, *Leptograptus* and *Bryograptus* here occur together on the same slabs, and he explains this unusual association by concluding that in Victoria *Bryograptus* ranges up to the Ordovician.

1898.

Elles,

"The Graptolite Fauna of the Skiddaw Slates," 'Quart. Journ. Geol. Soc.,' vol. liv.

In 1898 Elles published a revision of the Graptolite fauna of the Skiddaw Slates of the Lake District, the forms found in these rocks being re-described, and particular attention being paid to details of structure.

The development of *Bryograptus* is worked out in *B. Kjerulfi, B. cf. Callavei,* and *B. ramosus*; and a new variety, var. **cumbrensis,** is described. The following species are described: *Clonograptus flexilis, C. cf. tenellus, Clonograptus* sp.; *Loganograptus Logani; Trichograptus fragilis; Temnograptus multiplex; Trochograptus diffusus; Schizograptus reticulatus, S.* **tardifurcatus,** sp. nov.; *Pleurograptus vagans; Pterograptus* sp.; *Dichograptus octobrachiatus, D.* **separatus,** sp. nov.; *Tetragraptus quadribrachiatus, T. Headi, T. crucifer, T. Bigsbyi, T. serra,* and two new species, *T.* **pendens** and *T.* **Postlethwaitii;** *Phyllograptus ilicifolius* var. **grandis,** nov., *P. Anna, P. typus, P. angustifolius; Didymograptus gibberulus.* The structure of this last species is worked out in detail, and a somewhat anomalous point of structure is noticed, namely, the apparent presence of a second connecting canal, uniting the second theca of the primary stipe with the first theca of the secondary stipe.

The following species of *Didymograptus* are re-described: *D. nitidus, D. Nicholsoni, D. affinis, D. extensus, D. patulus, D. gracilis, D. fasciculatus, D. vfractus,* var. *volucer, D. indentus,* var. *nanus, D. bifidus.*

The genus *Azygograptus* is considered by Elles to belong to the Dichograptidæ (comp. Wiman) rather than to the Nemagraptidæ, on account of the structure of the proximal end being similar to that in the Dichograptidæ. *A. Lapworthi, A. cœlebs, A. suecicus,* are re-described, together with *Leptograptus* sp.; *Dicellog. moffatensis; Diplog. dentatus, D. cf. teretiusculus, D. appendiculatus; Climacog. Scharenbergi; Cryptograptus? antennarius, C. Hopkinsoni; Glossograptus fimbriatus, G. cf. Hincksii, G. armatus; Trigonograptus ensiformis, T. lanceolatus; Thamnograptus Doveri.* Some of the above-mentioned species are figured.

The range and distribution of the various species of Graptolites are given; the Skiddaw Slates are divided into zones and compared with similar beds in South Wales and Sweden, the sub-divisions of the Skiddaw Slates agreeing closely with those given by Marr in 1894.

As regards the phylogenetic relationships of the Skiddaw Slates Graptolites, Elles agrees with Marr and Nicholson in the main, namely, that (1) the resemblances between species of different genera are of genetic origin, and therefore (2) of systematic value; (3) in any natural group the forms with relatively fewer branches were developed from the more complex forms, and therefore (4) the so-called "genera" are far more of a chronological than of a zoological significance. She considers, however, that the various forms are "most probably the result of development along certain special lines."

According to her, therefore, there is a "Group relationship"; for example, " all the 'tuning forks' *Didymograpti* have been derived from what may be termed the *fruticosus* type of *Tetragraptus*, though not all from *T. fruticosus* itself."

She divides them into two main groups, (1) those derived from *Bryograptus*, (2) those derived from *Clonograptus*.

In the first group there are five sub-groups:

(*a*) Group containing *Bryograptus ramosus* var. *cumbrensis*, *Tetrag. pendens*, and *Didymog. indentus*.

(*b*) Group containing *Bryograptus ramosus* var. *cumbrensis*, *Tetrag. fruticosus*, and *Didymog. furcillatus*.

(*c*) Group containing *Bryograptus ramosus* var. *cumbrensis*, *Tetrag. Postlethwaitii*, and *Didymog. bifidus*.

(*d*) Group containing forms derived from *Tetragraptus Bigsbyi*.

(*e*) Group containing forms derived from *Bryograptus Callavei*.

In the second group there are three sub-groups:

(*a*) Group containing *Dichograptus octonarius*, *Tetrag. serra*, and *Didymograptus arcuatus*.

(*b*) Group containing *Loganograptus Logani*, *Didymograptus octobrachiatus*, *D. extensus*, and *Tetragraptus quadribrachiatus*.

(*c*) Group containing *Tetragraptus Headi* and *Didymograptus patulus*.

Elles does not regard the angle of divergence of the branches as of phylogenetic importance; the mode of development has been simply in the direction of "failure in dichotomous division."

1899.
Perner,
" Études sur les Grap-
tolites de Boheme,"
Prague, part iii,
sect. *b.*

In 1898 Perner published the second section of the third part of his monograph on the Graptolites of Bohemia, completing the descriptive part of his work. It is devoted to a description of the Graptolites of the upper part of Stage E.

Description of Species.—In the group of the Opisopodes of the genus *Monograptus*, he describes and figures the well-known forms: *Monog. priodon*, *M. riccartonensis*, *M. latus*, *M. sartorius*, and *M. vesiculosus*, and the new species and varieties *M. priodon* var. **rimatus**, var. **validus**, *M.* **Jaekeli**, *M.* **unguiferus**, and *M.* **Suessi.**

In the group Leptopodes Perner re-describes *M. Nilssoni* of Barrande and clears up the previous confusion as to the exact identity of this species, showing that Barrande had originally included three different species under this name.

The group Prosopodes includes *Monog. Roemeri, M. dubius, M. colonus, M. chimæra, M. testis, M. bohemicus*, and the following new species : *M.* **Kayseri**, *M.* **hercynius**, *M.* **gotlandicus**, *M.* **subcolonus**, *M.* **largus**, *M.* **transgrediens**, *M.* **vicinus**, *M.* **ultimus**, *M.* **clavulus**, *M.* **Fritschi**, *M.* **bohemicus** var. **rarus**.

The group Helicopodes includes only *Monog. spiralis* var. *subconicus*.

The group Orthopodes contains *M. crenulatus* and *M. vomerinus*.

The genus *Cyrtograptus* is represented by *Cyrtog. flaccidus, C. Lundgreni, C. Murchisoni, C. Carruthersi*, and the new species *C.* **tubuliferus**.

Under the genus *Retiolites* Perner describes *Retiolites Geinitzianus, R. (Gothograptus) nassa*, and *Stomatograptus grandis*.

1899. *Peach and Horne,* "The Silurian Rocks of Britain," vol. i, Scotland.	A memoir of very great importance as regards its bearing on the range and zonal value of the Graptolites was published in 1899. This was H. M. Geological Survey Memoir on the "Silurian Rocks of Scotland." The officers of the Survey confirm Lapworth's conclusions respecting the zonal distri-

bution of Graptolites in the rocks of the Southern Uplands, and employing these fossils as zone indices, they work out and illustrate in detail the geology of the districts in which they occur in the course of their description of the entire Upland Sequence.

1899. *Törnquist,* "The Monograptidæ of the Scanian *Rastrites* Beds," 'Acta Univ. Lund.,' vol. xxxv.	In 1899 Törnquist published the second part of his Monograph on the "Graptolites of the Scanian *Rastrites* Beds." This is devoted to the study of the Monograptidæ.

Törnquist employs throughout the terminology adopted by him in the first part of this work.

The following synopsis is given of the species of Monograptidæ described, the grouping being based mainly on the form of the polypary and the character of the sicula and thecæ.

A. All the thecæ of the same type ; each wholly adnate to the proximal wall of the theca next succeeding.

(*a*) Sicula attaining a length of more than 4 mm ; rhabdosome curved. *M. gregarius* and *M.* **acinaces**.

(*b*) Sicula not exceeding 2 mm. in length ; rhabdosome stout, straight. *M. leptotheca, M. jaculum, M. nudus, M.* **regularis**, n. s.

(*c*) Sicula not exceeding 2 mm. in length ; rhabdosome stout, proximally incurved. *M.* **inopinus**, n. s.

(*d*) Sicula not exceeding 2 mm. in length ; rhabdosome arcuate, gradually widening from the proximal extremity. *M. tenuis*.

(*e*) Sicula not exceeding 2 mm. in length; rhabdosome slender, distally straight, or irregularly bent. *M.* **incommodus**, n. s.

B. Thecæ dimorphous : the distal ones of the same type as that characteristic of the preceding section, each of the proximal thecæ wholly, or at least distally, free from the succeeding theca.

(*a*) Rhabdosome gradually widening. *M. revolutus* var. **austerus**, nov., *M.* **difformis**, n. s., *M. cf. cyphus.*

(*b*) Rhabdosome abruptly acquiring its normal width. *M. limatulus.*

C. All the thecæ of the same type ; each being wholly, or at least distally, free from the theca next in advance.

(*a*) Rhabdosome stout and straight, or only having the sicular portion bent backward. *M. runcinatus, M. priodon, M. Sedgwickii, M.* **harpago**, n. s.

(*b*) Rhabdosome slender, proximally arcuate, but not enrolled, distally straight or irregularly bent. *M.* **elongatus**, n. s.

(*c*) Proximal portion of the rhabdosome forming a more or less complete flat or sub-conical spiral, or at least showing a tendency to form such a figure ; prolific side convex. *M.* **denticulatus**, n. s., *M. fimbriatus, M. triangulatus, M.* **nobilis**, n. s., *M.* **decipiens**, n. s., *M. convolutus, M. subconicus.*

(*d*) Rhabdosome coiled up in an elongated conical helix bearing thecæ on the convex margin. *M. turriculatus.*

(*e*) Rhabdosome enrolled in a conical spiral-bearing theca on the concave margin. *M. proteus.*

(*f*) Rhabdosome forming a narrow flat spiral, bearing thecæ on the concave margin. *M. discus.*

(*g*) Rhabdosome fish-hook-shaped ; prolific side concave. *M. exiguus.*

Törnquist points out that he does not imagine that the above arrangement is " in every respect a natural one, though he is of opinion that several of the sub-divisions may be found to coincide with true natural groups."

1899.
Törnquist,
" Några anteckningar om Vestergötlands öfversiluriska Grapto-litskiffrar," 'Geol. Fören. Förh.,' bd. 21.

In a short stratigraphical paper published the same year, Törnquist noticed the various localities where the Upper Silurian Graptolitic zones are recognisable in Vestrogothland.

1899.
Hall, T. S.,
" The Graptolite-bearing Rocks of Victoria, Australia," 'Geol. Mag.,' dec. 4, vol. vi, no. x.

In 1899 Hall published a general account of the Graptolite-bearing beds of Victoria, and their divisions, comparing them with those of Europe.

Hall describes and figures three new species of Graptolites, namely *Tetra-graptus* **projectus**, *Goniograptus* **macer** and *Trigonograptus* **Wilkinsoni,** and a species of *Monograptus* which he does not name ; he re-describes *Didymog. gracilis* and *Dichog. octonarius,* and re-figures *Bryograptus Victoriæ,* and *Leptograptus antiquus.*

He recognises four main divisions in the Lower Ordovician of Victoria : (1) Lancefield Series, (2) Bendigo Series, (3) Castlemaine Series, and (4) Darri-will Series. The Graptolites characteristic of each group are fully given. The Upper Ordovician is represented in a few localities, but the Silurian contains only one or two species of Graptolites.

A special point dwelt upon in this paper is the apparent want of harmony between the Graptolite zones of Australia and of Europe. Thus, in addition to the occurrence of *Bryograptus* and *Leptograptus* in association, Hall considers that *Loganograptus* appears very high up in the series, and never in association with *Phyllograptus typus,* while *Didymog. bifidus,* which is characteristic of an Upper Arenig Fauna in Europe, here disappears before *Phyllograptus typus.*

1900.
Elles,
" The Zonal Classifica-tion of the Wenlock Shales of the Welsh Borderland," 'Quart. Journ. Geol. Soc.,' vol. lvi.

In 1900 Elles published a paper on the "Zones of the Wenlock Shales," demonstrating the systematic chronological arrangement of the various species of Graptolites in the Wenlock Rocks of Britain. A few new species were described, and many already well-known forms, all with special regard to the structure of the proximal end.

Under the species *Monograptus Flemingii* Elles recognises four varieties, a, β, γ, and δ, which are valuable zonally, as they are practically confined to certain definite horizons. The same is the case with *M. vomerinus,* of which she recognises three varieties; *M.* **flexilis,** *M.* **irfonensis,** and *M. testis* var. **inornatus,** are new forms. A new species of *Cyrtograptus,* *C.* **symmetricus,** is also described.

The Wenlock Shales are worked out in (1) the Builth district, (2) the Long Mountain, and (3) the Dee Valley. Elles finds that in the Builth district they are capable of division into six zones characterised by species of *Cyrtograptus,* and that the majority of these Wenlock zones are also to be found in the other areas. The evidences adduced in this paper prove for the first time in Britain, that the Wenlock Shales are as capable of Graptolitic zonal division as are the Birkhill Shales, and that these British Wenlock zones run parallel to the Wenlock zones already suggested by Tullberg for Scandinavia.

1900.
Wood,
" The Lower Ludlow Formation, and its Graptolite Fauna," ' Quart. Journ. Geol. Soc.,' vol. lvi.

In the same year Wood worked out the Graptolitic fauna of the Lower Ludlow Shales, and proved that these were equally capable of division into Graptolite zones.

Range and Distribution.—In this paper the distribution of the Ludlow Graptolites is first worked out in the typical Ludlow District, and four zones are recognised. Similar zones are afterwards shown (with the addition of a fifth zone) to hold in the main in the Builth District and Long Mountain area of the Welsh Borderland. Brief notes are also given of the Ludlow graptolitic fauna in the Dee Valley, the Lake District, South Scotland, Dudley, and the Abberley Hills.

Description of Species.—The following species, which had been already named by previous observers, are re-described and re-figured:

Group I.—*M. dubius, M. tumescens* var. *minor, M. gotlandicus, M. ultimus.*

Group II.—*M. colonus* var. *ludensis, M. Roemeri.*

Group III.—*M. chimæra* var. *Salweyi, M. leintwardinensis.*

Group IV.—*M. uncinatus* var. *micropoma.*

Group V.—*M. scanicus.*

Group VI.—*Gothograptus nassa, M. Nilssoni, M. bohemicus.*

The following new species are figured and described: *M.* **vulgaris**, var. **a**, var. **b**, *M.* **tumescens**, *M.* **comis**, *M. colonus* var. **compactus**, *M.* **varians**, var. **a**, var. **b**, var. **pumilus**, *M. chimæra* var. **a**, *M. leintwardinensis* var. **incipiens**, *M. uncinatus* var. **orbatus**, *M.* **crinitus**, *Retiolites* **spinosus**.

1900.

Lapworth, H., "The Silurian Sequence of Rhayader," 'Quart. Journ. Geol. Soc.,' vol. lvi.

A third paper of especial stratigraphical importance was published during this year by H. Lapworth on the "Silurian Sequence of Rhayader."

He shows that in the district of Rhayader there occur representatives of all the Llandovery-Birkhill Graptolitic zones as well as of the Tarannon; these zones are mapped by him in detail.

He describes and figures three new species: *Climacograptus* **parvulus**, *C.* **extremus**, and *Diplograptus* **magnus**, and gives a description of *Diplog. modestus*, which had previously been figured only.

1900.

Hall, T. S., " On a Collection of Graptolites from Mandurama," 'Records Geol. Survey, N. S. Wales,' vol. vii, pt. 1.

In the year 1900 Hall published the results of his examination of a collection of Graptolites made by Mr. E. F. Pittman from Mandurama.

Hall describes and figures three forms which he considers are new, but only names two, namely, *Climacograptus* **affinis** and *Diplograptus* **manduramæ**.

1901.

Newton, " Note on Graptolites from Peru," 'Geol. Mag.,' dec. 4, vol. viii, no. v

In 1901 Newton described and figured a slab of Graptolites collected by Mr. Jessop from the province of Carabaya, Peru. The species to which these belong is uncertain, but they closely resemble *Diplograptus truncatus*, Lapw. They are "indicative of beds near the uppermost part of the Lower Silurian."

In the year 1901 Törnquist published the results of his researches on the *Phyllo-Tetragraptus* beds of Scania and Vestrogothland, and recognised

1901.

Törnquist,
" Graptolites of the
Lower Zones of the
Scanian and Vestro-
gothian *Phyllo-Tetra-
graptus* Beds," ' Acta
Univ. Lund.,' vol.
xxxvii, pt. 2, no. 5.

in these beds for the first time the existence of five distinct zones :

    (*a*) Zone of *Tetragraptus phyllograptoides.*

    (*b*) Zone of *Didymograptus balticus.*

    (*c*) Zone of *Phyllograptus densus.*

    (*d*) Zone of *Isograptus gibberulus.*

    (*e*) Zone of *Phyllograptus cf. typus.*

The Orthoceras Limestone intervenes between the last two zones, and therefore the contrast between the faunas of these zones appears to be greater than between the others.

Before proceeding to the description of species, Törnquist suggests and explains certain terms which he employs throughout this paper.

The branch which "issues on the same side of the sicula as the first theca," he calls the " primordial stipe," the other he names the " complemental stipe."

He also distinguishes between the various parts of the first theca, and designates them: (1) The " initial portion," (2) " ramifying portion," and (3) " apertural portion," or true theca.

Törnquist again points out that there has been considerable confusion in the employment of the term " connecting canal," and that it can no longer be applied to " that part of the complemental stipe which crosses the sicula " in *Didymograptus*, etc., and he suggests the new name of " crossing canal."

In his figures he adheres to the old method of drawing them with the apex of the sicula directed downwards, though he " by no means under-rates the motives which may have prevailed upon some authors to figure these fossils in a different position."

Törnquist recognises the two genera established by Moberg as *Isograptus* and *Mæandrograptus*, though he is undecided whether to regard them as sub-genera of *Didymograptus* or as distinct genera.

Under the genus *Didymograptus, sensu latiori*, he describes and figures : *Didymog. suecicus, D. patulus* (=*D. hirundo*), *D. extensus, D. constrictus, D. balticus, D. vacillans, D. filiformis, D.* **flagellifer**, Tullb. MS., and the new species *D.* **undulatus**, *D.* **demissus**, *D.* **geometricus**, *D.* **Holmi**, *D.* **prænuntius**, *D.* **validus**, *D.* **Kurcki**, and *D.* **Mobergi**.

The genus *Isograptus* includes the one species *I. gibberulus*, and Törnquist adds a few additional notes on the structure of the initial end.

The genus *Mæandrograptus* also comprises only one species : *M. Schmalenseei ;* the structure of this peculiar form is well brought out by the illustrations.

1901.

Moberg,
" *Pterograptus scanicus,*
n. sp.," ' Geol. Fören.
Förh.,' bd. 23.

In 1901 Moberg described and figured a new species of *Pterograptus* under the name of *P.* **scanicus**, and he compares it with Holm's species *P. elegans.* This species occurs at Fågelsang associated with a *Climacograptus* and *Didymograptus geminus.*

1901.
Strandmark, J. E.,
" Undre Graptolit-
skiffer vid Fågelsång,"
' Geol. Fören. Förh.,'
vol. 23, pp. 548–556,
pl. xvii.

The author gives a description of the so-called " Lower Graptolite Shales " of Fågelsång, noting the lithological and palæontological sequence. He figures (taf. 17) the species *Tetragraptus Bigsbyi*, Hall, and *Tetragraptus phyllograptoides*, Linnrs., and also a remarkable new form, **Phyllograptus cor** Strandmark, and discusses their probable phylogenetic inter-relationships.

1901.
Elles and Wood,
" Monograph of British
Graptolites." (Edited
by Chas. Lapworth.)
pt. 1, pp. 1–54.
Palæontographical
Society, 1901.

The Monograph opens with a brief Introduction by the Editor, giving an account of the origin, object, and plan of the work and of the mode of illustration. The First Part is devoted to the description and figuring of the British forms assigned to the Genus *Didymograptus* of the family of the Dichograptidæ. The illustrations include quarto plates (I—IV), in which the species are represented on the natural scale, and Text-figures (1—35), in which minor details are given on a scale of five times the natural size. Some twenty-eight species are described, of which the following are noted as new : *Didymograptus* **uniformis,** *D.* **similans,** *D.* **deflexus,** *D.* **amplus,** *D.* **artus,** *D.* **stabilis,** and *D.* **acutidens** (Lapw. MS.).

The Descriptive Section is prefaced by short definitions of the morphological terms employed, and a synopsis of the several groups of *Didymograptus* recognised. At the close of the description of each such group the individual characteristics of its component forms are tabulated and compared.

1901.
Malaise, C.,
" Etat actuel de nos
connaissances sur le
Silurien de la
Belgique," ' Annales
Soc. Geol. du Nord,'
vol. xxx, pp. 188–190.

The author summarises the results of his long-extended and successful researches into the geology and fossils of the Cambrian, Ordovician, and Silurian deposits of the Massif of Brabant, and the Band of Sambre and Meuse, and gives a generalised scheme of the sequence in each area. Especial attention is directed to the close parallelism between the British and Belgian palæontological succession, and to the recognised presence of the same characteristic forms of Graptolites in several corresponding horizons in the two countries.

1901.
Kerforne, F.,
" Silurique de la
presqu'ile de Crozon
(Finistere)," Rennes,
1901, pp. 1–230.

An account of the author's study of the Silurian Succession in the peninsula of Crozon, with lists of the fossils detected. Seven successive Graptolite Zones are recognised, ranging from the equivalents of the British Upper Llandovery to those of the Lower Ludlow, both inclusive. Some nineteen forms of Graptolites are distinguished as present, and their horizons indicated.

1901.
Eisel, R.,
" Zonenfolge ost
Thüringischer und
Vogtländischer
Graptolithenschiefer,"
' Jahrb. Gesell. Freund.
Naturwissenschaften
in Gera,' year 1900.

This Memoir is remarkable as being the first to demonstrate the fact that the zonal succession of Graptolite species in Central Germany corresponds closely with that already established in Britain and North-West Europe. It is the fruit of the enthusiastic researches of the author carried on for many years into the chronological succession of Graptolite species in the Thuringian areas, long previously rendered classic by the works of Geinitz and Richter.

The author makes known the presence of several successive Graptolite zones in the sequence described, which ranges from Middle Llandovery to Lower Ludlow, and gives the names and ranges of the forms collected by himself personally. The zones described are distinguished by the author by names derived from their characteristic local species, and the following forms are given as new : *Monograptus* **priodon,** var. **reductus**; *M.* **veles**; *M.* **Sedgwickii,** var. **Voglandicus**; *Diplograptus* **binodosus**; *D.* **juncus**; *D.* **Thuringiacus**; *Retiolites* **præcursor**; *Climacogrgptus* **citrocrescens**; *Dimorphograptus* **Lapworthi.** (See also Törnquist, 1887.)

1901.
Ruedemann, R.,
" Hudson River Beds
near Albany, and their
Taxonomic Equiva-
lents," ' New York
State Mus.,' Bull. 42.

A summary of the author's discoveries among the complicated Graptolite-bearing strata of the Hudson River Valley.

CHAPTER V.

1902 TO 1913.

1902.
Moberg, J. C.,
" Didymograptus
Skiffer," ' Geol. Fören.
Förh.,' vol. 24,
pp. 44–48.

THE author proposes to unite under the single title " Didymograptus Skiffer " all the Scandinavian Graptolite-bearing strata ranging from the horizon where *Didymograptus* makes its earliest appearance to the top of the *Didymograptus geminus* beds.

1902.
Elles and Wood,
" Monograph of British
Graptolites," pt. 2,
Palæontographical
Society, 1902.

This Second Part of the British Monograph opens with the first two chapters of the promised " General Section." These chapters (pages i to xxviii) treat of the History of Research among the Graptolites in general from the times of Bromell and Linnæus (1727) to those of Barrande (1850) and Scharenberg (1851). References are made to the titles and dates of publication, etc., of the several Memoirs cited, and a synopsis is given of the more important facts and deductions brought forward by each author.

This " General Section " is followed (pp. 55–102) by the continuation from Part I. of the " Descriptive Section." It treats of the whole of the British species assigned to the family of the Dichograptidæ, with the exception of those belonging to the genus Didymograptus (already described in Part I). Twelve British genera are recognised (including Phyllograptus, previously regarded by many authorities as constituting a distinct family, and Azygograptus, hitherto assigned to the Leptograptidæ). Four new species are described and figured: Tetragraptus reclinatus, T. **Postlethwaitii**, T. **Amii** (Lapw. MS.), Holograptus **Deani** (Lapw. MS.), and Bryograptus **divergens**.

1902.
Kerforne, F.,
" Gothlandien inférieur du Massif Armoricain," ' Comptes Rendus, Acad. Sciences,' July 15, 1902.

An outline description of the lithological divisions of the Llandovery and Tarannon strata of Anjou, etc., with a note on their included Graptolites.

1902.
Ruedemann, R.,
" Graptolite Facies of the Beekmantown Formation in Renesslaer County," " New York State Palæontol. Ann. Rept.," 'New York State Mus.,' Bull. 52, p. 546.

A registration of the progress and results of the author's researches and discoveries among the Graptolite-bearing Lower Palæozoic Strata of the State of New York. (See also " Upper Cambrian Horizon of Dictyonema flabelliforme in New York," " New York State Palæontol. Ann. Rept.," ' New York State Mus., Bull. 69, p. 934, 1903.)

1902.
Ruedemann, R.,
" Growth and Development of *Goniograptus Thureaui*," ' Bull. 52, New York State Museum,' pp. 576, etc.

An account of the discovery of two forms of Goniograptus (Goniograptus Thureaui, M'Coy, and Goniograptus sp.) in the Deep Kill Graptolite Beds of New York, with a discussion of the bearing of the structural details and successive ontogenetic stages in Goniograptus upon the phylogeny of the Graptolites in general.

1904.
Ami, H. M.,
" Preliminary List of Fossils collected by Prof. L. M. Bailey from localities in New Brunswick," ' Summary Rept., Geol. Surv., Canada,' 1904.

Identification of genera and species of Graptolites collected from the Lower Palæozoic Strata of New Brunswick, with notes of correlation of the containing beds.

1902.
Hall, T. S.,
" Reports on Graptolites," ' Records Geol. Surv. Victoria,' vol. i, pp. 33–35.

A revision of the Graptolite species from Eastern Victoria. A list of all the forms recognised up to date is included with localities. Two new species, Didymograptus **ovatus** and Glossograptus **Hermani**, are described and figured.

1903.
Elles and Wood,
" Monograph of British
Graptolites," pt. 3.
Palæontographical
Society, 1903.

The General Section of this Third Part of the British Monograph continues the History of Research from 1851 to 1865 ; and the Descriptive Section is devoted to the British forms of the family of the Leptograptidæ. Four British genera—namely, *Leptograptus, Pleurograptus, Amphigraptus,* and *Nemagraptus (Cænograptus)*—are recognised, embracing eleven species with several varieties. Four of the species are given as new, viz. *Leptograptus* **latus,** *L.* **sacendens,** *L.* **validus** (Lapw. MS.), *L.* **grandis** (Lapw. MS.). In this, as in other parts of the work, the detailed descriptions are preceded by a summary of the general characteristics of the family or group under notice.

1904.
Noel, M. C.,
" Faune des Lydiennes
du grés Vosgien,"
' Comptes Rendus,
Acad. Sciences,'
June 13, 1904.

A list of the Graptolites discovered by Dr. Blücher (1898) and the author (1898—1915) in the pebbles of the Conglomerates of the ' Grés des Vosges ' of the higher parts of the basin of the Meuse and the Moselle. Sixteen forms are noted, ranging from Llandovery to Lower Ludlow inclusive.

1904.
Elles, G. L.,
" Some Graptolite
Zones in the Arenig
Rocks of Wales,"
' Geol. Mag.' dec. v,
vol. i, pp. 199-211.

The author summarises the results of her personal studies in the field and in the laboratory of the Vertical Distribution of the recognisable species of Graptolites occurring in Sedgwick's " Arenig " Series of his type area and their equivalents in other parts of Wales. The rock-successions and Graptolite-sequences, as developed in the Arenig District, in Cærnarvon, in the Lleyn Peninsula, at St. David's, Abereiddy, Whitesand Bay, and other areas in South Wales, are described. These are correlated with the corresponding successions in Shropshire, the Lake District, and Scania. She concludes that three successive Graptolite Zones are distinguishable in these British Arenig strata, namely, the Zones of (1) *Didymograptus extensus,* Hall ; (2) *Didymo. hirundo,* Salter ; and (3) *Didymo. bifidus,* Hall. The oldest of these zones follows in order of time the *Dichograptus* Beds of the Lake District, whilst the newest graduates upwards into the sub-formation usually termed the Lower Llandeilo, which is characterised by the presence of *Didymograptus Murchisoni.* The '*Murchisoni* Zone' and the underlying '*Bifidus* Zone' are both marked by the dominance of "tuning-fork" Graptolites, and constitute the Llanvirn Series of Dr. Hicks.

1904.
Elles and Wood,
" Monograph of British
Graptolites," pt. 4.
' Palæontographical
Society,' 1904.

In this Fourth Part of the British Monograph the Historical Section is carried on from 1865 to 1871. The Descriptive Section deals with the British forms assigned to the family of the Dicranograptidæ. Of these only two genera are recognised, *Dicellograptus,* Hopkinson, and *Dicranograptus,* Hall. Thirteen British species are assigned to the former genus and ten to the latter.

The species described as new are *Dicellograptus* **angulatus,** *Dicranograptus* **brevicaulis,** *D.* **celticus,** *D.* **cyathiformis,** and *D.* **tardiusculus** (Lapw. MS.).

1904.
Hall, T. S.,
" Reports on Grapto-
lites," 'Records Geol.
Surv. Victoria,' vol. i,
pp. 217–219.

Identifications and lists of Graptolite species from several localities in Victoria.

1904.
Törnquist, S. L.,
" Graptolites of the
Lower Zones of the
Scanian and Vestro-
gothian *Phyllo-Tetra-
graptus* Beds," pt. 2,
'Lunds Universitets
Årsskrift,' vol. 40,
pp. 1–29, pls. i–iv.

In this paper, of which the first part was published in 1901, Prof. Törnquist completes his figures and descriptions of the Graptolites of the Swedish strata corresponding broadly to the Arenig strata of Britain. The paper is marked throughout by the author's usual modesty of presentation, his care for accuracy of detail, and the beauty and clearness of his figures. There are four quarto plates, and an appendix showing the vertical distribution of the Graptolites described in the two parts of the Memoir. The Swedish species noted are about forty in number and are grouped in four successive zones named by the author. One genus is given as new (**Anthograptus**), and eight species—namely, *Bryograptus* **simplex**, *Tetragraptus* **Vestrogothus**, *Dichograptus* **regularis**, *Anthograptus* **crinitus** (Moberg, MS.), and *Azyograptus* **validus** (Moberg, MS.).

In a note in the body of the work Törnquist treats of the extreme difficulties in the classification of the so-called genera and species of the compound forms of the Dichograptidæ in general, and gives a provisional classification of those referred to in his own Memoirs. He expresses his agreement with Wiman and Ruedemann, that there can be " no fundamental difference between the 'dichotomous' and 'lateral' mode of division of the stipes," but notes at the same time the remarkable constancy of the one or the other mode in certain forms.

1904.
Törnquist, S. L.,
" Sundry Geological
Notes," nos. 2 and 3,
'Geol. Fören. För-
handl.,' vol. 28,
pp. 497—515.

In the second of these notes the author expresses his agreement with W. C. Brögger (1896) in regarding the Swedish Ordovician as commencing with the *Dictyonema* Shales (Tremadoc), and embracing three main members: (1) *Dictyonema* Shales; (2) *Didymograptus* Shales; and (3) *Dicellographus* Shales. In the third note he discusses the synonymy of *Didymograptus gibberulus*, Nicholson; *Didymo. patulus*, Hall, and *D. constrictus*, Hall.

1904.
Ruedemann, R.,
" Graptolites of New
York, Part I;
Graptolites of the
Lower Beds." 'New
York State Museum,
Memoir 7.'

This Memoir constitutes the first volume of a comprehensive Monograph by the author dealing with the Graptolites of New York State and their relations to those of the equivalent Graptolite-bearing regions elsewhere in the United States, in Canada, Europe, Britain, and Australia.

This first volume includes not only a description and illustration of all the forms of Graptolites hitherto collected from the "Upper Cambrian" (Tremadoc) and "Lower Ordovicic" (Arenig) of New York State (the majority of which were the fruits of the author's long-extended personal field researches), but embraces also a detailed account and discussion of Graptolitic literature, knowledge, and speculation up to the date of publication.

The volume may be said to be divisible into two main sections: a first (pp. 498—577), treating of the Graptolites in general, and a second (pp. 577—783), devoted to the description of the New York forms.

The General Section opens with an outline Bibliography ranging from 1874 to 1903. This is followed by paragraphs dealing with the History of Research among the Graptolites, in which the difficulties of the various branches of the subject are pointed out, and the past stages of advance in knowledge and speculation adduced and frankly discussed. In sub-section 3 the author, after passing in review the methods of previous observers, describes the modes of illustration adopted, which on the whole are similar to those followed in the present work.

In sub-section 4, where Ruedemann is dealing with Terminology, he notes his adoption of Törnquist's term "rhabdosome" instead of polypary as the equivalent of the term "hydrosome" or the whole colony in the terminology of the "Hydrozoans," and suggests the new title "synrhabdosome" for each of those stellate assemblages which have been interpreted by himself as colonies of colonies. He calls attention to Lapworth's use (1897) of the term "graptotheca" as the equivalent of "hydrotheca," but follows the general procedure in using the simple term "theca" without qualification. He regards the Graptolites as being separable into the two orders of the Dendroidea (Nich.) and the Graptoloidea (Lapw.) and the latter as being divisible into the two sub-orders, viz. Graptoloidea-Axonolipa (Frech. Ruedemann, em.) and Graptoloidea-Axonophora (Frech.).

He draws a sharp distinction between the filiform tubular structure in the Axonolipa, termed by Lapworth the "nema" or "nemacaulus," and the solid axis, rod or "virgula" of the Axonophora, agreeing with those who hold that the latter is wanting as such in the Axonolipa, but is apparently present inside the nemacaulus in the polypary of the Diplograptidæ, etc. He employs Hall's term "funicle" for the common base of the component nemacauluses in the stellate forms.

After describing and illustrating the sicula and its relations, as worked out by the Swedish authorities, he expresses his full accord with those who hold that notwithstanding its resemblance to an ordinary serial theca it is always appropriate to designate by the special title "sicula" alone.

The various classifications of the Graptoloidea are passed in review, and that of Lapworth (1879) accepted, with the additional families due to subsequent discoveries. As respects the phylogeny of the Graptoloidea, the views of Marr and Nicholson (1895, as extended by Elles 1898) are adopted, and a comprehensive table is appended, illustrating the author's "Suggested Phylogeny of the American Graptolite Axonolipa."

In dealing with the geological range of the Graptolites Ruedemann summarises in brief the far-reaching results already arrived at on both sides of the Atlantic by previous graptolithologists, and combines them with the main results of his own researches in a general "Correlation Table" of the Upper Cambrian and Lower Ordovician formations of Scandinavia, Great Britain, Canada, and New York. In calling special attention to the long-accepted generalisation of the almost world-wide distribution of the same (or a large percentage of the same) characteristic genera and species in corresponding zones, he argues that while there appear to be some evidences of a local nature accordant more or less with the four theoretic Lower Palæozoic "provinces" of Frech and others (deduced from the geographical distribution of the Trilobita, Brachiopoda, etc.), yet there are other facts entirely at variance with the distribution of the land and water areas as they can be constructed for the Lower Siluric (Ordovician) age from the study of the littoral faunas. His remarks are illustrated by a world-chart.

As respects the mode of existence of the Graptolites, Ruedemann agrees with those who regard the majority of the genera of the Dendroidea as being permanently attached to rocks or growing sea-weeds in the littoral regions, thus forming a part of the marine *benthos* of their time; while the majority of the Graptoloidea were probably attached to sea-weeds alone, and shared their fate, whether remaining permanently moored to rocks or the sea-floor off-shore, or drifting (like the modern Sargassum) as *pseudo-plankton* far and wide over the surface of the deep seas. Some forms, however, and notably those arranged in stellate groups Rudemann considers, had possibly attained a holoplanktonic stage of existence corresponding with that of the Siphonophora of the present day.

In the Descriptive Section of the volume Ruedemann gives diagnoses of some forty species of Graptolites occurring in the New York formations dealt with. These are arranged by him in six families, viz. 1, Dendrograptidæ (Roemer); 2, Dichograptidæ (Lapw.); 3, Cœnograptidæ (nov.); 4, Phyllograptidæ (Lapw.); 5, Diplograptidæ (Lapw.); and 6, Climacograptidæ (Frech); and seventeen genera, of which two, **Sigmagraptus** and **Strophograptus,** are new. The following are the new species described: *Dendrograptus* (?) **succulentus,** *D.* **fluitans,** *Ptilograptus* **tenuissimus,** *Dictyonema* **furciferum,** *D.* **rectilineatum,** *Desmograptus* **intricatus,** *Goniograptus* **gometricus,** *G.* **perflexilis,** *Temnograptus* **Noveboracensis,** *Bryograptus* **Lapworthi,** *B.* **pusillus,** *Tetragraptus* **Clarkei,** *T.* **Woodi,** *T.* **taraxicum,** *T.* **pygmæus,** *T.* **lentus,** *Didymograptus* **cuspidatus,** *D.* **Ellesi,** *D.* **Tornquisti,** *D.* **spinosus,** *D.* **forcipiformis,** *D.* **incertus,** *Sigmagraptus* **præcursor,** *Strophograptus* **trichomanes,** *Diplograptus* **laxus,** *D.* **longicaudatus,** *Glossograptus* **hystrix,** *G.* **echinatus,** *Climacograptus* **spongens**

These are illustrated in seventeen excellent quarto plates, in which the majority of the forms are given in natural size and detail.

The letterpress of the work is enriched by 106 Text-figures, most of them on a common scale of five times the original size.

1905.

Flamande, B. M.,
"Existence de schistes à Graptolithes à Haci-El-Khenig (Sahara Central)," 'Comptes Rendus Acad. Sciences,' vol. cxl, pp. 954–957.

Note of the discovery of Graptolites of Silurian Age in the Sahara, by Captain Cottenest and Mons. M. Foureau, in 1902, with a description of the position and lithology of the containing rocks.

1905.

Gentil, L.,
" Sur la présence de schistes à Graptolithes dans le Haut-Atlas Marocain," 'Comptes Rendus Acad. Sciences,' vol. cxl, p. 1659.

A brief account of the author's discovery of Silurian (Llandovery) Graptolites in the strata of the High Atlas, Morocco, with lists of species present and notes upon the forms previously obtained from Silurian strata in the Sahara. (See also Munier Chalmas, " Notice sur les travaux scientifiques," Lille, 1903, p. 94, and " Documents de la Mission Saharienne," by MM. Foureau, Gentil and Haug, 1905, pp. 755–756.

1905.

Fearnsides, W. G.,
"The Geology of Arenig Fawr and Moel Llyfnant," 'Quart. Journ. Geol. Soc.,' vol. lxi, pp. 608–640.

A detailed description of the geological succession ranging from the base of the Upper Cambrian to the Middle Bala in the classic Arenig region. Some twenty-five species of Graptolites are named as present in the succession, and their relative horizons and ranges are carefully noted.

1905–6.

Lapworth, C.,
" On the Graptolites from Bratland, Gausdal, Norway," 'Norges Geol. Undersogelse,' no. 39.

A description of the species of Graptolites collected by Herr Bjorlekke, of the Geological Survey of Norway, in the Ordovician Shales of Bratland, Gudbrandsdal, in 1890. Ten distinct forms are noted, and illustrated on a plate. Two forms are given as new : *Dicellograptus* **laxatus** and *Didymograptus euodus* (Lapworth) var. **Bjorlekki.** A table is given (p. 5) showing the distribution of the Bratland forms in Norway, Sweden, Great Britain, and North America, and the containing strata are assigned to the higher Arenig rocks of the British Succession (Zones of *Didymograptus extensus* and *D. hirundo*).

1905.

Pirie, J. H.,
" Graptolite-bearing Rocks of the South Orkneys, 'Proc. Roy. Soc. Edinburgh,' vol. xxv, pp. 463–470.

A note of the discovery of the genus *Pleurograptus* in the sedimentary rocks of the South Orkney Islands, 800 miles south-east of Cape Horn.

1905.
Hall, T. S.,
" Victorian Grapto-
lites," pt. 3, ' Proc.
Roy. Soc. Victoria,'
vol. xviii, pp. 20–24,
pl. vi.

Identification of several species of Graptolites collected by Mr. Thiele from the Upper Ordovician Strata of Victoria. Two new species are figured—*Diplograptus* **Thielei** and *Dicranograptus* **hians.**

1905.
Törnquist, S. L.,
" Paleontologiska med-
delanden," 'Geol. Fören.
Förhandl.,' vol. 27,
pp. 452–457.

A correction of the specific nomenclature of several Swedish forms.

1905.
Schepotieff, A.
" Ueber Stellung der
Graptolithen im Zoolo-
gischen System,"
' Neues Jahrbuch für
Mineralogie,' 1905,
vol. 2, pp. 79–98.

The author commences by pointing out (with references) how the majority of earlier investigators (1865—1895) were in accord in assigning the Graptolites to the Hydroidea, while the latest researchers (1895—1901) have been gradually led to infer that the Graptolites either (1) constitute a special class of Cœlenterata, having a very distant relationship to the Hydroidea, or (2) they cannot be compared with any of the accepted groups of recent animals, but must be regarded simply as Invertebrata of unknown systematic position.

The paper itself is devoted to a comparison of the many points of resemblance between the structure of the polypary in the Graptolites (especially the Mono-graptidæ) and that of the polypary in the remarkable recent marine organism *Rhabdopleura Normani,* Allman, some of which were indicated by Allman himself as early as 1872. At that time *Rhabdopleura* was assigned to the Polyzoa (Bryozoa); but the subsequent progress of zoological research has placed it in company with the equally remarkable recent genus, *Cephalodiscus,* in a special class, Pterobranchiata (Ray Lankester, 1884), of the subphylum, Adelochorda, some members of which also suggest affinities with the Echinodermata on the one hand and with the Chordata on the other.

Schepotieff's residence in Norway, from whose deeper sea-waters the majority of the known examples of *Rhabdopleura* have been dredged, gave him special advantages on the study of its organisation. He describes his results in detail, and compares them with those of his microscopic study of sections of Bohemian specimens of *Monograptus* preserved in limestone.

He describes the main element or stipe of *Rhabdopleura* as a creeping, longitudinal tube, attached throughout by its flattened dorsal surface to some foreign body. In reality this longitudinal tube is composed of the basal portions of a uniserial succession of dwelling chambers inhabited by the individual zooids of the colony. Each such chamber may be described as divisible into two regions directed approximately at right angles to each other—namely, a basal, or proximal region, forming a constituent part of the longitudinal tube; and a distal, lateral, and free region, having a mouth at its outer extremity. Thus the stipe in *Rhabdopleura* presents an appearance similar to that in some forms of the Monograptidæ (especially *Rastrites*); the longitudinal tube in *Rhabdopleura* answering in form and relation to the common canal in *Rastrites*, and the free portions of the successive chambers to the so-called "isolate thecæ" in that genus. In *Rastrites*, however, while the successive thecæ intercommunicate freely by means of the common longitudinal canal, in *Rhabdopleura* they intercommunicate only in their earliest stages, and later on in life are completely shut off from each other.

Imbedded in the dorsal wall of the longitudinal tube in *Rhabdopleura*, there runs from end to end an intensely black filament or cord, denominated by Schepotieff as the *stolo* (a title which we here retain in this place for convenience of reference). This stolo is composed of four members, namely, (1) an exceedingly minute axial rod running continuously down its centre and surrounded by three concentric tubular structures, (2) an inner cell-string, (3) a middle branching cell-layer, and (4) an outer cover of intense blackness.

Schepotieff agrees emphatically with those of his predecessors who have pointed out the agreement in position and relation of this stolo in *Rhabdopleura* with the virgula in the Monograptidæ, and he fortifies his view by the results of his own microscopic studies and those of others into the structure of *Monograptus*. He also brings forward other points of similarity in structural detail between *Rhabdopleura* and *Monograptus*, etc., such as the common presence of minute growth-lines or transverse rings which meet at an angle on the ventral side of the thecæ, the presence of periodic projections on the solid axis, etc.

We are wholly unable to agree with Schepotieff's identification of the growing portion in *Rhabdopleura* with the sicula in *Monograptus*, but his paper (which is illustrated by several good Text-figures) should be read by all students of Graptolite literature.

1906.
Turnquist, S. L.,
"Sundry Geological and Palæontological Notes," 'Geol. Fören. Förhandl.,' vol. 28, pp. 497–515.

In a first of these notes the author expresses his agreement with the views of those who regard the *Dictyonema* and *Ceratopyge* beds of Sweden as constituting in combination the first member of the Scandinavian Ordovician. In a second he discusses at length the synonomy of the forms which have been referred to *Isograptus gibberulus* (Nich.), *Didymograptus patulus* (Hall), and *Didymograptus constrictus* (Hall).

1906.
Moberg, J. C., and Segerberg, C. O.,
"Bidrag till kanndomen om Ceratopygeregionen," 'Lund's Geol. Fältklubb.,' ser. *b*, no. 2, pp. 1–110, pls. i–vii.

A comprehensive essay on the *Dictyonema* and *Ceratopyge* beds of Sweden, regarded as constituting collectively a single formation. This is divided by the authors into two main sections, namely, a lower or *Dictyograptus* section, and an upper or *Ceratopyge* section, each of these sections being composed of a lower and an upper zone. All the fossil forms —Brachiopoda, Trilobites, and Graptolites—recognised by the authors as afforded by the formation, are figured and described, and an admirable account is given of the history of previous research and opinion respecting the formation as a whole and its extra-Scanian equivalents. Some eight forms of Graptolites are recognised by the authors as characteristic of the formation. Of these the following are given as new: *Clonograptus tenellus* (Linnarsson) var. **hians** (Moberg), *Clonogr.* **heres** (Westergård), and *Bryograptus* **Hunnebergensis** (Moberg).

1906.

Wood, E. M. R., " Graptolites from Bolivia," 'Quart. Journ. Geol. Soc.,' vol. lxii, pp. 431, 432.

A brief note on the Graptolites collected by Dr. J. W. Evans during his expedition to Bolivia in 1901—1902. Seven species of Bolivian Graptolites are recognised; and on the evidence afforded by the collective assemblage the containing strata are paralleled with the Upper Arenig Formation of Britain.

1906.

Wood, E. M. R., " Tarannon Series of Tarannon," ' Quart. Journ. Geol. Soc.,' vol. lxii, pp. 644–701.

This comprehensive paper is devoted to the description and the discussion of the stratigraphy and palæontology of the Tarannon Formation as worked out in detail by the authoress in the typical area of central Wales. The great vertical extent of the formation is demonstrated and its conformable relations to the Llandovery Formation below and the Wenlock above. It is shown to consist of four recognisable sub-formations, marked by characteristic graptolitic sub-faunas. These are arranged in four zones, namely, those of (1) *Monograptus turriculatus*, (2) *Mono. crispus*, (3) *Mono. Griestonensis*, (4) *Mono. crenulatus*.

The immediately overlying Wenlock and underlying Llandovery local Graptolite zones are also worked out, and their containing species noted. Some thirty-eight forms of Graptolites are quoted from the Tarannon Series itself, of which seven are survivals from the Llandovery beds and five range upwards into the Wenlock Series.

The paper is illustrated by tables showing (i) the correlation of the local Tarannon sub-formations with their representatives elsewhere in Britain, Sweden, and Bohemia, and (ii) the geological and geographical distribution of the species of Graptolites identified in the typical Tarannon area. Upon one of these tables the Graptolites are arranged in the order of their chronological appearance.

1906.

Elles and Wood, " Monograph of British Graptolites," pt. 2, Palæontographical Society, 1906.

In this Fifth Part of the British Monograph the Historical Section embraces the period 1871—1880. The Descriptive Section opens with an account of the structure and mode of development of the family of the Diplograptidæ (Lapworth) in general. The remainder of the Part is devoted to the diagnoses and illustration of the British forms belonging to the genus *Climacograptus*, Hall (which is assigned to that family) and its constituent forms arranged under five different types. Nineteen British species of *Climacograptus* are recognised, the following being described as new: *Climaco-*

graptus **Tornquisti,** *Cl.* **brevis,** *Cl.* **supernus,** and *C.* **latus,** together with several varieties.

1906. *Hall, T. S.,* "Reports on Grapto- lites," 'Records Geol. Surv. Victoria,' vol. i, pp. 266–278, pl. xxxiv.	Annotated lists of Graptolites identified by the author from many localities in Victoria, with figures and descriptions of three new species: *Climacograptus* **mensuris,** *C.* **Baragawanathi,** and *Diplograptus* **ingens.**

1906. *Evans, D. C.,* "Ordovician Rocks of Western Caermarthen- shire," 'Quart. Journ. Geol. Soc.,' vol. lxii, pp. 597–643.	An excellent summary of the results of the author's field-researches in a most complicated district. The paper is illustrated by a geological map of the area described and tables of fossils, etc. The stratigraphical succession, which extends from Middle Arenig (*Didymograptus bifidus* zone) to the Lower Llandeilo (Llandovery) inclusive, is rich in Graptolites on several horizons. Fifty-four species are

recognised by the author, and their localities and ranges given.

1907. *Fearnsides, W. G.,* *Elles, G., and Smith, B.,* "The Lower Palæozoic Rocks of Pomeroy," 'Proc. Roy. Irish Acad.,' vol. xxvi, pp. 97–128.	A description of the results of a combined field-study of the strata and fossil contents of the Lower Palæozoic Rocks of Portlock's classic district of Pomeroy, North Ireland. Especial attention is paid to the zonal sequence of the included forms of Graptolites, of which thirty-four species and varieties are named from the Pomeroy formations, and their sequence and correlation given in an accompanying table.

1907. *Vinassa de Regny, P.,* "Graptoliti Carniche," 'Congresso dei Natu- raliste Italiani,' 1906, pp. 1–27, pl. i.	The author gives a table of some twenty-five species and varieties of Graptolites collected from the Lower Palæozoic rocks on the Italian versant of the main range of the Carnic Alps, and assigns the containing beds in part to the Llandovery and in part to the Wenlock-Ludlow. The majority of the forms are referred to well-known British species, but

three are noted as new: *Dendrograptus* (?) **carnicus,** *Desmograptus* **italicus,** *M.* **colonus,** (Barr.) var. **intermedius.** A brief bibliography introduces the work. A description of each species and an accompanying plate allows of the identification of most of the forms represented.

1907. *Moberg, J. C.,* "Skånes *Dicellograptus* Skiffer," etc, 'Geol. Fören. Förhandl.,' vol. 29, pp. 75–83, pl. i.	The author noted the discovery of *Pleurograptus linearis* (Carr.) in the *Dicellograptus* Beds of Scania, and discusses and tabulates in zonal form the whole of the Ordovician of Scania, parallels the Trilobite-facies and Graptolite-facies, and recognises under distinct names some fourteen Graptolite Zones in the general sequence.

Founding mainly upon collections made by himself in Sweden and Central Europe, supplemented by examples collected by Herr Eisel of Gera, Thuringia,

Törnquist revises, describes, and figures the chief recognisable forms of Monograptidæ assignable to the group *Rastrites* of Barrande. He discusses and accepts the generic or sub-generic value of the group, and in the main body of the paper deals with some thirteen species and varieties of *Rastrites*, among which the following are given as new: *R. peregrinus* (Barr.) var. **pecten** and var. **socialis**, *R. approximatus* (Perner) var. **Geinitzii**. The remainder of the paper is devoted to a critical description of those forms of *Monograptus* proper which are most closely allied to *Rastrites*. Of these one new form is given, *Monograptus* **amphibolus** (Törnquist).

1907.
Törnquist, S. L.,
" On the genus Rastrites and some allied species of *Monograptus*," ' Lund's Univ. Årsskrift,' vol. 3, pp. 1–22.

The Historical Section of this Sixth Part embraces the period 1881—1895. The Descriptive Section deals with the forms collectively assigned to the genus *Diplograptus* proper. This section commences with an account of the general characteristics of the genus, the nomenclature employed, and the grouping adopted. The British species and varieties are arranged in four main groups. One of these receives a new name, **Mesograptus**. A minor group is also novel, viz., **Amplexograptus** (Lapw. MS.). Some twenty-five British species of *Diplograptus* are described. The new forms noted are *Diplograptus* **mutabilis**, *D.* **serratus**, *D.* **multidens**, *D.* **artus**, and *D.* **Pageanus** (Lapw. MS.).

1907.
Elles and Wood,
" Monograph of British Graptolites,' pt. 2, Palæontographical Society, 1907.

Dr. Hind describes and figures two forms of Graptolites discovered by Mr. Tate in the Pendleside Series of the British Carboniferous, viz. *Callograptus* **carboniferus**, and *Desmograptus* **Monensis**. These forms are of especial interest to geologists and palæontologists, as no Graptolites had previously been detected in British strata of more recent geological date than the Silurian.

1907.
Hind, W.
" On the Occurrence of Dendroid Graptolites in British Carboniferous Rocks," ' Proc. Yorks. Geol. Soc.,' vol. xvi, pp. 155–157, pl. xviii.

This second volume, completing the author's Monograph of the Graptolites of New York State, sustains the high reputation of the first volume in respect both to matter and illustrations. It treats of the Graptolites which occur in the New York strata of later geological age than those of the Chazy Formation, commencing with those of the famous Ordovician " Norman's Kill Zone," and ending with those of the American Carboniferous. Like the first volume, it falls into two sections— a General Section, dealing with the literature and geological and geographical distribution of the genera and species concerned, their morphology and zoological relationships, etc.; and a Descriptive Section, in which the forms themselves are diagnosed and figured.

1908.
Ruedemann, R.,
" Graptolites of New York. Pt. 2: Graptolites of the Higher Beds," ' New York State Museum, Memoir XI,' pp. 1–487, pls. i–xxxi.

Full justice is done to the results arrived at by previous observers, and especial attention is very properly called throughout to the memoirs and conclusions of Dr. Gurley on American Graptolites. Several of his MS. diagnoses, figures, and descriptions are included, and their sources acknowledged in the body of the work.

The General Section of the volume is illustrated by tables of local distribution, zonal range, and classification. The earlier and later nomenclatures of the Graptolite-bearing formations are given and discussed. Five successive Graptolite zones are distinguished in the Upper Ordovician, namely, those of (1) *Nemagraptus gracilis*, (2) *Diplograptus amplexicaulis*, (3) *Glossograptus quadrimucronatus*, (4) *Diplograptus peosta*, (5) *Dicellograptus complanatus*, embracing the American strata ranging from the base of the Black River Formation to the top of the New York Ordovician.

The notes on morphology, etc., given in the first volume are here supplemented and extended. The known facts and varied views respecting spines, basal discs, vesicles, virgula, nema, retiolite-structure, etc., are adduced and discussed, and illustrated by abundant text-figures, and several new observations respecting the polypary and thecæ in the Graptoloidea are brought forward. Attention ought also to be called to Ruedemann's notes, descriptions, and figures of the so-called genera which he classes together under the title of 'Genera incertæ sedis.' Of these, the most important constitute the group *Corynoides* of Nicholson, which Hopkinson and Lapworth (1875) erected into a distinct family under the title of Corynograptidæ. The author changes the name to Corynoidæ, and gives several instructive figures of the New York forms. These he identifies with the known British species, but recognises several varieties.

The Descriptive Section of the volume is devoted to the diagnoses and figures of the Graptolite forms themselves. One hundred and seventeen species are recognised by the author, of which fifty-one are assigned to the Dendroidea and sixty-six to the Graptoloidea. Three new genera, viz. **Ptiograptus** and **Mastiograptus**, are classed among the Dendroidea. The new species of Dendroidea include *Dendrograptus* **rectus**, *Ptiograptus* **Poctai**, *P.* **Hartnageli**, *Odontaculis* **hepaticus**, *Ptiograptus* **percorrugatus**, *Desmograptus* **tenuiramosus**, *D.* **cadens**, *D.* **Vandelooi**, *Cactograptus* **crassus**, *Palæodictyota* **anastomotica**, *P.* **Clintonensis**, *Mastigograptus* **circinalis**, together with the following MS. forms of Dr. Gurley: *Dictyonema* **polymorphum**, *D.* **Leroyense**, *D.* **Areyi**, *D.* **megadictyon**, *D.* **perradiatum**.

The new forms of Graptoloidea embrace *Azyograptus* **simplex**, *Syndograptus* **pecten**, *Dicellograptus* **mensurans**, *D.* **Smithi**, *C.* **Mississippiensis**, *C.* **modestus**, *Cyrtograptus* **Ulrichi**, and the following MS. species of Lapworth: *Azyograptus* **Walcotti**, *Nemagraptus* **exilis**, *Dicranograptus* **Gurleyi**.

This second volume of Ruedemann's Monograph is clearly the work of an enthusiast in the subject; but one who nowhere poses as an authority, or disguises either the defects in our knowledge or the inevitability of diverse interpretations.

He takes it for granted throughout that the reader wishes to know in brief the facts already discovered, and by whom; and the various inferences already drawn, and why. That the reader, like himself, may be in a position to arrive at his own unbiased conclusions, he supplies him with a wealth of illustration and detail that are certain to render the monograph of especial value, not only to the field-geologist, but also to the beginner in Graptolithology.

1908.
Moberg, J. C.,
" Nomenklaturen för våra Paleozoiska bild-ningar," 'Geol. Fören. Förhandl.,' vol. 30, pp. 343–351.

After giving a carefully written but brief review and discussion of the history of the successive past improvements in the nomenclature of the Swedish Lower Palæozoic Rocks, the author expresses his agreement with the plan advocated by De Lapparent in 1900, namely, the retention of Murchison's term Silurian for the whole of the Palæozoic Formations regarded collectively, and the employment of the names Cambrian (Sedgwick), Ordovician (Lapworth, 1879), and Gothlandien (De Lapparent, 1900) for its three component divisions and their respective faunas. This paper, which had already been published by the author in the 'Geological Magazine' for 1907 (December 5, vol. vi, pp. 273—279), is of especial importance from the historical point of view, as the nomenclature advocated in its pages became subsequently adopted in the Geological Guides and Memoirs issued in connection with the International Geological Congress held in Stockholm in 1910, and in the majority of those Swedish Graptolite memoirs and papers which have been published since that date.

1908.
(Mrs. Shakespear)
Wood, E. M. R.,
" On some New Zealand Graptolites," 'Geol. Mag. [5] vol. v, pp. 145–148.

An identification and discussion of twenty-one forms of Graptolites collected by Mr. Isaacson from Slaty Creek, New Zealand, for the British Museum (Nat. Hist.). It is pointed out that all the species distinguished in the collection are either identical with, or closely allied to, previously named British and North American forms whose association and geological range are already fairly well established in the Northern Hemisphere. Two Graptolitic sub-zones, distinguishable by lithological and palæontological characters, are noted by the author as represented in the collection. A table is given, showing the distribution of these New Zealand species in America and Great Britain, and the conclusion is drawn that, as the association of species in this collection is practically the same as their association in the Northern Hemisphere, it may be anticipated that further work will result in proving that the zonal succession of Graptolites, well established in the Northern Hemisphere, prevails in New Zealand also.

1908.
Hall, T. S.,
" Reports on Grapto-lites," 'Records Geol. Surv. Victoria,' vol. ii, pp. 221–227.

Lists of Graptolites collected by the officers of the Geological Survey in several Victorian localities.

1908.
Elles and Wood,
" Monograph of British
Graptolites," pt. 7,
Palæontographical
Society, 1908.
In this Part the Historical Section is devoted to the Literature of Graptolite Research during the period 1895—1901. In the Descriptive Section the British forms of Diplograptidæ assigned to the sub-genus, *Petalograptus,* and also those of the genera, *Cryptograptus* and *Trigonograptus,* are treated of. The family of the Glossograptidæ is next taken up, then the Retiolitidæ, and finally the Dimorphograptidæ.

A remarkable new form, *Petalograptus* (?) *phylloides* is provisionally assigned to the Diplograptidæ. Two previously named but undescribed sub-genera, **Hallograptus** (Carruthers, MS.) and **Neurograptus** (Lapworth) are classed with the Glossograptidæ, together with two new sub-genera, **Thysanograptus** and **Nymphograptus,** and **Plegmatograptus** is assigned to the Retiolitidæ. The new species described in this part include *Petalograptus* **altissimus,** *P.* (?) **phylloides,** *Glossograptus* **acanthus,** *Nymphograptus* **velatus,** and *Plegmatograptus* **nebula.**

1908.
Kiær, J.,
' Das Obersilur im
Kristianiagebiete,
Christiania,' pp. 1–595.
A brilliant monograph dealing with the author's researches into the lithological and palæontological sequence in the Silurian (Llandovery to Ludlow) of the classical Norwegian district extending from Skien to the head of Lake Mjösen.

The work is illustrated by maps, plates, sections, abundant text-figures, tables of zonal distribution, and a special chapter on correlation. Four successive formations are recognised, parallel broadly with the British Lower Llandovery. Upper Llandovery (inclusive of the Tarannon), Wenlock and Ludlow respectively, the whole being divided into nineteen zones. In the palæontological parts of the work, although the chief attention is naturally directed to the extraordinarily prolific Brachiopod, Trilobite, and Coral faunas, the Graptolites are by no means neglected, some twenty-six distinct species being recognised, and their geographical and geological ranges in the districts described carefully indicated.

1908.
Hall, T. S.,
" Graptolite Beds of
Dalesford," 'Proc. Roy.
Soc. Victoria,' vol. ii,
pp. 271–284.
Lists of the Arenig Graptolites from some thirty-four different Victorian localities.

1908.
Hall, T. S.,
" Reports on Grapto-
lites," ' Records Geol.
Surv. Victoria,' vol. ii,
pp. 137–143, pl. xv.
Lists of Graptolites collected from Victorian localities, with figures and descriptions of two new species, *Diplograptus* **tardus** and *Didymograptus* **latus.**

1909.
Westergård, A. H.,
" Studier ofver Dicho-
graptus Skiffern in
Skåne, etc.," ' Lund's
Geol. Fältklubb.,' ser. B,
no. 4, pp. 1–79, pls. i–v.

Under the collective title of Didymograptus Skiffer, Westergard includes the Lower and shalier division of Angelin's *Ceratopyge* Region in contradistinction to the more compact and calcareous Upper Division, embracing the *Shumardia* beds and *Ceratopyge* Kalk proper, the Upper Division being distinguished further by a fauna consisting mainly of Trilobites and Brachiopods, while the Lower Division under description is marked by the presence of Graptolites. The present paper summarises the main facts and conclusions respecting the strata of this Lower Division, regarded by Scandinavian geologists as the basal member of the Swedish Ordovician, and of the beds which are in immediate contact with the sub-formation above and below. The memoir is introduced by a full bibliography of the history of previous research, followed by a detailed account of the results of the author's personal field-studies of the succession in Scania, Westergötland, Östergötland, Öland, etc.

The author arranges the strata of the *Dictyograptus* Skiffer in three sub-zones, characterised respectively by : (1) *Dictyograptus flabelliformis* (Eichw.). (2) Idem *Forma typica.* (2) *Clonograptus tenellus* (Linnrs.). (3) *Dictyograptus flabelliformis* var. *Norvegica* (Kjerulf.).

All the known fossils, so far as they are represented in the Lund Museum, are described in the body of the paper, and figured in the accompanying plates.

Especial attention is devoted to the Graptolites present. Nine distinct species and varieties are recognised. Two of these are noted as new, *Dictyograptus flabelliformis* (Eichw.) var. **confertus** (Linn. MS.) and *Clonograptus tenellus* (Linn.) var. **grandis.** All the Graptolite forms noted are admirably figured in the plates, some from striking photographs by Professor Moberg, and others from drawings by Mauda Broman and the author.

As respects that large group of Graptolites generally united by palæontologists under the common generic title of *Dictyonema*, Hall, Westergard expresses his opinion that the group is separable into two distinct sections or genera : (1) a section typified by *Dictyonema flabelliforme* (Eichw.), and (2) a section typified by *Dictyonema cervicorne* (Holm.), etc. To the first of these sections he restricts the generic title, *Dictyograptus* (Hopkinson and Lapw.), and for the second he proposes a new generic title—**Dictyodendron.** *Dictyograptus* he assigns to the Graptoloidea and *Dictyodendron* to the Dendroidea.

1909.
Elles, G. L.,
" Ordovician and
Silurian Rocks of
Conway," ' Quart.
Journ. Geol. Soc.,'
vol. lxv, pp. 169–194.

A description of the detailed results of the authoress' study of the lithological and palæontological succession in the rocks of the Conway District, ranging from the Llandeilian to Middle Wenlockian. Particular attention is paid to the Graptolites, some fourteen separate zones being recognised and named, and paralleled with their equivalents elsewhere in Great Britain.

1909.
Jones, O. T.,
"The Hartfell-Valentian Succession around Plynlimon and Pont Erwyd," 'Quart. Journ. Geol. Soc.,' vol. lxv, pp. 463-537.

An admirably detailed description of the local members and distribution of the lithological and palæontological succession, as worked out by the author, in the Plynlymmon District, Central Wales, through strata ranging from Upper Bala to Middle Tarannon (zone of *Dicellograptus anceps* to zone of *Monograptus Griestoniensis*). The strata described are locally rich in Graptolites. More than 100 distinct forms are recognised as present by the author; their distribution in the several local formations, stages, and zones is tabulated, and paralleled with their known arrangement in equivalents elsewhere in Britain. Two new species of Graptolites, *Monograptus* **atavus** and *M.* **Rheidolensis,** are figured and described.

1908-9.
Moberg, J. C., and Törnquist, S. L.,
"Retioloidea Skånes Colonusskiffer,"
, Sveriges Geol. Undersökning,'Årsbok 2, no. 5.

The authors record the occurrence of **Plectograptus macilentus** (Törnquist), *Retiolites spinosus* (Wood), *Gothograptus nassa* (Holm), in the Scanian-*Colonus* Shales. The structure of the first named, which is made the type of a new genus, is described and figured in detail. For the central rod-like body, or "filiform organ," figured by Holm in his *Gothograptus nassa,* and provisionally termed "Virgula" by Wiman—with the reservation, however, that it is not apparently morphologically identical with the virgula of the Diplograptidæ—the authors of this paper propose the neutral title of "fulcell" (Latin "fulcimen," a prop or stay).

1909.
Hall, T. S.,
"Notes on Graptolites from Tallong, New South Wales,"
'Records Geol. Surv. N.S. Wales,' vol. viii, pp. 339-341, pl. lv.

A list of twelve forms of Graptolites collected by the officers of the New South Wales Geological Survey. The majority are identified with previously known species, but one variety (*Dicranograptus hians* (T. S. Hall) var. **apertus**) is noted as new.

1910.
Moberg, J. C.,
"Guide for the Principal Silurian Districts of Scania," 'Geol. Fören. Förhandl.,' vol. 32, pp. 45-194.

In this valuable memoir the author gives a minutely detailed description of the most important Cambrian, Ordovician, and Silurian fossiliferous localities in Scania including the famous Graptolite-bearing areas of Fögelsång, Jerrestad, Röstånga, Tösterup, etc., illustrating the work by many maps, tables of local zonal sequence, and by a bibliography ranging from 1827 to 1909 inclusive. The long extended, detailed, and successful researches of Prof. Moberg in the Lower Palæozoic rocks and fossils of Scania give an especial authority to this memoir, and it is certain long to remain the standard field guide to the region.

1910.
Elles and Wood.
" Monograph of British
Graptolites," pt. 2,
Palæontographical
Society, 1910.

This Part of the Monograph is wholly Descriptive. It is introduced by a section dealing with the general characteristics of the family Monograptidæ and of the genus *Monograptus*, followed by diagnoses and illustrations of British species belonging to the first three of the seven component groups of Monograpti recognised by the authors. None of the species described are named as new, but several forms previously regarded as species by other authors are here classed as varieties.

1910.
Fricke, M.
" Die Silurischen Ablagerungen am Südrande des Zwickauer Kohlenbeckens,"
pp. 1–53.
Zwickau, 1910.

A privately printed geological and palæontological memoir on the Silurian strata south of Zwickau, Saxony, with special reference to the Graptolite fauna. The work is somewhat popular in character, but apparently the result of several years' enthusiastic research. It is introduced by a discussion of the results obtained by previous researchers in the neighbouring regions of Thuringia, etc., and embraces a brief description, illustrated by many good text-figures, of the typical forms and structures of the Graptolites in general. A list of the local Graptolites of the region, including some eighty species and varieties, is carefully tabulated, the forms cited being grouped for eight different localities, and in the zones and sub-zones previously defined by Eisel, to whose long extended labours and successful results in the equivalent Graptolitic deposits of Thuringia appropriate references are made.

1910.
Törnquist, S. L.,
" *Cyrtograptus*-arter från Thüringen, etc.,"
' Geol. Fören. För-handl.,' vol. 32,
pp. 1559–1575, pl. lxii.

The author describes and figures Thuringian examples of his species, *Cyrtograptus radians* (Törnquist) (see bd. 9, p. 491) and a remarkable new form, *Cyrtograptus* **multiramis.** He appends sundry historical and critical observations upon the detailed grouping of the Monograptidæ in general, and of several of the forms referred to by Jaekel (1899) and Frech (1897) in particular.

1910.
Lapworth, C.
" Graptolites," ' Encyclopædia Britannica,'
11th ed., vol. xii,
pp. 365–367, figs. 1–28.

A brief summary of the existent state of knowledge and opinion respecting the Graptolites in general, their structure, development, classification, systematic position, and geological and geographical range. Two main sections—viz. Graptoloidea and Dendroidea—are recognised, united under the collective title of **Graptolithina.** The article is illustrated by figures of the adult polypary in several of the more characteristic genera, and of the sicula and the early parts of the polypary in the best known species.

1911.
Hadding, A.,
" Svenska Arterna af Släktet Pterograptus, Holm," ' Geol. Fören. Förhandl.,' vol. 33,
pp. 487–494, pl. vii.

A welcome summary, discussion, and extension of previous knowledge and opinion respecting the geological distribution, range, and alliances of the genus *Pterograptus*, Holm, illustrated by clear and instructive figures.

1911.
Törnquist, S. L.,
" Graptolitologiska
bidrag," III–VII,
' Geol. Fören. För-
handl.,' vol. 33,
pp. 421–438, pls. v, vi.

Törnquist describes his discovery of *Lasiograptus* (*Hallo-graptus*) *mucronatus*, Hall (= var. *bimucronatus*, Nich.), in the Flag-kalk of the Siljan region, together with *Glyptograptus teretiusculus*, His., and discusses the geological horizon of the containing beds. He treats of the synonymy of two previously known forms in the upper *Didymograptus* Shales of Fögelsång, and classes them as *Didymograptus bifidus*, Hall, and *D.* **lentus**, Törnquist, sp. nov. He indicates the resemblance of *Cyrtograptus Ulrichi*, Ruede-mann, to *C. multiramis*, Törnq., and describes and figures three forms from the *Phyllograptus densus* zone of Flagebro, viz. an example of the form *Chætoides*, Gurley, and two of *Clonograptus*. A specimen of *Dendrograptus* conf. *serpens*, Hopk., is also described by him from Bornholm.

1911.
Horn, E.,
" Eine Graptolithen-
kolonie aus Westergöt-
land," ' Geol. Fören.
Förhandl.,' vol. 33,
pp. 237–239.

The author figures and describes a radiating assemblage of examples of *Climacograptus scalaris*, from the *Rastrites* beds of Westrogothia, of the type of one of Ruedemann's " synrhabdosomes."

1911.
Wade, A.,
" Llandovery and
Associated Rocks of
N.E. Montgomery-
shire," ' Quart. Journ.
Geol. Soc.,' vol. lxvii,
pp. 415–459.

A summary of the results of the author's field work in the Welshpool district, the sequence described ranging from Llandeilo-Caradoc (Glenkilk-Hartfell) to Lower Ludlow in-clusive. Twenty-eight Graptolite species are noted in the succession, and their stratigraphical and distributional ar-rangement described and discussed.

1911.
*Watney, G. R., and
Welch, E. G.*,
" Zonal Classification
of Salopian Rocks,
Cautley and Raven-
stonedale," ' Quart.
Journ. Geol. Soc.,' vol.
lxvii, pp. 215–237.

An account of the detailed local sequence in the Wenlock and Lower Ludlow beds of a large area in Westmoreland, especial attention being devoted to the Graptolites present. Six distinct Graptolite zones are distinguished and named. Thirty-four distinct forms are recognised, and their localities and ranges given and discussed.

1912.
Törnquist, S. L.,
"Graptolitologiska
bidrag," VIII, IX, X,
' Geol. Fören. För-
handl.,' vol. 34,
pp. 603–622, pl. viii.

In the first of these contributions Törnquist gives a minutely detailed critical review of the history of discovery and opinion respecting the species *Monograptus spiralis*, Geinitz, illustrating it by a plate, including not only Geinitz's original species, but also those of examples collected by Eisel from the actual locality whence Geinitz obtained his original specimen. In the second contribution Törnquist discusses the asserted synonymy of *Monograptus discus*, Törnq., and *M. veles*, Richter. In the third he makes known the presence of the British zonal forms, *Diplograptus acuminatus*, Nich., and *D. vesiculosus*, Nich., in the Lower *Rastrites* Shales of Röstånga.

1912.
Ruedemann, R.,
" Lower Siluric Shales
of the Mohawk Valley,"
' New York State
Museum Bull.,' clxii,
pp. 1–151.

In this paper the author gives a detailed description of the local succession, lithology, and fauna of the " Utica Slate " and associated sub-formations outcropping in the broad Shale Belt, extending from Albany north-westwards up the valley of the Upper Hudson River to Saratoga and Glenfalls, and westwards up the valley of the Mohawk to Littlefalls and Utica. The memoir, which is illustrated by several good plates and by local lists of the Graptolites, etc., collected *in situ*, goes far to satisfy a long-felt want in respect to the vexed question of the relations of the so-called Utica Slate in its typical localities to the neighbouring sub-formations, ranging from the Trenton on the one hand to the Loraine on the other.

Among the Graptolites cited in the four successive local groups recognised by the author, the following forms are given as new : *Dictyonema* **multiramosum,** *Dicranograptus Nicholsoni* (Hopkinson) var. **parvulus,** *Diplograptus* (*Mesograptus*) **Mohawkensis,** *D.* (*Amplexograptus*) **macer.**

The example of *D. Nicholsoni* figured and described (Fig. 17, p. 79) is especially interesting, each of its branches showing a virgula-like axis prolonged distally as a naked rod or fibre well beyond the theca-bearing parts of the branch.

1912.
Elles and Wood,
" Monograph of British
Graptolites," pt. 2,
Palæontographical
Society, 1912.

This Ninth Part of the British Monograph is wholly descriptive. Dia es and figures of some fifty-three British species of Monc tidæ are given, among which are many forms not previ recorded from British strata. Four species are noted new, namely, *Monograptus* **remotus,** *M.* **undulates,** *M.* **ensis,** *M.* **delicatulus.**

1913.
Ulrich, E. O.,
" The Ordovician-
Silurian Boundary,"
' Compte-Rendu
Congrès Géologique
International, Canada
(Toronto),' 1913,
pp. 593–669.

A detailed discussion, mainly from the palæontological point of view, of the evidences relating to the probable local positions of the North American (United States and Canada) stratigraphical horizon answering to that usually accepted in Britain and Europe as marking the upper limits of the Bala and the lower limits of the Llandovery. The bearing of the Graptolite species upon the subject is carefully noted, and the discovery of typical Birkhill Graptolites in the United States (Arkansas) is for the first time made known.[1]

1913.
Hadding, A.,
" Undre Dicellograp-
tuskiffern i Skåne, etc.,"
' Lund's Universitets
Årskrift,' vol. 9, no. 15,
pp. 1–39, pls. iii, viii.

A well-illustrated memoir on the sequence and fossils of the Swedish Graptolite-bearing strata which follow at once upon the "*Didymograptus geminus* strata," and answer more or less to the British Llandeilo Flags and Upper Llandeilo. The series is divided by the author into four successive zones, namely, the zones of (1) *Glossograptus Hincksii* ; (2) *Climacograptus putillus* ; (3) *Nemagraptus gracilis* ; and (4) a zone in which

[1] Compare also Ulrich's "Revision of the Paleozoic Systems," 'Bull. Geol. Soc. of America,' vol. xxii, 1911, pp. 481–680; and Ulrich and C. Schuchert, "Palæozoic Seas and Barriers in Eastern North America," ' New York State Museum Bull.,' 1902, lii, pp. 633–663.

Graptolites are wanting. Twenty-six forms of Graptolites are described, and well illustrated by figures drawn by the author. Of these forms twelve are given as new—*Azygograptus* **Mobergi**, *Glossograptus* **Scanicus**, *Cryptograptus* **lanceolatus**, *Thysanograptus* **spinatus**, *Diplograptus* **Törnquisti**, *D.* **notabilis**, *D.* **propinquus**, *Dicranograptus* **irregularis**, *Dicellograptus* **vagus**, *D.* **minimus**, *Nemagraptus* **subtilis**, and *Desmograptus* **Tullbergi**. Founding upon specimens of which parts are preserved in relief, the author discusses the probable form and structure of the polypary and thecæ in the genus *Glossograptus*.

1913.
La Touche, T. H. D.,
"Geology of the Northern Shan States, Burma," 'Mem. Geol. Surv. India,' vol. xxxix, pt. 2.

A description of the geology of a wide region of Northern Burma, surveyed by the author ; with notes on the characteristic fossils and a correlation of the containing formations with their Asiatic, European, and American equivalents.

In this work there is made known for the first time the discovery of Graptolites in the Lower Palæozoic rocks of South-East Asia. A list of the forms collected by the author and Mr. J. Coggin Brown, and identified by Miss G. Elles, is given. All the forms named are referred to well-known British species. (See also Coggin Brown, 'Records of Geological Survey of India.')

1913.
Davies, A. M., and Pringle, J.,
"Deep Borings at Calvert, Buckinghamshire," 'Quart. Journ. Geol. Soc.,' vol. lxix, pp. 308–342.

Notable from the point of view of British Graptolitic literature as recording the earliest discovery of Graptolites (*Clonograptus*) of Tremadoc Age in deep borings in Central England.

1913.
Törnquist, S. L.,
"Några anmärkningar om indelningar inom Sveriges Kambrosilur," 'Geol. Fören. Förhandl.,' vol. 35, pp. 407–438.

In this paper Törnquist gives an historical and critical summary and review of the discoveries and opinions respecting the distribution of those species of Graptolites which are of zonal significance in the various recognised formations and sub-formations of the Lower Palæozoic strata of Sweden, Norway, Denmark and Bohemia, and Britain. References to the papers cited are given in footnotes, and the most recent zonal nomenclatures for the Swedish Graptolite-bearing strata are given at appropriate places in the text.

1913.
Elles and Wood,
'Monograph of British Graptolites," pt. 10, pp. 487–526, pls. l–lii, Palæontographical Society, 1913.

This Tenth Part of the British Monograph is composed of three divisions. The first division deals with those forms of the Monograptidæ which the authors arrange in their seventh group *Rastrites* (*auctorum*). Of these only one species is noted as new, *Monograptus* (*Rastrites*) **setiger.**

The second division is devoted to the genus *Cyrtograptus* (Carruthers).

The third division deals with the Zonal Range of all the forms of the British Graptoloidea definitely recognised by the authors in the Descriptive sections of the Monograph. These forms are named as amounting to 372 in number, and are grouped in the order of the consecutive families and genera described in the body of the work. Some thirty-six British Graptolite zones are recognised and named by the authors A first table (Table A, pp. 516–525) entitled " The Zonal Distribution of the British Graptoloidea," shows the presence, so far as known, of each species and variety in the several British zones recognised. A second table (Table B, p. 526) entitled " The Vertical Range of the Zones of British Graptoloidea," exhibits the approximate relative position of each of the Graptolite zones named in the ascending Succession of the British Lower Palæozoic formations and sub-formations generally.

A MONOGRAPH

OF

BRITISH GRAPTOLITES.

BY

GERTRUDE L. ELLES, Sc.D.,

LATE GEOFFREY FELLOW, NEWNHAM COLLEGE, CAMBRIDGE;

AND

ETHEL M. R. WOOD, D.Sc.

[MRS. SHAKESPEAR],

OF NEWNHAM COLLEGE, CAMBRIDGE; AND THE UNIVERSITY OF BIRMINGHAM.

EDITED BY

CHARLES LAPWORTH, LL.D., F.R.S.,

LATE PROFESSOR OF GEOLOGY IN THE UNIVERSITY OF BIRMINGHAM.

PLATES.

LONDON:

PRINTED FOR THE PALÆONTOGRAPHICAL SOCIETY.

1901—1918.

PLATE I.

Genus **Didymograptus**, M'Coy.

E. M. R. Wood, del.   *Bemrose, Ltd., Derby.*

DIDYMOGRAPTUS.

PLATE II.

Didymograptus—*continued.*

E. M. R. Wood, del.

Bemrose, Ltd., Derby.

DIDYMOGRAPTUS.

PLATE III.

Didymograptus—*continued*.

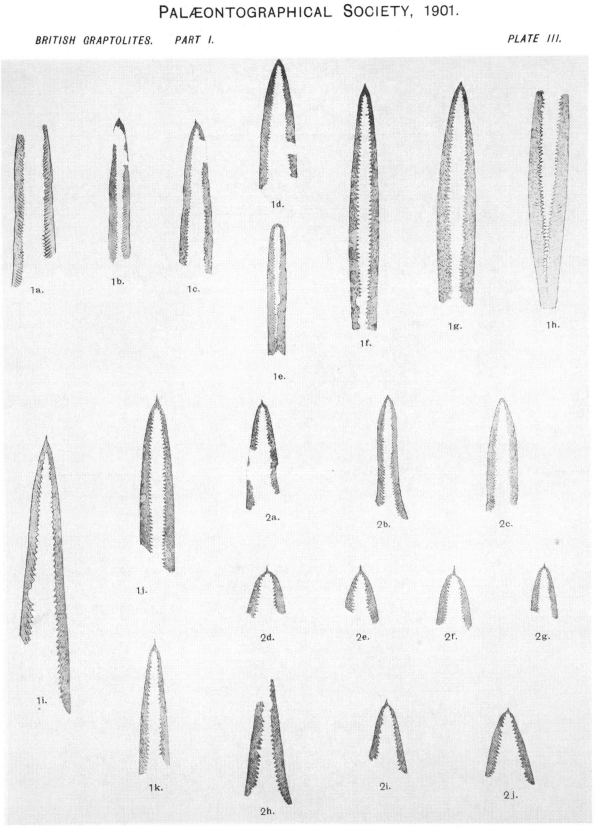

E. M. R. Wood, del.   *Bemrose, Ltd., Derby.*

DIDYMOGRAPTUS.

PLATE IV.

Didymograptus—*continued.*

1 *a—f.—Didymograptus bifidus*, Hall.

    1 *a.* Preserved in black shale. Figured by Hopkinson and Lapworth, Quart. Journ. Geol. Soc., 1875, pl. xxxiii, fig. 8 *a.* Porth-hayog, Ramsey I. Up. Arenig. Woodwardian Museum.

    1 *b.* Ibid., fig. 8 *d.*

    1 *c.* Larger specimen in relief. Pont Seiont. Up. Arenig. Woodwardian Museum.

    1 *d.* Ibid.

    1 *e.* Large specimen, preserved partly in relief, partly as a cast (labelled *D. Murchisoni*). 1½ miles S.E. of Duleek, co. Meath. Museum of Geol. Survey, Ireland.

    1 *f.* Widely divergent stipes. Pont Seiont. Up. Arenig. Woodwardian Museum.

2.—*Didymograptus stabilis*, Elles and Wood, sp. nov.

    Specimen redeveloped. Originally figured by Hopkinson as *D. indentus*, Quart. Journ. Geol. Soc., 1875, pl. xxxiii, fig. 7 *a.* Porth-hayog, Ramsey I. Up. Arenig. Woodwardian Museum.

3 *a—c.—Didymograptus amplus*, Elles and Wood, sp. nov.

    3 *a.* Obverse view. Preserved in black shale. Abereiddy Bay. Llandeilo (Up. Llanvirn, Hicks). G. L. Elles' Collection.

    3 *b.* Large specimen, affected by cleavage. Abereiddy Bay. E. M. R. Wood's Collection.

    3 *c.* Specimen not much cleaved. Abereiddy Bay. Woodwardian Museum.

4 *a—c.—Didymograptus*, cf. *indentus*, Hall.

    4 *a.* Badly preserved example. Mungrisedale, Glenderamakin Valley, Lake district. Upper Arenig. Woodwardian Museum.

    4 *b.* Ibid.

    4 *c.* Poorly preserved. Figured by Nicholson as *D. geminus*, Ann. Mag. Nat. Hist., 1870, p. 346, fig. 6 *a.* Thornship Beck, Lake district. Up. Arenig. Woodwardian Museum.

5 *a—h.—Didymograptus nanus*, Lapw.

    5 *a.* Poorly preserved specimen affected by cleavage. Figured by Hopkinson and Lapworth, Quart. Journ. Geol. Soc., 1875, pl. xxxv, fig. 4 *a.* Abereiddy Bay. Llandeilo (Up. Llanvirn, Hicks). Woodwardian Museum.

    5 *b.* Ibid., fig. 4 *b.*

    5 *c.* Quarry N.W. of Llanvirn. Lower Llanvirn (Hicks). Woodwardian Museum.

    5 *d.* Outerside, Keswick. Upper Skiddaw Slates. British Museum (Nat. Hist.), S. Kensington.

    5 *e.* Reverse view. Quarry N.W. of Llanvirn, Lower Llanvirn (Hicks). E. M. R. Wood's Collection.

    5 *f.* Preserved partly in relief, partly as a cast. Ellergill, Lake District. Lapworth's Collection.

    5 *g.* Ibid.

    5 *h.* Poorly preserved specimen. ? Figured Salter, Quart. Journ. Geol. Soc., vol. xix, p. 137, fig. 13 *c.* Eggbeck, Ullswater. Upper Skiddaw Slates. Museum of Practical Geology, Jermyn Street.

6 *a—d.—Didymograptus artus*, Elles and Wood, sp. nov.

    6 *a.* Large specimen. Thornship Beck. Upper Skiddaw Slates. Woodwardian Museum.

    6 *b.* Two specimens showing effect of cleavage. Porth-hayog, Ramsey I. Upper Arenig (Hicks). G. L. Elles' Collection.

    6 *c.* Porth-hayog. Woodwardian Museum.

    6 *d.* Obverse view. Porth-hayog. G. L. Elles' Collection.

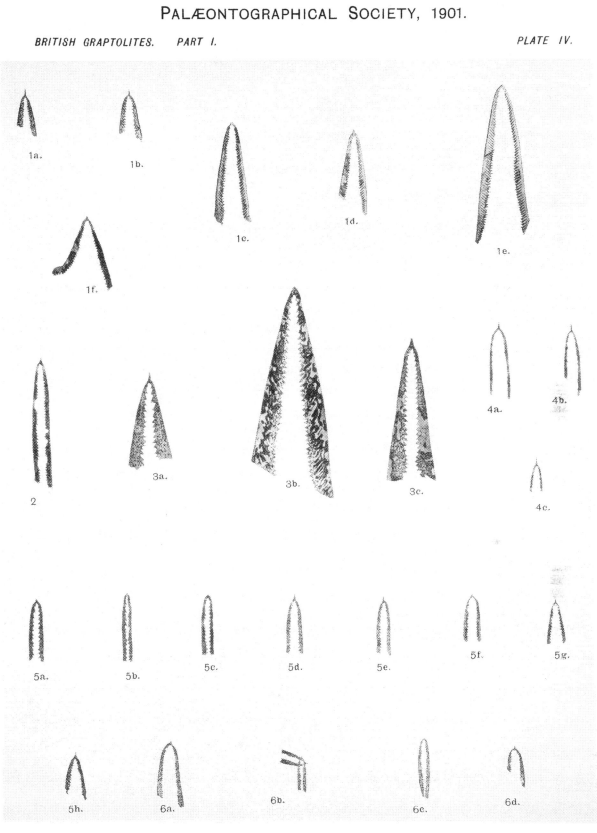

E. M. R. Wood, del.   *Bemrose, Ltd., Derby.*

DIDYMOGRAPTUS.

PLATE V.

Genus **Tetragraptus,** Salter.

E. M. R. Wood, del.   *Bemrose, Ltd., Derby.*

TETRAGRAPTUS.

PLATE VI.

Tetragraptus (*continued*) and Schizograptus, Nicholson.

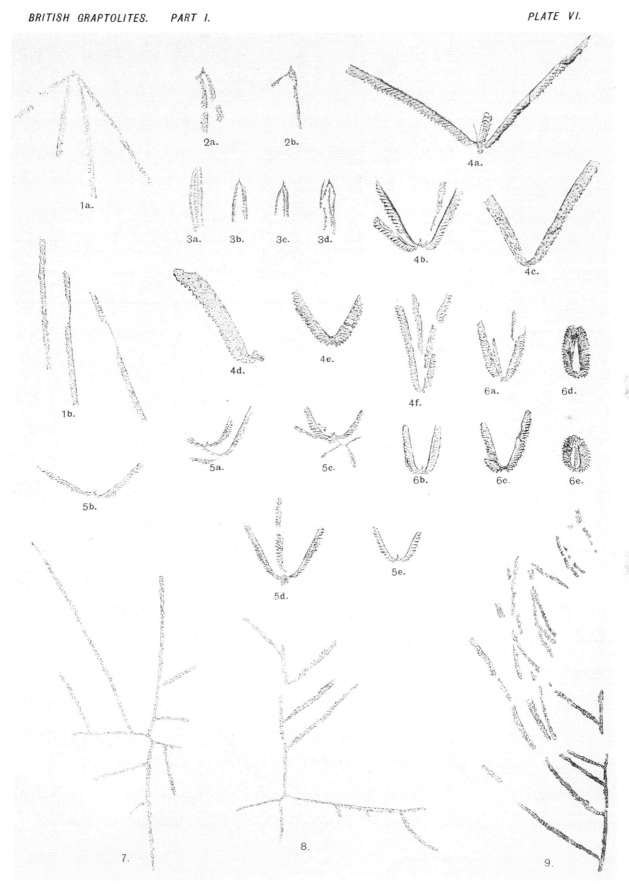

E. M. R. Wood, del.　　　　　　　　　　　　　　　　*Bemrose, Ltd., Derby.*

TETRAGRAPTUS AND SCHIZOGRAPTUS.

PLATE VII.

Genus **Trochograptus**, Holm.

FIG.

1.—*Trochograptus diffusus*, Holm.

Specimen showing disc. Figured, Elles, Quart. Journ. Geol. Soc., vol. liv, pl. xxvii. New Quarry, Scow Gill. Middle Skiddaw Slates. Fitz-Park Museum, Keswick.

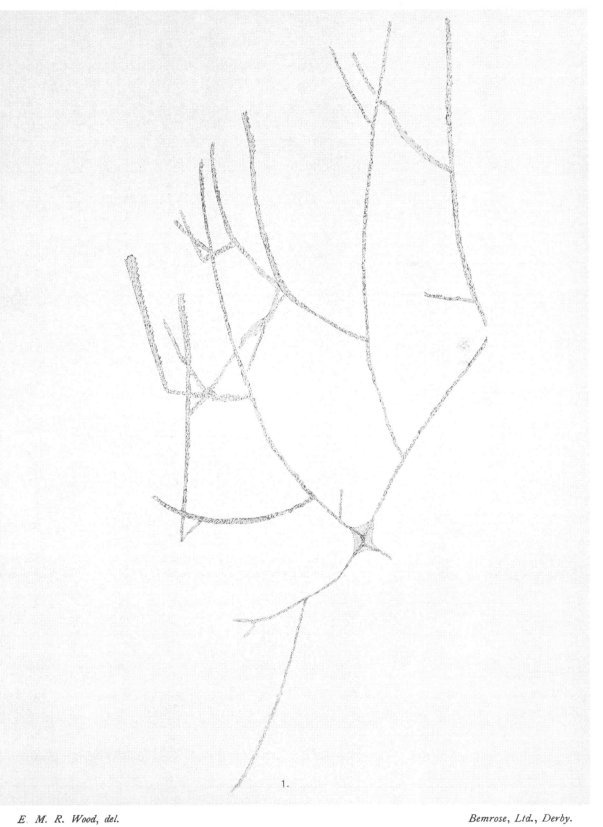

1.

E. M. R. Wood, del.   *Bemrose, Ltd., Derby.*

TROCHOGRAPTUS.

PLATE VIII.

Genus **Trochograptus** (*continued*) and **Holograptus**, Holm.

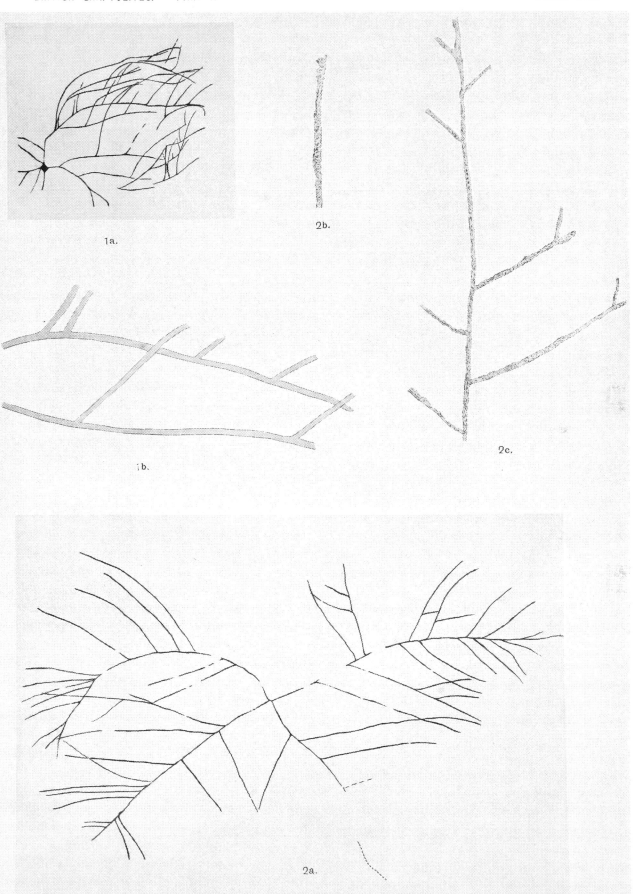

1a.

2b.

;b.

2c.

2a.

E. M. R. Wood, del.

Bemrose, Ltd., Derby.

TROCHOGRAPTUS AND HOLOGRAPTUS.

PLATE IX.

Genus **Dichograptus**, Salter.

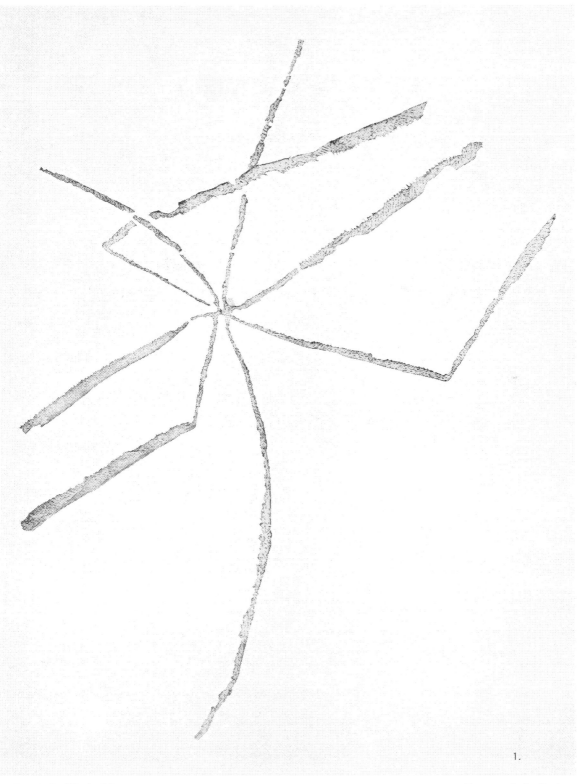

1.

E. M. R. Wood, del.

Bemrose, Ltd., Derby.

DICHOGRAPTUS.

PLATE X.

Dichograptus—(*continued*).

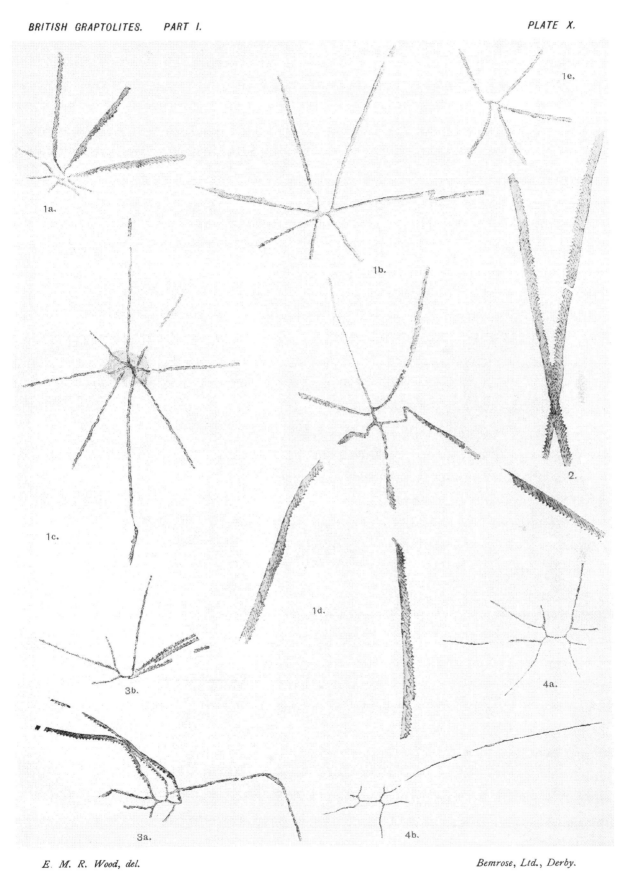

E. M. R. Wood, del.

Bemrose, Ltd., Derby.

DICHOGRAPTUS.

PLATE XI.

Loganograptus, Hall, and Clonograptus, Hall.

FIGS.

1 a—g.—*Loganograptus Logani*, Hall.

 1 a. Large specimen. Figured by Nicholson, Monograph of British Graptolites, 1872, p. 109, fig. 52 c. ? Skiddaw Slates. British Museum (Natural History), S. Kensington.

 1 b. Fifteen-stiped form, distorted by cleavage. Barf. Skiddaw Slates. Woodwardian Museum.

 1 c. Ten-stiped form. Randal Crag. Skiddaw Slates. Woodwardian Museum.

 1 d. Twelve-stiped form. Gate Gill, Blencathra. Skiddaw Slates. Fitz-Park Museum, Keswick.

 1 e. Nine-stiped form. Randal Crag. Skiddaw Slates. Woodwardian Museum.

 1 f. Seven-stiped form. Randal Crag. Skiddaw Slates. Woodwardian Museum.

 1 g. Eight-stiped form, broader than usual, poorly preserved. Ibid.

2 a—c.—*Clonograptus tenellus* (Linnarsson).

 2 a. Specimen preserved in low relief. Mary Dingle. Shineton Shales. H. M. Geological Survey Collection.

 2 b. Ibid.

 2 c. Fragment of distal stipes. Skiddaw Slates. Fitz-Park Museum, Keswick.

3 a—c.—*Clonograptus*, var. *Callavei*, Lapworth.

 3 a. Specimen showing form of thecæ. Mary Dingle. Shineton Shales. H. M. Geological Survey Collection.

 3 b. Specimen showing stipes of the sixth order. Ibid.

 3 c. Fragmentary example. Mary Dingle. Shineton Shales. Callaway's Collection.

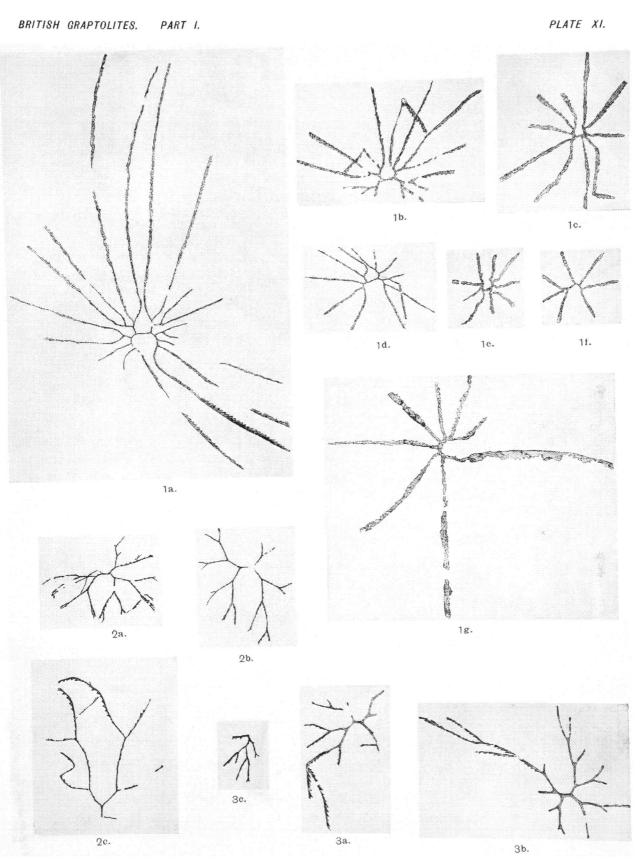

E. M. R. Wood, del.

Bemrose, Ltd., Derby.

LOGANOGRAPTUS AND CLONOGRAPTUS.

PLATE XII.

Temnograptus, Nicholson ; **Bryograptus**, Lapworth ; and **Trichograptus**, Nicholson.

FIGS.

1.—*Temnograptus multiplex*, Nicholson.

Two specimens on same slab. Right-hand one figured by Nicholson, Quart. Journ. Geol. Soc., 1868, vol. xxiv, pl. vi, figs. 1—3 ; left-hand specimen figured by Elles, Quart. Journ. Geol. Soc., vol. liv, p. 477, fig. 6. Peelwyke, Bassenthwaite. Skiddaw Slates. Christopherson's Collection.

2.—*Bryograptus divergens*, Elles and Wood.

Specimen preserved partially in relief. Figured by Marr as *Bryograptus Callavei* (?), Geol. Mag., 1894, p. 130, fig. 6. Barf. Lower Skiddaw Slates. Woodwardian Museum.

3 *a, b.*—*Bryograptus Kjerulfi*, Lapworth.

3 *a.* Specimen figured by Salter as *Dichograpsus*, sp., Quart. Journ. Geol. Soc., 1863, vol. xix, p. 137, fig. 12. Keswick. Skiddaw Slates. Museum of Practical Geology, Jermyn Street.

3 *b.* Large slab, showing specimens of *B. Kjerulfi* associated with its variety v. *cumbrensis*. Figured, Marr, Geol. Mag., 1894, p. 130, fig. 1. Barf. Skiddaw Slates. Postlethwaite's Collection.

4 *a—c.*—*Bryograptus*, var. *cumbrensis*, Elles.

4 *a.* Three specimens in association, the largest showing conspicuous sicula with its nema. Figured, Marr, Geol. Mag., 1894, p. 130, fig. 3. Barf. Lower Skiddaw Slates. Woodwardian Museum.

4 *b.* Small specimen on same slab as 4 *a.* Figured, Marr, *op. cit.*, figs. 4, 5.

4 *c.* Specimen from Barf. Lower Skiddaw Slates. Lapworth's Collection.

5.—*Trichograptus fragilis*, Nicholson.

Type specimen. Figured, Nicholson, Ann. and Mag. Nat. Hist. (4), vol. iv, pl. xi, figs. 1—3. Thornship Beck, near Shap. Upper Skiddaw Slates. Natural History Museum, S. Kensington.

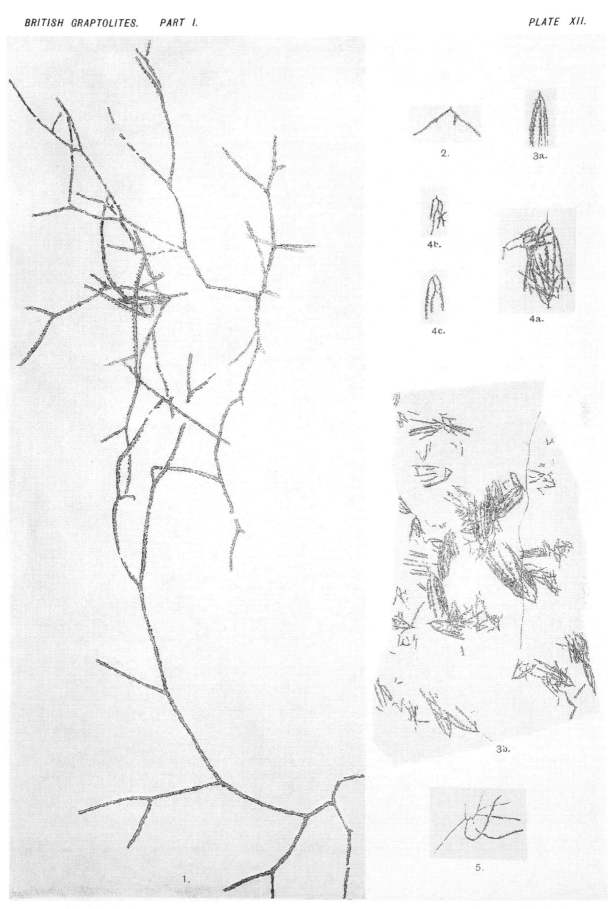

E. M. R. Wood, del.

Bemrose, Ltd., Derby.

TEMNOGRAPTUS, DRYOGRAPTUS, AND TRICHOGRAPTUS.

PLATE XIII.

Azygograptus, Nicholson, and Phyllograptus, Hall.

E. M. R. Wood, del.

Bemrose, Ltd., Derby.

AZYGOGRAPTUS AND PHYLLOGRAPTUS.

PLATE XIV.

Genus **Leptograptus**, Lapworth.

1 *a—g.*—*Leptograptus flaccidus* (Hall).

    1 *a.* Typical form. Hartfell Spa, Moffat. Hartfell Shales (zone of *Pleurog. linearis*). Lapworth's Collection.

    1 *b.* Typical form, preserved in low relief. Mount Benger Burn. Hartfell Shales. Geological Survey of Scotland, Edinburgh Museum.

    1 *c.* Smaller specimen, with a slightly more irregular curvature of the stipes. Hartfell Spa. Hartfell Shales (zone of *Dicranog. Clingani*). Lapworth's Collection.

    1 *d.* Larger specimen with typical curvature. On same slab as fig. 1 *c.*

    1 *e.* Centribrachiate form. Hartfell Spa. Hartfell Shales. Lapworth's Collection.

    1 *f.* Ibid.

    1 *g.* Centribrachiate form, showing four extra branches. On same slab as fig. 1 *e.*

2 *a—c.*—*Leptograptus flaccidus*, var. *spinifer*, Elles and Wood, nov.

    2 *a.* Typical specimen, well preserved in low relief. Hartfell Spa. Hartfell Shales. Lapworth's Collection.

    2 *b.* Specimen with well-developed spines on the proximal thecæ. Ibid.

    2 *c.* Broader form with strongly curved stipes, doubtfully referable to this variety. Ibid.

3 *a—c.*—*Leptograptus flaccidus*, var. *macilentus*, Lapworth, MS.

    3 *a.* Typical form, with somewhat rigid stipes. Hartfell Spa. Hartfell Shales (zone of *Pleurog. linearis*). Lapworth's Collection.

    3 *b.* Stipes more flexed. On same slab as fig. 3 *a.*

    3 *c.* Specimen showing complete sicula. On same slab as figs. 3 *a*, 3 *b.*

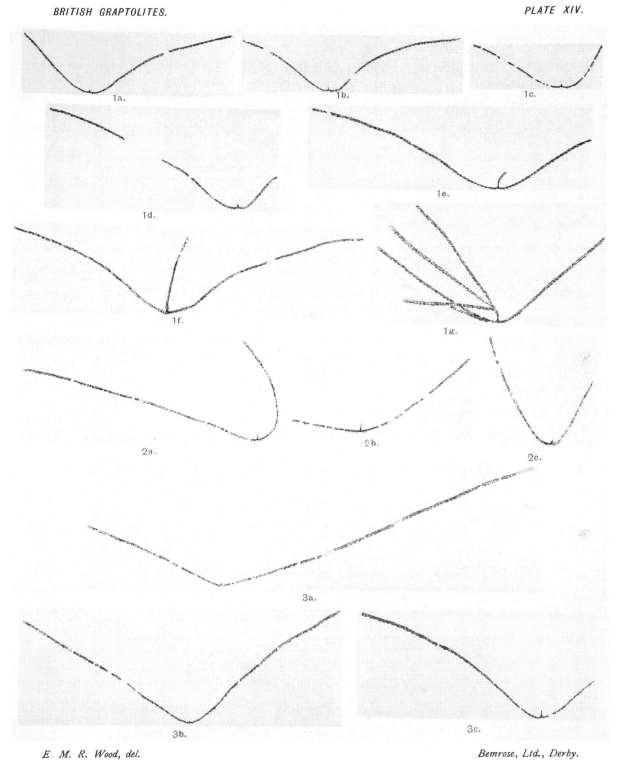

E. M. R. Wood, del.

Bemrose, Ltd., Derby.

LEPTOGRAPTUS.

PLATE XV.

Leptograptus—*continued.*

1 *a*—*c*.—*Leptograptus flaccidus*, cf. var. *macilentus*, Lapworth, MS.

    1 *a*. Large, strongly curved specimen. Hartfell Spa, Moffat. Hartfell Shales. Lapworth's Collection.

    1 *b*. Ibid.

    1 *c*. Straighter form with conspicuous sicula. On same slab as fig. 1 *b*.

2 *a*—*i*.—*Leptograptus flaccidus*, var. *macer*, Elles and Wood, nov.

    2 *a*. Portion of a slab showing specimens in association. Hartfell Spa. Hartfell Shales. Elles' Collection.

    2 *b*. Two specimens in association. Belcraig Burn. Hartfell Shales (zone of *Pleurog. linearis*). Wood's Collection.

    2 *c*. More flexed form with prominent sicula. Hartfell Spa. Hartfell Shales. Wood's Collection.

    2 *d*. Isolated stipe. On same slab as fig. 2 *c*.

    2 *e*. Well-preserved specimen. Hartfell Spa. Hartfell Shales. Wood's Collection.

    2 *f*. Centribrachiate form. Hartfell Spa. Hartfell Shales. Lapworth's Collection.

    2 *g*. Flexed form. Belcraig Burn. Hartfell Shales. Wood's Collection.

    2 *h*. Double centribrachiate form. Hartfell Spa. Hartfell Shales. Wood's Collection.

    2 *i*. Ditto ?, or *Amphigraptus ?* On same slab as fig. 2 *h*.

3 *a*—*c*.—*Leptograptus flaccidus*, var. *arcuatus*, Elles and Wood, nov.

    3 *a*. Symmetrically curved form. Hartfell Spa. Hartfell Shales. Lapworth's Collection.

    3 *b*. Typical form with sigmoid curvature. On same slab as fig. 3 *a*.

    3 *c*. Ditto. On same slab as figs. 3 *a* and 3 *b*.

4 *a*—*d*.—*Leptograptus capillaris* (Carruthers).

    4 *a*. Type specimen. Figured, Carruthers, Geol. Mag., 1868, vol. v, pl. v, fig. 7 *a*. Hartfell Spa ? Hartfell Shales. British Museum (Natural History), S. Kensington.

    4 *b*. Part of a slab, showing general habit of the species. Hartfell Spa. Hartfell Shales. Lapworth's Collection.

    4 *c*. Single stipe, showing curvature. Ibid.

    4 *d*. Centribrachiate form. Belcraig Burn. Hartfell Shales (zone of *Pleurog. linearis*). Elles' Collection.

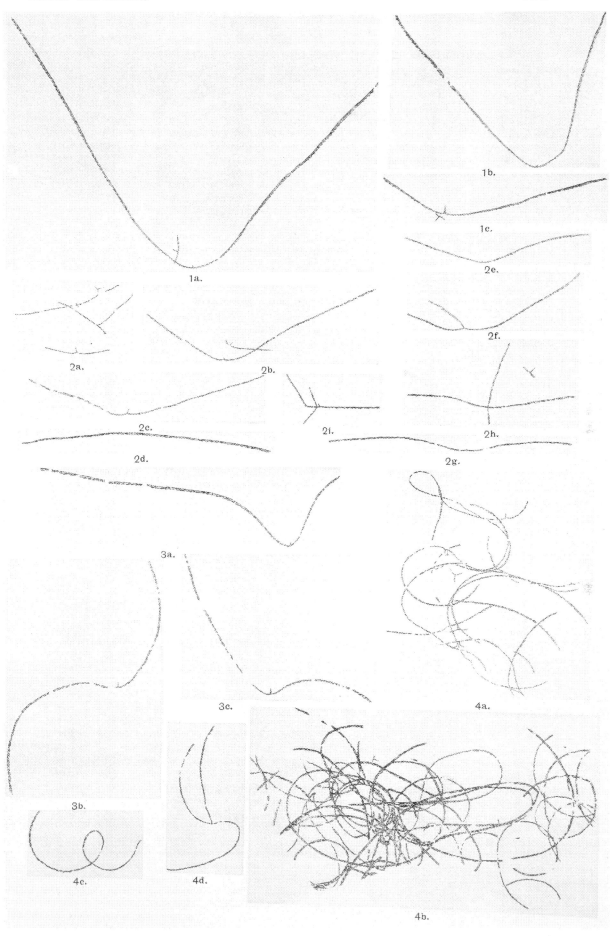

LEPTOGRAPTUS.

PLATE XVI.

Leptograptus—*continued*—and Pleurograptus, Nicholson.

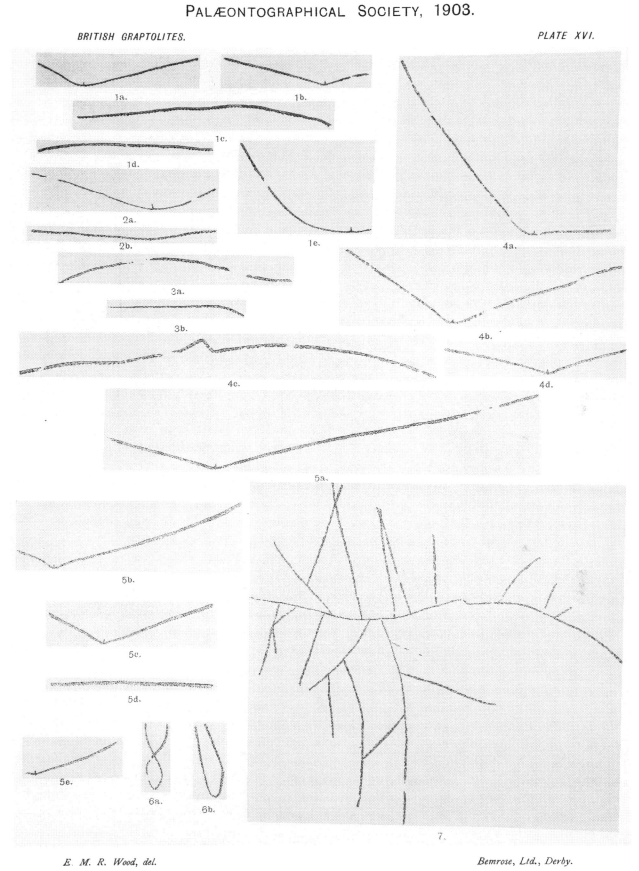

E. M. R. Wood, del.

Bemrose, Ltd., Derby.

LEPTOGRAPTUS AND PLEUROGRAPTUS.

PLATE XVII.

Pleurograptus—*continued.*

1.

2.

E. M. R. Wood, del.   *Bemrose, Ltd., Derby.*

PLEUROGRAPTUS.

PLATE XVIII.

Genus **Amphigraptus**, Lapworth.

1.—*Amphigraptus divergens* (Hall).

Large specimen, well preserved. Mount Benger Burn, Selkirkshire. Hartfell Shales. Lapworth's Collection.

2 *a—c.*—*Amphigraptus divergens*, var. *radiatus*, Lapworth.

2 *a*. Type specimen. Figured, Lapworth, Cat. West. Scott. Fossils, 1876, pl. iii, fig. 71. Hartfell Spa, Moffat. Hartfell Shales (zone of *Pleurog. linearis*). Lapworth's Collection.

2 *b*. Specimen with only four branches. Ibid.

2 *c*. Narrow specimen, poorly preserved. Hartfell Spa. Hartfell Shales. Geological Survey of Scotland, Edinburgh Museum.

3.—*Amphigraptus distans*, Elles and Wood, sp. nov.

Type specimen. Barskeoch Burn, near St. John's, Dalry, Kirkcud-brightshire. Hartfell Shales. Geological Survey of Scotland, Edinburgh Museum.

4.—*Amphigraptus?* sp. Mount Benger Burn?, Selkirkshire. Hartfell Shales. Lapworth's Collection.

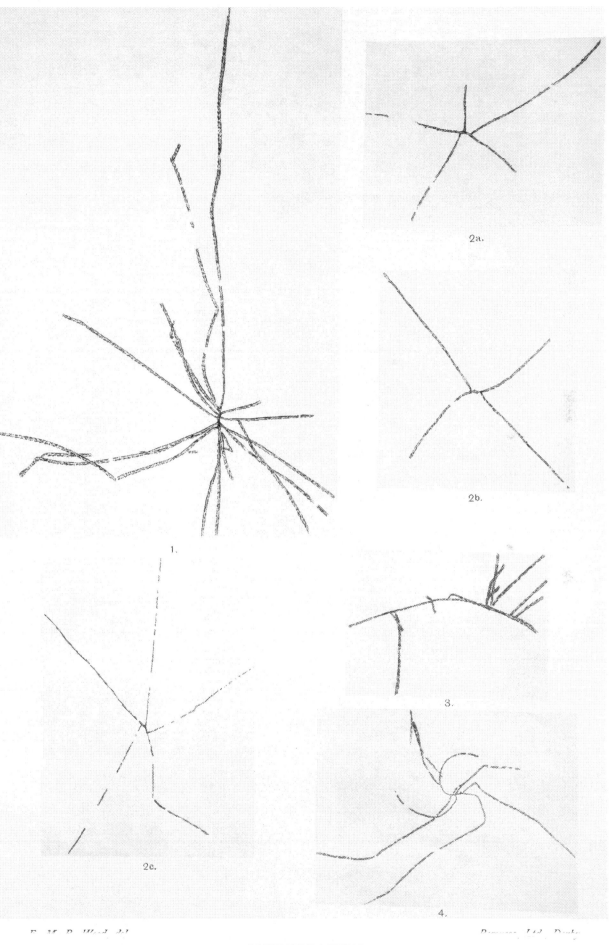

AMPHIGRAPTUS.

PLATE XIX.

Genus **Nemagraptus** (*Cænograptus*, Hall), Emmons.

1 *a—f.—Nemagraptus gracilis* (Hall).

    1 *a.* Very large specimen, well preserved, partly in low relief. Bail Hill, Sanquhar. Glenkiln Shales. Geological Survey of Scotland, Edinburgh Museum.

    1 *b.* Typical form. Figured, Lapworth, Cat. West. Scott. Foss., 1876, pl. iii, fig. 65. Glenkiln Burn. Glenkiln Shales. Lapworth's Collection.

    1 *c.* Specimen showing sicula. Berrybush Burn, St. Mary's Loch, Selkirkshire. Glenkiln Shales. Geological Survey of Scotland, Edinburgh Museum.

    1 *d.* Specimen preserved in low relief. Spy Burn, Shropshire. Rorrington Flags. H.M. Geological Survey Collection.

    1 *e.* Portion of large specimen. Belcraig Burn. Glenkiln Shales. Elles' Collection.

    1 *f.* Specimen preserved in low relief. Spy Burn, Shropshire. Rorrington Flags. H.M. Geological Survey Collection.

2 *a—d.—Nemagraptus gracilis*, var. *surcularis* (Hall).

    2 *a.* Typical form. Figured, Lapworth, Cat. West. Scott. Foss., 1876, pl. iii, fig. 64. Cairn Hill, Euchan Water, nr. Sanquhar. Glenkiln Shales. Lapworth's Collection.

    2 *b.* Two specimens in close association. Ibid.

    2 *c.* Small specimen, showing the apertural spine of the sicula. Glenkiln Burn. Glenkiln Shales. Lapworth's Collection.

    2 *d.* Elongated specimen. Belcraig Burn. Glenkiln Shales. Lapworth's Collection.

3 *a—h.—Nemagraptus gracilis*, var. *remotus*, Elles and Wood, nov.

    3 *a.* Part of a slab, showing numerous specimens in association. Rein Gill, Wandel Water, Lanarkshire. Glenkiln Shales. Geological Survey of Scotland, Edinburgh Museum.

    3 *b.* Specimen on same slab as fig. 3 *a.*

    3 *c.* Ditto? On same slab as figs. 3 *a* and 3 *b.*

    3 *d.* Characteristic form, well preserved. Cairn Hill? Glenkiln Shales. Lapworth's Collection.

    3 *e.* Specimen showing general form, poorly preserved. Belcraig Burn. Glenkiln Shales. Wood's Collection.

    3 *f.* Fragment of larger specimen. Ballygrot, co. Down. Glenkiln Shales. Lapworth's Collection.

    3 *g.* Large specimen, but with few branches. Morrach Bay, Portpatrick. Glenkiln Shales. Geological Survey of Scotland, Edinburgh Museum.

    3 *h.* Specimen showing only one branch. Morrach Bay, Portpatrick. Glenkiln Shales. Geological Survey of Scotland, Edinburgh Museum.

4 *a—d.—Nemagraptus gracilis*, var. *nitidulus* (Lapworth).

    4 *a.* Portion of a slab, showing three specimens in association. Cairn Hill, Euchan Water. Glenkiln Shales. Lapworth's Collection.

    4 *b.* ? Type specimen. ? Figured, Lapworth, Cat. West. Scott. Foss., 1876, pl. iii, fig. 66. Ibid.

    4 *c.* Small specimen. Belcraig Burn. Glenkiln Shales. Lapworth's Collection.

    4 *d.* Large specimen, poorly preserved. Birnock Water, Abington. Glenkiln Shales. Lapworth's Collection.

5.—*Nemagraptus, sp.*

    Well-preserved specimen, showing pendent form of the stipes. Craiglure Lodge, Head of Stinchar. Glenkiln Shales. Geological Survey of Scotland, Edinburgh Museum.

6 *a—c.—Nemagraptus explanatus* (Lapworth).

    6 *a.* Type specimen. Figured, Lapworth, Cat. West. Scott. Foss., 1876, pl. iii, fig. 68. Glenkiln Burn. Glenkiln Shales. Lapworth's Collection.

    6 *b.* Specimen showing the sicula. Cairn Hill? Glenkiln Shales. Lapworth's Collection.

    6 *c.* Fragment with two branches. On same slab as fig. 2 *c.*

7 *a—f.—Nemagraptus explanatus*, var. *pertenuis* (Lapworth).

    7 *a.* Type specimen. Figured, Lapworth, Cat. West. Scott. Foss., 1876, pl. iii, fig. 67. Birnock Water, Abington. Glenkiln Shales. Lapworth's Collection.

    7 *b.* Fragmentary stipe. Ibid.

    7 *c.* Specimen showing sicula. Cairn Hill. Glenkiln Shales. Lapworth's Collection.

    7 *d.* Single straight stipe with sicula well shown. Glenkiln Burn? Glenkiln Shales. Lapworth's Collection.

    7 *e.* Ditto. On same slab as fig. 2 *a.*

    7 *f.* Two specimens strongly flexed. Rein Gill, Wandel Water, Lanarkshire. Glenkiln Shales. Geological Survey of Scotland, Edinburgh Museum.

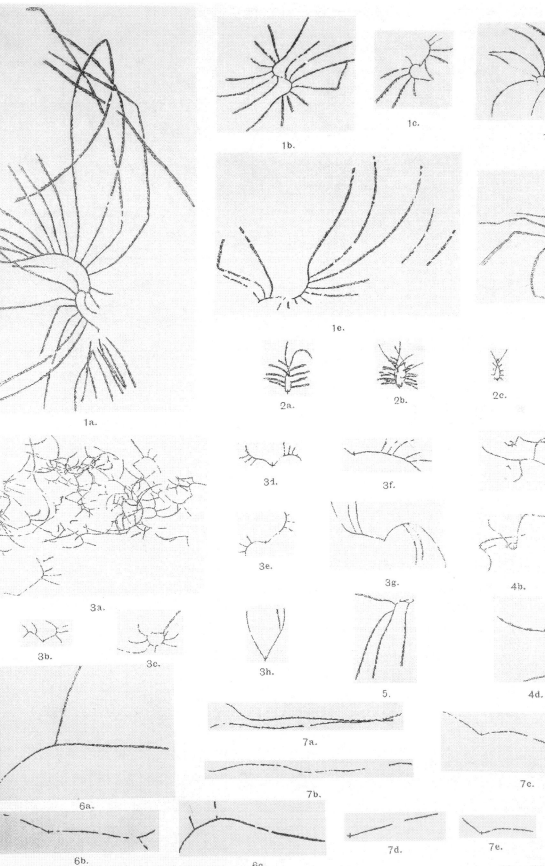

E. M. R. Wood, del.

Bemrose, Ltd., Derby.

NEMAGRAPTUS (COENOGRAPTUS).

PLATE XX.

Genus **Dicellograptus,** Hopkinson.

FIGS.

1 *a*—*d*.—*Dicellograptus complanatus*, Lapworth. (Page 139.)
  1 *a*. Typical form. Dobb's Linn, Moffat. Upper Hartfell Shales (base of " Barren Mudstones "). Lapworth's Collection.
  1 *b*. Typical form, showing "initial" and "lateral" spines. Same slab as 1 *a*.
  1 *c*. More convergent form. Dobb's Linn, Moffat. Upper Hartfell Shales. Wood's Collection.
  1 *d*. Form with slightly curved stipes. On same slab as figs. 1 *a* and 1 *b*.

2 *a*—*c*.—*Dicellograptus complanatus*, var. *ornatus*, Elles and Wood, nov. (Page 140.)
  2 *a*. Typical specimen, showing apex of sicula. Dobb's Linn, Moffat. Upper Hartfell Shales (zone of *D. anceps*). Sedgwick Museum.
  2 *b*. Specimen with extremely inconspicuous "lateral" spines. Dobb's Linn, Moffat. Upper Hartfell Shales. Elles' Collection.
  2 *c*. Small specimen with more divergent stipes. Dobb's Linn, Moffat. Upper Hartfell Shales. Elles' Collection.

3 *a*—*e*.—*Dicellograptus anceps*, Nicholson. (Page 141.)
  3 *a*. Typical form in low relief. Dobb's Linn, Moffat. Upper Hartfell Shales (zone of *D. anceps*). Wood's Collection.
  3 *b*. Smaller specimen, showing apex of sicula. Ibid. Lapworth's Collection.
  3 *c*. Very long specimen, but indifferently preserved. Ibid.
  3 *d*. Broad specimen, somewhat distorted, with indications of a " web " in the axil. Ettrick-bridge-end. Upper Hartfell Shales. Geological Survey of Scotland, Edinburgh Museum.
  3 *e*. Small, but well-preserved specimen in low relief showing mesial spines. Dobb's Linn, Moffat. Upper Hartfell Shales. Wood's Collection.

4 *a*—*f*.—*Dicellograptus intortus*, Lapworth. (Page 146.)
  4 *a*. Typical form. Birnock Water, S. Scotland. Glenkiln Shales. Lapworth's Collection.
  4 *b*. Somewhat narrow specimen with stipes crossing. Carco, Crawick Water. Glenkiln Shales. Geological Survey of Scotland, Edinburgh Museum.
  4 *c*. Large, but incomplete specimen. Gwernyfed, Builth Road, Radnorshire. " Cœnograptus " Beds, Glenkiln. Dr. Fraser's Collection, Wolverhampton.
  4 *d*. Specimen showing crossing of stipes near proximal extremity. Ibid.
  4 *e*. Specimen showing parallelism of stipes near proximal extremity. Ibid.
  4 *f*. Specimen showing proximal extremity. Same slab as fig. 4 *d*.

5 *a, b*.—*Dicellograptus*, cf. *divaricatus*, Hall. (Page 143.)
  5 *a*. Wide form showing sudden increase in breadth of stipes. Wanlock Head, S. Scotland. Glenkiln Shales. Lapworth's Collection.
  5 *b*. Very poorly preserved specimen. Figured as *D. moffatensis*, Hopkinson and Lapworth, Quart. Journ. Geol. Soc., 1875, vol. xxxi, pl. xxxv, fig. 5 *b*. Abereiddy Bay, S. Wales. Middle Llandeilo (Hicks). Sedgwick Museum.

6 *a*—*e*.—*Dicellograptus divaricatus*, var. *rigidus*, Lapworth. (Page 144.)
  6 *a*. Typical form without " web " in axil. Birnock Water, S. Scotland. Glenkiln Shales. Lapworth's Collection.
  6 *b*. Typical form with " web " in axil. Ibid.
  6 *c*. Two narrower specimens in association, probably referable to this variety. Dobb's Linn, Moffat. Lower Hartfell Shales. Sedgwick Museum.
  6 *d*. Specimen similar to fig. 6 *b*.
  6 *e*. Specimen similar to fig. 6 *a*.

7 *a*—*e*.—*Dicellograptus divaricatus*, var. *salopiensis*, Elles and Wood. (Page 145).
  7 *a*. Typical form. Spy Burn, Shropshire. Rorrington Flags (" Cœnograptus " Beds). Professor T. McK. Hughes' Collection, Sedgwick Museum.
  7 *b*. Ibid.
  7 *c*. Specimen with somewhat more widely divergent stipes. Birnock Water, S. Scotland. Glenkiln Shales. Lapworth's Collection.
  7 *d*. Specimen showing sicula and "lateral" spines. Meggat Water, S. Scotland. Glenkiln Shales. Lapworth's Collection.
  7 *e*. Typical specimen. Birnock Water. Glenkiln Shales. Lapworth's Collection.

E. M. R. Wood, del.   *Bemrose, Ltd., Derby.*

DICELLOGRAPTUS.

PLATE XXI.

Dicellograptus—(continued).

FIGS.

1 *a—e.*—*Dicellograptus sextans*, Hall. (Page 153.)

      1 *a.* Type specimen of Hopkinson's *Dicranograptus formosus*, Geol. Mag., 1870, vol. vii, pl. xvi, figs. 2 *b*, 2 *c*, 2 *d.* Belcraig Burn. Glenkiln Shales. Sedgwick Museum.

      1 *b.* Ibid. Fig. 2 *a.*

      1 *c.* More widely divergent form. Belcraig Burn. Glenkiln Shales. Lapworth's Collection.

      1 *d.* Typical form. Belcraig Burn. Glenkiln Shales. Wood's Collection.

      1 *e.* Ditto. Lapworth's Collection.

2 *a—d.*—*Dicellograptus sextans*, .var. *exilis*, Elles and Wood, nov. (Page 155.)

      2 *a.* Typical form. Dobb's Linn, Moffat. Upper Glenkiln Shales. Wood's Collection.

      2 *b.* Ditto. Birnock, S. Scotland. Glenkiln Shales. Lapworth's Collection.

      2 *c.* Ditto. Wandel Water, Abington. Glenkiln Shales. Geological Survey of Scotland, Edinburgh Museum.

      2 *d.* Ditto. On same slab as fig. 2 *c.*

3 *a—f.*—*Dicellograptus pumilus*, Lapworth. (Page 149.)

      3 *a.* Large typical specimen. Hartfell Spa. Hartfell Shales. Lapworth's Collection.

      3 *b.* Imperfect specimen, but with typical form of axil. Hartfell Spa, Hartfell Shales. Lapworth's Collection.

      3 *c.* Well-preserved specimen showing sicula. Ibid.

      3 *d.* Incomplete specimen. Ibid.

      3 *e.* Specimen showing sicula. Ibid.

      3 *f.* Ditto. Conway Railway Cutting, N. Wales. Hartfell Shales. Lapworth's Collection.

4.—*Dicellograptus angulatus*, Elles and Wood, sp. nov. (Page 149.)

      4. Portion of slab with examples showing typical habit. Morroch Bay. Upper Glenkiln to Lower Hartfell Shales. Geological Survey of Scotlaand, Edinburgh Museum.

5 *a—e.*—*Dicellograptus patulosus*, Lapworth. (Page 147.)

      5 *a.* Typical form. Craigmichan Scaurs. Glenkiln Shales. Lapworth's Collection.

      5 *b.* Typical form with long sicula. Same slab as fig. 5 *a.*

      5 *c.* Somewhat more slender form. Builth. Glenkiln Shales. Lapworth's Collection.

      5 *d.* Specimen widened by compression, showing prominent " initial " and " lateral " spines. Belcraig Burn. Glenkiln Shales. Wood's Collection.

      5 *e.* Slender specimen. Glenkiln Burn. Glenkiln Shales. Lapworth's Collection.

6 *a—d.*—*Dicellograptus Morrisi*, Hopkinson. (Page 155.)

      6 *a.* Young specimen, doubtfully referable to this species. Figured, Hopkinson, Geol. Mag., 1871, vol. viii, pl. i, fig. 2 *e.* Dobb's Linn. Hartfell Shales. Sedgwick Museum.

      6 *b.* Typical form. Glenkiln Burn. Hartfell Shales. Lapworth's Collection.

      6 *c.* Ditto. Dobb's Linn. Hartfell Shales. Lapworth's Collection.

      6 *d.* Ditto. Hartfell Spa. Wood's Collection.

E. M. R. Wood, del.

Bemrose, Ltd., Derby.

DICELLOGRAPTUS.

PLATE XXII.

Dicellograptus—*continued.*

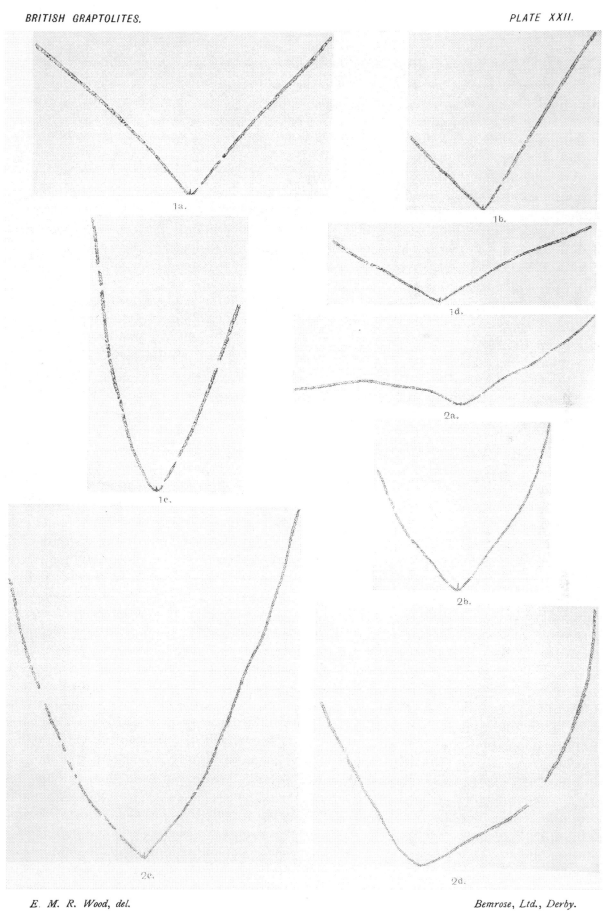

E. M. R. Wood, del.   *Bemrose, Ltd., Derby.*

DICELLOGRAPTUS.

PLATE XXIII.

Dicellograptus—*continued.*

1 *a—f.—Dicellograptus moffatensis*, Carruthers. (Page 157.)

    1 *a*. ? Type specimen. Figured, Carruthers, Trans. Roy. Phys. Soc. Edinb., p. 469, fig. 3. Hartfell Spa. Hartfell Shales. Museum of Natural History, S. Kensington.

    1 *b*. Typical form. Glenkiln Burn. Hartfell Shales (zone of *Dicranog. Clingani*). Lapworth's Collection.

    1 *c*. Specimen with continuously divergent stipes. On same slab as fig. 1 *b*.

    1 *d*. Widely divergent form, doubtfully referable to this species. Craigmichan Scaurs, Selcoth Burn. Hartfell Shales. Lapworth's Collection.

    1 *e*. Specimen figured, Hopkinson, Geol. Mag., 1871, vol. viii, pl. 1, figs. 4 *a, b*. Hartfell Spa. Hartfell Shales. Sedgwick Museum.

    1 *f*. Poorly preserved specimen. Figured, Hopkinson and Lapworth, Quart. Journ. Geol. Soc., 1875, vol. xxxi, pl. xxxv, fig. 5 *a*.

2 *a—e.—Dicellograptus elegans*, Carruthers. (Page 159.)

    2 *a*. Type specimen. Figured, Carruthers, Geol. Mag., 1868, vol. v, pl. v, fig. 8 *d*. Dobb's Linn. Hartfell Shales. Museum of Natural History, S. Kensington.

    2 *b*. Small specimen. Figured, Hopkinson, Geol. Mag., 1871, vol. viii, pl. i, figs. 3 *a, b*. Dobb's Linn. Hartfell Shales. Sedgwick Museum.

    2 *c*. Ibid., figs. 3 *c, d*.

    2 *d*. Typical form. Dobb's Linn. Hartfell Shales (zone of *Pleurog. linearis*). Lapworth's Collection.

    2 *e*. Typical form, showing prominent " lateral " spines. Hartfell Spa, Hartfell Shales. Lapworth's Collection.

3.—*Dicellograptus elegans*, var. *rigens*, Lapworth, MS. (Page 161.)

    3. Typical rigid form. Mount Benger Burn, Selkirkshire. Hartfell Shales. Lapworth's Collection.

4 *a—c.—Dicellograptus caduceus*, Lapworth. (Page 161.)

    4 *a*. Typical form. Dobb's Linn. Hartfell Shales. Lapworth's Collection.

    4 *b*. Large specimen. Hartfell Spa. Hartfell Shales. Lapworth's Collection.

    4 *c*. Specimen with stipes crossing only once. Ibid.

E. M. R. Wood, del.

Bemrose, Ltd., Derby.

DICELLOGRAPTUS.

PLATE XXIV.

Genus **Dicranograptus**, Hall.

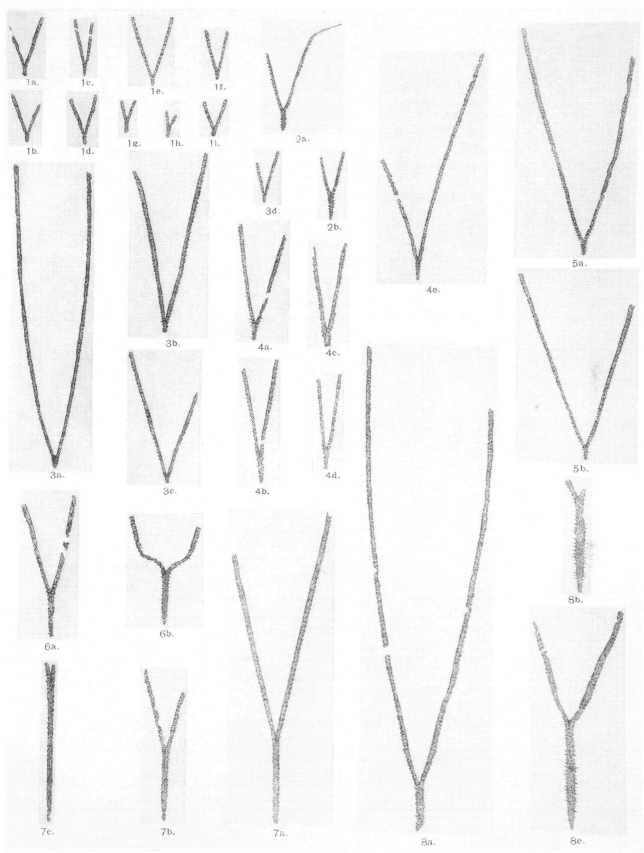

E. M. R. Wood, del.

Bemrose, Ltd., Derby.

DICRANOGRAPTUS.

PLATE XXV.

Dicranograptus—(continued).

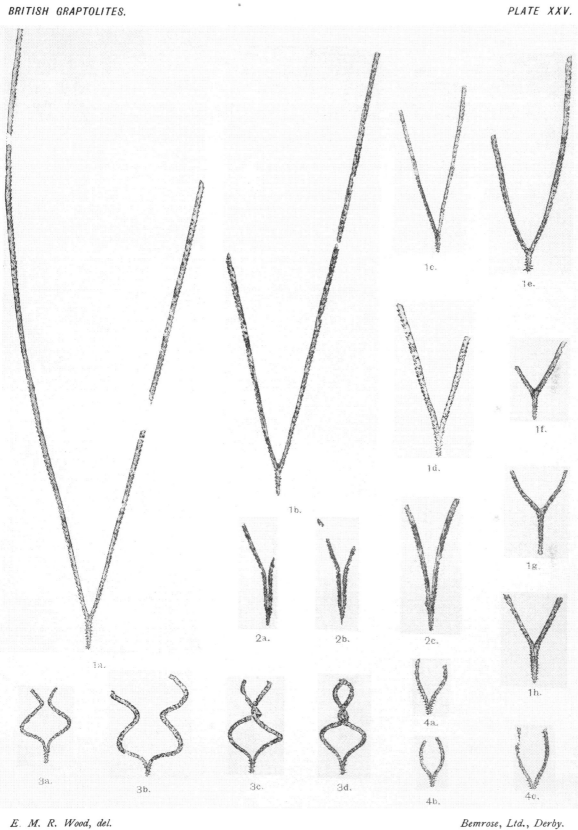

E. M. R. Wood, del.   *Bemrose, Ltd., Derby.*

DICRANOGRAPTUS.

PLATE XXVI.

Genus **Climacograptus.**

1 *a—c.*—*Climacograptus scalaris* (Hisinger). (Page 184.)
    1 *a.* Specimen showing proximal end. Belcraig Burn, S. Scotland. Upper Birkhill Shales. Hopkinson Collection, Sedgwick Museum.
    1 *b.* Incomplete specimen. Ibid. Elles' Collection.
    1 *c.* Larger specimen. Donaghadee, Ireland. Birkhill Shales. Elles' Collection.

2 *a—g.*—*Climacograptus scalaris,* var. *normalis,* Lapworth. (Page 186.)
    2 *a.* Type specimen. Figured, Lapworth, Graptolites of Co. Down, 1877, pl. vi, fig. 31. Lower Birkhill Shales (zone of *Cephalog. acuminatus*). Dobb's Linn. Lapworth's Collection.
    2 *b.* Narrow specimen showing sicula. Ibid. Elles' Collection.
    2 *c.* Reverse view. Ibid.
    2 *d.* Long specimen. Ibid. H. Lapworth's Collection.
    2 *e.* Specimen with long distal virgula. Ibid.
    2 *f.* Obverse view. Ibid.
    2 *g.* Specimen in low relief. North Cliff, Dobb's Linn. Lapworth's Collection.

3 *a—h.*—*Climacograptus scalaris,* var. *miserabilis,* Elles and Wood, nov. (Page 186.)
    3 *a.* Complete specimen. Dobb's Linn. Upper Hartfell Shales (zone of *Dicellog. complanatus*). Lapworth's Collection.
    3 *b.* Incomplete specimen. Ibid.
    3 *c.* Specimen showing proximal end. Ibid.
    3 *d.* Ditto. Ibid.
    3 *e.* Ibid.
    3 *f.* Narrow form (?) Upper Hartfell Shales (zone of *Dicellog. anceps*). Dobb's Linn. Wood's Collection.
    3 *g.* Slender parallel-sided form, doubtfully referable to this variety. Ibid. Elles' Collection.
    3 *h.* Ibid.

4 *a—f.*—*Climacograptus medius,* Törnquist. (Page 189.)
    4 *a.* Complete and typical specimen. Lower Birkhill Shales (zone of *Diplog. vesiculosus*). Dobb's Linn. Sedgwick Museum.
    4 *b.* Specimen which tapers more at the proximal end. Lower Birkhill Shales (zone of *Diplog. vesiculosus*). Main Cliff, Dobb's Linn. Elles' Collection.
    4 *c.* Distal fragment. On same slab as fig. 4 *a.*
    4 *d.* Young specimen preserved as a cast, showing incomplete septum. Fachdre Beds (zone of *Dimorphograptus*), Plas Pennant, Llanbrynmair. Wood's Collection.
    4 *e.* Young specimen, showing long virgella. Same slab as figs. 4 *a* and 4 *c.*
    4 *f.* Ditto, showing two long proximal spines. Ibid.

5 *a—e.*—*Climacograptus rectangularis* (M‘Coy). (Page 187.)
    5 *a.* Type specimen in relief. Figured, M‘Coy, British Palæozoic Fossils, pl. i B, fig. 8 *a.* Birkhill Shales. Moffat. Sedgwick Museum.
    5 *b.* Cast of same. Fig. 8.
    5 *c.* Longer specimen, preserved in scalariform view. Ibid. Fig. 9.
    5 *d.* Small incomplete specimen. Birkhill Shales. Moffat. Sedgwick Museum.
    5 *e.* Ibid.

6 *a—f.*—*Climacograptus Törnquisti,* Elles and Wood, sp. nov. (Page 190.)
    6 *a.* ? Specimen figured, Lapworth, as *C. rectangularis,* Graptolites of Co. Down, 1877, pl. vi, fig. 32. Dobb's Linn. Birkhill Shales. Lapworth's Collection.
    6 *b.* Smaller specimen, showing reverse view, on same slab as fig. 6 *a.*
    6 *c.* Larger specimen, showing tubular virgella. Dobb's Linn. Birkhill Shales. Lapworth's Collection.
    6 *d.* Ditto, preserved in scalariform view. Ibid.
    6 *e.* Specimen with very long tubular virgella. Obverse view. Ibid.
    6 *f.* Small well-preserved specimen, reverse view. Waterfall, Long Cliff, Dobb's Linn. Birkhill Shales. Lapworth's Collection.

PLATE XXVI—*continued*.

FIGS.

7 *a—d.*—*Climacograptus tuberculatus*, Nicholson. (Page 213.)
    7 *a.* Complete specimen, poorly preserved. Dobb's Linn. Lower Birkhill Shales. British Museum (Natural History), S. Kensington.
    7 *b.* Ibid.
    7 *c.* Ibid.
    7 *d.* Ibid.?

8 *a—f.*—*Climacograptus bicornis* (Hall). (Page 193.)
    8 *a.* Specimen in full relief; obverse view. Dobb's Linn. Lower Hartfell Shales (zone of *Climacog. Wilsoni*). Elles' Collection.
    8 *b.* Slender specimen with small proximal spines. Kirkton Burn, Wanlock Head. Glenkiln Shales. Lapworth's Collection.
    8 *c.* Long specimen. Glenkiln Burn. Lower Hartfell Shales (zone of *Climacog. Wilsoni*). Wood's Collection.
    8 *d.* Scalariform view of specimen with stout spines. The Cornice, Hartfell. Hartfell Shales. Lapworth's Collection.
    8 *e.* Specimen with very long slender spines. The Cornice, Hartfell. Hartfell Shales (zone of *Climacog. Wilsoni*). Elles' Collection.
    8 *f.* Ibid.

9 *a—c.*—*Climacograptus bicornis*, var. *tridentatus*, Lapworth. (Page 195.)
    9 *a.* Specimen with short virgella. Glenkiln. Glenkiln Shales. Lapworth's Collection.
    9 *b.* Specimen with three equally prominent spines. Wanlock Water. Glenkiln Shales. Lapworth's Collection. (The reverse side of this specimen is in the Sedgwick Museum.)
    9 *c.* Specimen with shorter spines enclosed in prominent disc. Glenkiln. Glenkiln Shales. Lapworth's Collection.

10 *a—c.*—*Climacograptus bicornis*, var. *peltifer*, Lapworth. (Page 196.)
    10 *a.* Specimen with proximal spines enclosed in a small disc. Dobb's Linn. Upper Glenkiln Shales. Wood's Collection.
    10 *b.* Small specimen with large disc enveloping proximal end of polypary. Tiddyndicwm, N. Wales. Upper Glenkiln Shales. Sedgwick Museum.
    10 *c.* Typical specimen. Dobb's Linn. Glenkiln Shales. Lapworth's Collection.

11 *a—d.*—*Climacograptus supernus*, Elles and Wood, sp. nov. (Page 196.)
    11 *a.* Typical specimen. Main Cliff, Dobb's Linn. Upper Hartfell Shales (zone of *Dicellog. anceps*). Wood's Collection.
    11 *b.* Ibid. Elles' Collection.
    11 *c.* Ibid. Wood's Collection.
    11 *d.* Specimen showing virgella and short spines. Ibid.

12 *a—d.*—*Climacograptus Wilsoni*, Lapworth. (Page 197.)
    12 *a.* Typical specimen in partial relief. Main Cliff, Dobb's Linn. Lower Hartfell Shales (zone of *Climacog. Wilsoni*). Lapworth's Collection.
    12 *b.* Specimen showing distorted position of sac. Ibid.
    12 *c.* Specimen with sac in normal position. Ibid.
    12 *d.* Long specimen in partial relief, sac broken off. Ibid.

13.—*Climacograptus Wilsoni*, var. *tubularis*, Elles and Wood, nov. (Page 199.)
    13. Large specimen in relief showing long tubular virgella. Dobb's Linn. Lower Hartfell Shales (zone of *Climacog. Wilsoni*). Lapworth's Collection.

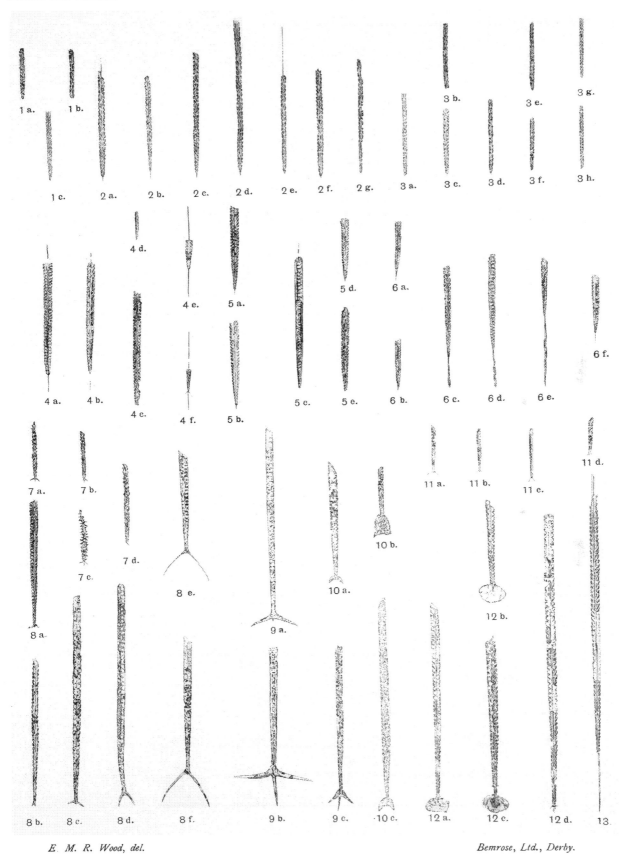

E. M. R. Wood, del.

Bemrose, Ltd., Derby.

PLATE XXVII.

Climacograptus—*continued.*

1 *a*—*g.*—*Climacograptus minimus* (Carruthers). (Page 191.)

    1 *a.* Complete specimen. Mount Benger Burn, S. Scotland. Hartfell Shales (zone of *Pleurog. linearis*). Lapworth's Collection.

    1 *b.* Smaller specimen. On same slab as fig. 1 *a.*

    1 *c.* Wider fragment. On same slab as figs. 1 *a* and 1 *b.*

    1 *d.* Distal fragment, showing virgular tube. Dobb's Linn (zone of *P. linearis*). Elles' Collection.

    1 *e.* Complete specimen, showing proximal end. Ibid.

    1 *f.* Long specimen. Ibid.

    1 *g.* Smaller specimen. Ibid.

2 *a*—*f.*—*Climacograptus brevis,* Elles and Wood, sp. nov. (Page 192.)

    2 *a.* Typical specimen, reverse view, showing sicula. Gwern-y-fed-fach, Builth. Llandeilo Beds. Sedgwick Museum.

    2 *b.* Specimen on same slab as fig. 2 *a.*

    2 *c.* Smaller specimen. Ibid.

    2 *d.* Complete specimen. Ibid.

    2 *e.* Young specimen, obverse view, showing sicula. Dobb's Linn. Hartfell Shales (zone of *Pleurog. linearis*). Elles' Collection.

    2 *f.* Dorsal view of wide specimen. Gwern-y-fed-fach, Builth. Llandeilo Beds. Elles' Collection.

3 *a*—*h.*—*Climacograptus latus,* Elles and Wood, sp. nov. (Page 204.)

    3 *a.* Typical specimen, obverse view. Main Cliff, Dobb's Linn. Upper Hartfell Shales (zone of *Dicellog. anceps*). Elles' Collection.

    3 *b.* Ibid. Wood's Collection.

    3 *c.* Small specimen. Ibid. Elles' Collection.

    3 *d.* Wide distal fragment. Ibid. Wood's Collection.

    3 *e.* Small specimen. Ibid.

    3 *f.* Larger specimen. Ibid.

    3 *g.* Small narrow specimen, reverse view. Ibid. Elles' Collection.

    3 *h.* Larger specimen, doubtfully referable to this species. Ibid. Wood's Collection.

4 *a*—*e.*—*Climacograptus antiquus,* Lapworth. (Page 199.)

    4 *a.* Typical specimen. Black Linn, Glenkiln. Glenkiln Shales. Lapworth's Collection.

    4 *b.* Large but incomplete specimen. Ibid.

    4 *c.* Narrower specimen. Kirkmichael Burn, S. Scotland. Glenkiln Shales. Lapworth's Collection.

    4 *d.* Specimen with prominent proximal spines. Oak Wood, Pontesford, Shropshire. Lower Bala. Lapworth's Collection.

    4 *e.* Ibid.

5 *a*—*f.*—*Climacograptus antiquus,* var. *lineatus,* Elles and Wood, nov. (Page 201.)

    5 *a.* Small but complete specimen. Craigmichan Scaurs, S. Scotland. Upper Glenkiln Shales (zone of *Dicellog. patulosus*). Lapworth's Collection.

    5 *b.* Distal fragment. Ibid.

    5 *c.* Long distal fragment, narrowed by compression. Llanystwmdwy. *Dicranograptus* Shales. Sedgwick Museum.

    5 *d.* Proximal end of specimen preserved in scalariform view. Ibid.

    5 *e.* Distal fragment. Ibid.

    5 *f.* Distal fragment, not distorted. Ibid.

6 *a*—*d.*—*Climacograptus antiquus,* var. *bursifer,* Elles and Wood, nov. (Page 201.)

    6 *a.* Typical specimen. Kirkmichael Burn, Dumfries. Glenkiln Shales. Lapworth's Collection.

    6 *b.* Smaller specimen, showing distal prolongation of the virgula. On same slab as fig. 6 *a.*

    6 *c.* Specimen showing proximal end and sac. On same slab as figs. 6 *a* and 6 *b.*

    6 *d.* Small specimen with large sac. Same locality as figs. 6 *a*—*c.*

PLATE XXVII.

Climacograptus—*continued.*

PLATE XXVII—*continued*.

7 *a*—*e.*—*Climacograptus caudatus*, Lapworth. (Page 202.)

    7 *a.*—Typical specimen with long virgella. Glenkiln Burn, Kirkmichael, Dumfries. Hartfell Shales. Lapworth's Collection.

    7 *b.* Small specimen showing virgella and distal virgula. Ibid.

    7 *c.* Specimen showing tubular virgella. Hartfell Spa. Ibid.

    7 *d.* Typical form. Same locality as figs. 7 *a* and 7 *b*.

    7 *e.* Specimen having a long proximal spine in addition to virgella. Same locality as fig. 7 *c*.

8 *a*—*d.*—*Climacograptus tubuliferus*, Lapworth. (Page 203.)

    8 *a.* Typical specimen. Hartfell Spa. Hartfell Shales. Lapworth's Collection.

    8 *b.* Ditto, showing broad virgular tube. On same slab as fig. 8 *a*.

    8 *c.* Small specimen showing proximal end. Ibid.

    8 *d.* Larger specimen. Ibid.

9 *a*—*e.*— *Climacograptus styloideus*, Lapworth. (Page 205)

    9 *a.* Typical specimen showing virgula with sac, scalariform view. Hartfell Spa. Hartfell Shales. Lapworth's Collection.

    9 *b.* Distal fragment, showing virgula and sac. Ibid.

    9 *c.* Proximal part. Ibid.

    9 *d.* Wide specimen. Ibid.

    9 *e.* Complete specimen, scalariform view. Ibid.

10 *a*—*e.*—*Climacograptus innotatus*, Nicholson. (Page 212.)

    10 *a.* Typical specimen. Dobb's Linn. Birkhill Shales (zone of *Monog. gregarius*). Elles' Collection.

    10 *b.* Small specimen showing thecal spines. Ibid.

    10 *c.* Long narrow specimen with few thecal spines. Dobb's Linn. Birkhill Shales. Lapworth's Collection.

    10 *d.* Small specimen, obverse view. Ibid. Elles' Collection.

    10 *e.* Ibid.

11 *a*—*e.*—*Climacograptus Hughesi* (Nicholson). (Page 208.)

    11 *a.* Typical specimen in full relief. Ambleside, Lake District. Skelgill Beds. British Museum (Natural History), S. Kensington.

    11 *b.* Compressed specimen, in relief. Skelgill. Skelgill Beds (zone of *Monog. fimbriatus*). Marr's Collection.

    11 *c.* Specimen in the flat. Branch Linn, Dobb's Linn. Birkhill Shales (zone of *M. gregarius*). Elles' Collection.

    11 *d.* Small specimen, cast. Skelgill. Skelgill Beds (zone of *M. argenteus*). Marr's Collection.

    11 *e.* Specimen preserved in the flat. Main Cliff, Dobb's Linn. Birkhill Shales. Elles' Collection.

12 *a*—*c.*—*Climacograptus minutus*, Carruthers. (Page 211.)

    12 *a.* Complete specimen. Frenchland Burn, S. Scotland. Birkhill Shales. British Museum (Natural History), S. Kensington.

    12 *b.* Ibid. On same slab as fig. 12 *a*.

    12 *c.* Distal fragment. Ibid.

13 *a. b.*—*Climacograptus extremus*, H. Lapworth. (Page 210.)

    13 *a.* Typical specimen in full relief, reverse view. Rhyd Hir Brook, Rhayader. Rhayader Pale Shales (zone of *Monog. crassus*). H. Lapworth's Collection.

    13 *b.* Ibid.

14 *a*—*e.*—*Climacograptus Scharenbergi*, Lapworth. (Page 206.)

    14 *a.* Typical specimen, full relief. Dobb's Linn. Lower Hartfell Shales (zone of *Climacog. Wilsoni*). Lapworth's Collection.

    14 *b.* Large specimen in relief. Ibid.

    14 *c.* Specimen preserved in the flat, showing no zig-zag septum. Hartfell Spa. Lower Hartfell Shales (zone of *Climacog. Wilsoni*). Lapworth's Collection.

    14 *d.* Small specimen preserved as a cast, showing very long virgula. Ibid.

    14 *e.* Small fragment. Pont Seiont, N. Wales. Upper Arenig Beds (zone of *Didymog. bifidus*). Sedgwick Museum.

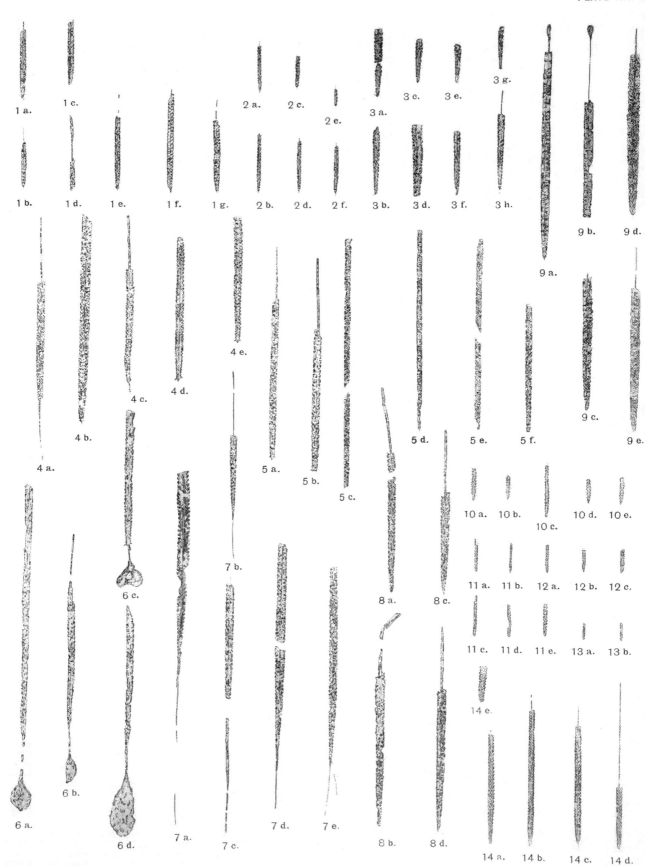

E. M. R. Wood, del.   *Bemrose, Ltd., Derby.*

PLATE XXVIII.

Sub-genus **Orthograptus**.

PLATE XXVIII (*continued*).

FIGS.

6 a—d.—*Orthograptus Whitfieldi* (Hall).   (Page 227.)

> 6 a. Incomplete, but typical, specimen. Rein Gill, Abington. Glenkiln Shales. Geological Survey of Scotland, Edinburgh.
>
> 6 b. Characteristic specimen, with long virgella. Berrybush Burn, St. Mary's Loch. Glenkiln Shales. Geological Survey of Scotland.
>
> 6 c. Broad distal fragment. Belcraig Burn. Glenkiln Shales. Geological Survey of Scotland.
>
> 6 d. Small proximal fragment. Tiddyndicwm, N. Wales. Llandeilo Beds. Sedgwick Museum.

7 a—c.—*Orthograptus insectiformis*, Nicholson.   (Page 228.)

> 7 a. Typical, but incomplete, specimen. Dobb's Linn. Birkhill Shales (zone of *Monog. gregarius*). Lapworth's Collection.
>
> 7 b. Somewhat broader distal fragment. Ibid.
>
> 7 c. Small, incomplete specimen. Ibid.

8 a—d.—*Orthograptus vesiculosus*, Nicholson.   (Page 229.)

> 8 a. Typical and complete specimen, "bi-scalariform" view. Coalpit Bay, Co. Down. Lower Birkhill Beds. Museum of Natural History Society, Belfast.
>
> 8 b. Small, complete specimen, showing sicula. Dobb's Linn. Birkhill Shales (zone of *Orthog. vesiculosus*). Lapworth's Collection.
>
> 8 c. Broad specimen, " bi-scalariform " view, obverse aspect. Ibid.
>
> 8 d. Well-preserved specimen, sub-scalariform view. North end of Clanyard Bay, Drummore, Wigtown. Lower Birkhill Shales. Geological Survey of Scotland.

9 a—c.—*Orthograptus vesiculosus*, var. *penna*, Hopkinson.   (Page 231.)

> 9 a. Type specimen, incomplete, figured Hopkinson, 'Geol. Mag.,' 1872, vol. ix, pl. xii, fig. 6. Frenchland Burn, Moffat. Lower Birkhill Shales. Sedgwick Museum.
>
> 9 b. Incomplete specimen. Keisley, Westmoreland. Llandovery. Fearnsides' Collection.
>
> 9 c. Complete typical specimen. Frenchland Burn ? Lower Birkhill Shales. Geological Survey of Scotland.

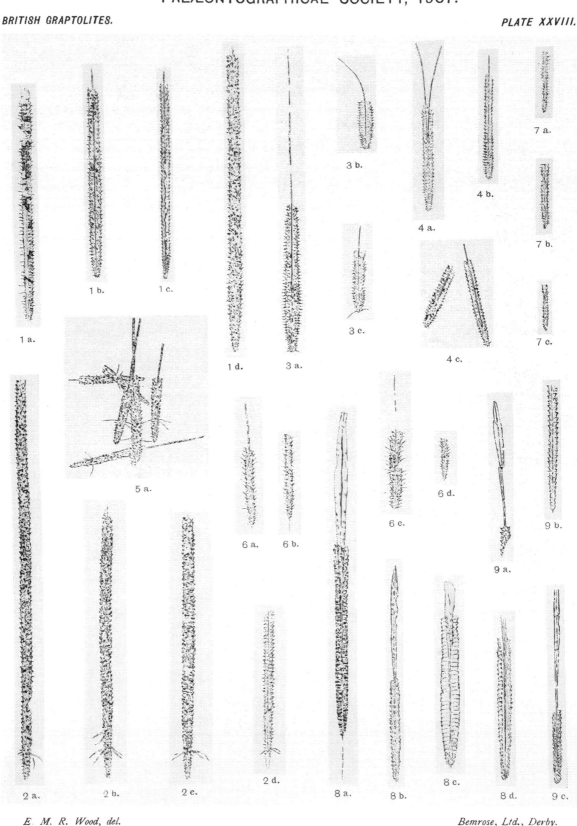

E. M. R. Wood, del.

Bemrose, Ltd., Derby.

ORTHOGRAPTUS.

PLATE XXIX.

Sub-genus Orthograptus.

Figs.

1 *a—d.—Orthograptus mutabilis*, Elles and Wood, nov.   (Page 232.)

    1 *a*. Large typical specimen, reverse aspect. Dobb's Linn. Birkhill Shales. Lapworth's Collection.

    1 *b*. Smaller specimen, obverse aspect. On same slab as fig. 1 *a*.

    1 *c*. Short, broad specimen, reverse aspect. Ibid.

    1 *d*. Distal portion, compressed, doubtfully referable to this species. Ibid.

2 *a—e.—Orthograptus bellulus*, Törnquist.   (Page 231.)

    2 *a*. Typical specimen, showing long virgella. Dobb's Linn. Upper Birkhill Shales (*Monog. Clingani* band). Geological Survey of Scotland.

    2 *b*. Narrower specimen. Ibid. Lapworth's Collection.

    2 *c*. Ibid.

    2 *d*. Well-preserved specimen. Plewlands Burn, Raehills, Moffat. Birkhill Shales. Geological Survey of Scotland.

    2 *e*. Small specimen in relief, reverse aspect. Skelgill, Ambleside. Stockdale Shales. British Museum (Natural History), S. Kensington.

3 *a—e.—Orthograptus truncatus*, Lapworth.   (Page 233.)

    3 *a*. Typical specimen. Mount Benger Burn. Hartfell Shales (zone of *Dicranog. Clingani*). Lapworth's Collection.

    3 *b*. Ditto. On same slab as fig. 3 *a*.

    3 *c*. Smaller form, showing more gradual increase in width. Dobb's Linn. Hartfell Shales (zone of *Dicranog. Clingani*). Geological Survey of Scotland.

    3 *d*. Characteristic form. Hartfell Spa. Hartfell Shales (zone of *Dicranog. Clingani*). Wood's Collection.

    3 *e*. Ditto. Elles' Collection.

4 *a—e.—Orthograptus truncatus*, var. *intermedius*, Elles and Wood, nov. (Page 236.)

    4 *a*. Short, but typical specimen. Hartfell. Lower Hartfell Shales (zone of *Climacog. Wilsoni*). Lapworth's Collection.

    4 *b*. Long distal portion. Ibid. Elles' Collection.

    4 *c*. Proximal portion, reverse aspect. On same slab as fig. 4 *b*.

    4 *d*. Very long, complete specimen. Hartfell Shales (zone of *Dicranog. Clingani*). Lapworth's Collection.

    4 *e*. Long distal portion. Hartfell Shales (zone of *Climacog. Wilsoni*). Wood's Collection.

PLATE XXIX (*continued*).

Figs.

5 *a*—*d*.—*Orthograptus truncatus*, var. *pauperatus*, Lapworth MS.   (Page 237.)

> 5 *a*. Small characteristic specimen, reverse aspect.   Hartfell Spa. Hartfell Shales (zone of *Dicranog. Clingani*).   Lapworth's Collection.
>
> 5 *b*. Ditto.   Dobb's Linn.   Hartfell Shales.   Dr. Fraser's Collection.
>
> 5 *c*. Smaller specimen.   Hartfell Spa.   Hartfell Shales.   Lapworth's Collection.
>
> 5 *d*. Ditto.   On same slab as fig. 5 *c*.

6 *a*—*e*.—*Orthograptus truncatus*, var. *abbreviatus*, Elles and Wood, nov.   (Page 235.)

> 6 *a*. Typical specimen, obverse aspect, well preserved.   Dobb's Linn. Upper Hartfell Shales (zone of *Dicellog. anceps*).   Elles' Collection.
>
> 6 *b*. Compressed specimen.   Belcraig Burn.   Hartfell Shales (zone of *Dicellog. anceps*).   Elles' Collection.
>
> 6 *c*. Distal portion.   Dobb's Linn.   Hartfell Shales (zone of *Dicellog. anceps*).   Wood's Collection.
>
> 6 *d*. Fairly complete specimen.   On same slab as fig. 6 *c*.
>
> 6 *e*. Long narrow specimen, probably referable to this variety.   Dobb's Linn.   Lower Birkhill Shales.   Wood's Collection.

7 *a*—*e*.—*Orthograptus truncatus*, var. *socialis*, Lapworth.   (Page 237.)

> 7 *a*. Typical specimen, obverse aspect.   Dobb's Linn.   Upper Hartfell Shales (zone of *Dicellog. complanatus*).   Lapworth's Collection.
>
> 7 *b*. Long specimen.   Ibid.
>
> 7 *c*. Small specimen.   Ibid.
>
> 7 *d*. Broad specimen.   Ibid.
>
> 7 *e*. Characteristic form.   Ibid.

8 *a*—*c*.—*Orthograptus cyperoides*, Törnquist.   (Page 238.)

> 8 *a*. Relatively large specimen, obverse aspect.   Dobb's Linn (Long Cliff).   Birkhill Shales (zone of *Monog gregarius*).   Elles' Collection.
>
> 8 *b*. Characteristic form, reverse aspect.   Ibid.
>
> 8 *c*. Small specimen, obverse aspect.   Ibid.

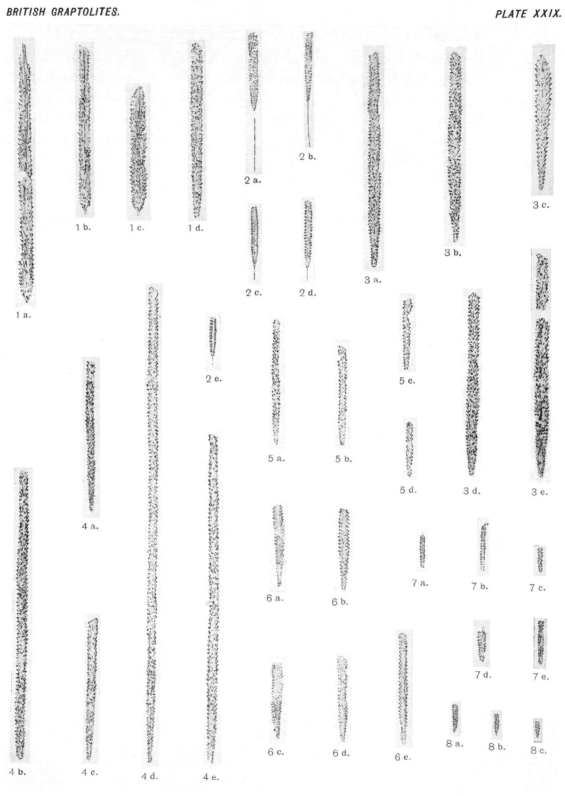

E. M. R. Wood, del.

Bemrose, Ltd., Derby.

ORTHOGRAPTUS.

PLATE XXX.

Sub-genera **Orthograptus** and **Glyptograptus**.

PLATE XXX—*continued.*

Figs.

6 *a—c.—Orthograptus calcaratus,* var. *priscus,* Elles and Wood, nov.   (Page 244.)

> 6 *a.* Small narrow specimen, compressed, reverse aspect.   Figured, Hopkinson and Lapworth, ' Quart. Journ. Geol. Soc.,' 1875, vol. xxxi, pl. xxxv, fig. 7 *b.*   Abereiddy Bay.   Llandeilo (zone of *Didymog. Murchisoni*).   Sedgwick Museum.
>
> 6 *b.* Long, incomplete specimen, showing characteristic view of thecæ. Ibid., fig 7 *c.*
>
> 6 *c.* Small broad specimen, compressed.   Ibid., fig. 7 *a.*

7 *a—d.—Orthograptus rugosus,* var. *apiculatus,* Elles and Wood, nov.   (Page 245.)

> 7 *a.* Typical specimen, preserved in low relief, obverse aspect.   Laggan Gill, Girvan.   Ardwell Beds.   Geological Survey of Scotland.
>
> 7 *b.* Smaller perfect specimen, reverse aspect.   Ibid.
>
> 7 *c.* Larger specimen, reverse aspect.   On same slab as 7 *b.*
>
> 7 *d.* Small specimen.   Ibid.   Lapworth's Collection.

8 *a—d.—Glyptograptus tamariscus,* Nicholson.   (Page 247.)

> 8 *a.* Small characteristic specimen.   Duffkinnel.   Birkhill Shales. Geological Survey of Scotland.
>
> 8 *b.* Broader and longer specimen.   Dobb's Linn.   Birkhill Shales. Wood's Collection.
>
> 8 *c.* Characteristic specimen.   Dobb's Linn.   Birkhill Shales.   Lapworth's Collection.
>
> 8 *d.* Specimen on cleaved rock, reverse aspect.   Belcraig Burn.   Birkhill Shales.   Elles' Collection.

9 *a—d.—Glyptograptus tamariscus,* var. *incertus,* Elles and Wood, nov.   (Page 249.)

> 9 *a.* Characteristic specimen, incomplete.   Dobb's Linn.   Upper Birkhill Shales.   Sedgwick Museum.
>
> 9 *b.* Ditto.   Dobb's Linn.   Upper Birkhill Shales.   Lapworth's Collection.
>
> 9 *c.* Somewhat distorted specimen.   Garple Linn, Moffat.   Birkhill Shales.   Lapworth's Collection.
>
> 9 *d.* " Bi-scalariform " view.   On same slab as Fig. 9 *c.*

10 *a—c.—Glyptograptus serratus,* Elles and Wood, sp. nov.   (Page 249.)

> 10 *a.* Typical specimen.   Belcraig Burn.   Birkhill Shales.   Wood's Collection.
>
> 10 *b.* Broader specimen showing virgula, obverse aspect.   Ibid.
>
> 10 *c.* Ditto, reverse aspect.   On same slab as fig. 10 *b.*

11 *a, b.—Glyptograptus serratus,* var. *barbatus,* Elles and Wood, nov.   (Page 250.)

> 11 *a.* Typical specimen.   Pary's Mount, Anglesea.   Birkhill Shales (zone of *Monog. Sedgwickii*).   Greenly's Collection.
>
> 11 *b.* Somewhat narrower specimen.   Ibid.

E. M. R. Wood, del.

Bemrose, Ltd., Derby.

ORTHOGRAPTUS AND GLYPTOGRAPTUS.

PLATE XXXI.

Sub-genera Glyptograptus, Mesograptus, and Amplexograptus.

PLATE XXXI—*continued.*

FIGS.

10 *a*—*c.*—*Mesograptus multidens*, var. *compactus*, Lapworth. (Page 262.)

> 10 *a.* Complete specimen, somewhat narrower than usual, reverse aspect. Dobb's Linn. Hartfell Shales (zone of *Dicranog. Clingani*). Lapworth's Collection.
>
> 10 *b.* Shorter but broader specimen, showing virgular tube. Ibid.
>
> 10 *c.* Specimen showing average width, reverse aspect. Ibid.

11 *a*—*e.*—*Mesograptus modestus*, Lapworth. (Page 263.)

> 11 *a.* Typical specimen, showing characteristic appearance. Dobb's Linn. Lower Birkhill Shales (zone of *Orthog. vesiculosus*). Lapworth's Collection.
>
> 11 *b.* Large, but incomplete, specimen. On same slab as fig. 11 *a.*
>
> 11 *c.* Specimen with long virgula, somewhat distorted. Ibid. Sedgwick Museum.
>
> 11 *d.* Narrower specimen. Ibid. On same slab as figs. 11 *a*, 11 *b.*
>
> 11 *e.* Part of narrow specimen in partial relief. Fuches-gau Farm, near Pont Erwyd. Lowest Llandovery (base of zone of *Cephalog. acuminatus*). O. T. Jones' Collection.

12 *a*—*d.*—*Mesograptus modestus*, var. *parvulus* (H. Lapworth). (Page 264.)

> 12 *a.* Type specimen, reverse aspect, but with surface removed, so as to show sicula. Figured, H. Lapworth, 'Quart. Journ. Geol. Soc.,' 1900, vol. lvi, p. 132, fig. 20 *b.* Gwastaden, Rhayader. Lower Gwastaden Series (zone of *Cephalog. acuminatus*). H. Lapworth's Collection.
>
> 12 *b.* Larger specimen, well preserved, obverse aspect. Fuches-gau Farm, near Pont Erwyd. Lowest Llandovery Beds. O. T. Jones' Collection.
>
> 12 *c.* Small, broad specimen, reverse aspect. Ibid.
>
> 12 *d.* Incomplete specimen, obverse aspect. Ibid.

13 *a*—*c.*—*Mesograptus modestus*, var. *diminutus*, Elles and Wood, nov. (Page 265.)

> 13 *a.* Typical specimen, reverse aspect. Skelgill, Ambleside. Skelgill Beds (*Dimorphograptus* band). Sedgwick Museum.
>
> 13 *b.* Complete specimen, bi-profile view. Dobb's Linn. Lower Birkhill Shales. Elles' Collection.
>
> 13 *c.* Complete specimen, bi-scalariform view. Dobb's Linn. Lower Birkhill Shales (*Dimorphograptus* band). Lapworth's Collection.

14 *a*—*c.*—*Mesograptus magnus*, H. Lapworth. (Page 266.)

> 14 *a.* Type specimen, part in relief, part as an impression. Figured, H. Lapworth, 'Quart. Journ. Geol. Soc.,' 1900, vol. lvi, p. 133, fig. 21 *b.* Ddôl Farm, Rhayader. Gwastaden Series (zone of *Monog. fimbriatus*). H. Lapworth's Collection.
>
> 14 *b.* Smaller specimen in low relief. Rheidol Gorge, below Pont Erwyd, Cardiganshire. Llandovery Beds. O. T. Jones' Collection.
>
> 14 *c.* Narrow specimen, probably referable to this species. Ibid.

15 *a*—*d.*—*Amplexograptus perexcavatus*, Lapworth. (Page 267.)

> 15 *a.* Typical, complete specimen, obverse aspect. Kirkmichael Burn, S. Scotland. Upper Glenkiln Shales. Lapworth's Collection.
>
> 15 *b.* Long specimen, obverse aspect. On same slab as fig. 15 *a.*
>
> 15 *c.* Complete specimen, intermediate view. Dobb's Linn. Upper Glenkiln Shales. Geological Survey of Scotland.
>
> 15 *d.* Complete specimen in partial relief, bi-profile view. Ibid. Lapworth's Collection.

16 *a*—*d.*—*Amplexograptus arctus*, Elles and Wood, sp. nov. (Page 271.)

> 16 *a.* Large, fairly complete specimen, long virgella. Spittal Railway Cutting, S. Wales. Lower *Dicranograptus* Shales. Geological Survey of England and Wales.
>
> 16 *b.* Complete specimen, showing characteristic form. Ibid.
>
> 16 *c.* Distal fragment, showing virgula. On same slab as fig. 16 *b.*
>
> 16 *d.* Specimen showing general form. Ibid.

17 *a*—*c.*—*Amplexograptus cœlatus* (Lapworth). (Page 270.)

> 17 *a.* Type specimen, sub-scalariform view. Figured, Lapworth, 'Quart. Journ. Geol. Soc.,' 1875, vol. xxxi, pl. xxxv, fig. 8 *c.* Abereiddy Bay, S. Wales. Llandeilo Beds (zone of *Didymog. Murchisoni*). Sedgwick Museum.
>
> 17 *b.* Incomplete specimen, sub-scalariform view. Ibid., fig. 8 *b.*
>
> 17 *c.* Distorted specimen, bi-profile view, showing *Diplograptus* appearance. Ibid., fig. 8 *a.*

18 *a*—*e.*—*Amplexograptus confertus* (Lapworth). (Page 269.)

> 18 *a.* Specimen much distorted and compressed. Figured, Lapworth, 'Quart. Journ. Geol. Soc.,' 1875, vol. xxxi, pl. xxxiv, fig. 2 *a.* Porth Hayog, Ramsey Island. Upper Arenig (zone of *Didymog. bifidus*). Sedgwick Museum.
>
> 18 *b.* Specimen elongated by compression. Ibid., fig. 2 *c.*
>
> 18 *c.* Typical specimen, preserved as an impression, obverse aspect. Upper Arenig (zone of *Didymog. bifidus*). Geological Survey of England and Wales.
>
> 18 *d.* Two small specimens in association. Near Ffairfach Railway Station, Llandeilo. Upper Arenig (zone of *Didymog. bifidus*). Geological Survey of England and Wales.
>
> 18 *e.* Specimen of unusual length, preserved as an impression. Roadside E. of Church, Lampeter Velfrey. Upper Arenig. Turnbull's Collection, Sedgwick Museum.

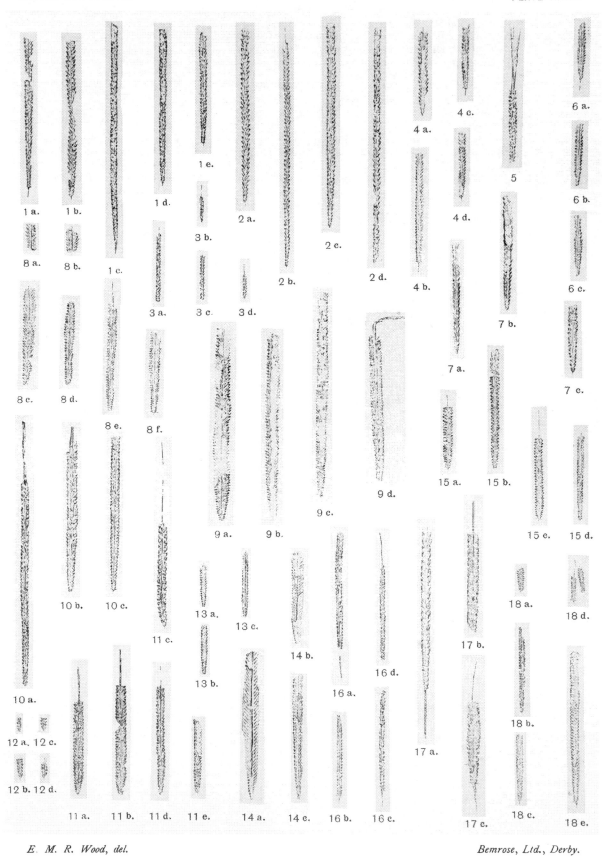

E. M. R. Wood, del.

Bemrose, Ltd., Derby.

GLYPTOGRAPTUS, MESOGRAPTUS AND AMPLEXOGRAPTUS.

PLATE XXXII.

Sub-genera **Petalograptus** and **Cephalograptus**; genus **Cryptograptus.**

FIGS.

1 *a—d.—Petalograptus palmeus* (Barrande). (Page 274.)

1 *a.* Typical specimen, with long virgular tube. Dobb's Linn, S. Scotland. Birkhill Shales. Lapworth's Collection.

1 *b.* Typical specimen, showing sicula. Dobb's Linn. Birkhill Shales (band with *Monog. Clingani*). Lapworth's Collection.

1 *c.* Wider specimen, obverse aspect. Ibid.

1 *d.* Wide specimen, doubtfully referable to this species. Dobb's Linn. Birkhill Shales (zone of *Monog. gregarius*). Geological Survey of Scotland, Edinburgh.

2 *a—f.—Petalograptus palmeus*, var. *latus* (Barrande). (Page 275.)

2 *a.* Typical specimen, obverse aspect. Dobb's Linn. Birkhill Shales. Lapworth's Collection.

2 *b.* Typical specimen, reverse aspect. Ibid.

2 *c.* Shorter specimen, somewhat compressed. Dobb's Linn. Birkhill Shales. Sedgwick Museum.

2 *d.* Small specimen, figured Elles, 'Quart. Journ. Geol. Soc.,' 1897, vol. liii, pl. xiv, fig. 3. Garple Linn, near Beattock. Birkhill Shales (zone of *Monog. gregarius*). Sedgwick Museum.

2 *e.* Characteristic small specimen, showing virgella, reverse aspect. Long Linn, Dobb's Linn. Birkhill Shales (zone of *Monog. gregarius*). Elles' Collection.

2 *f.* Small specimen, obverse aspect. Ibid.

3 *a—d.—Petalograptus palmeus*, var. *tenuis* (Barrande). (Page 276.)

3 *a.* Typical specimen, cast, showing fine growth-lines. Dobb's Linn?, Birkhill Shales. Lapworth's Collection.

3 *b.* Typical specimen, showing sicula, figured Elles, 'Quart. Journ. Geol. Soc.,' 1897, vol. liii, pl. xiv, fig. 9. Pull Beck, Lake District. Browgill Beds (zone of *Monog. crispus*). Sedgwick Museum.

3 *c.* Small specimen in full relief, showing sicula, ? figured Elles, Ibid., fig. 10. Dobb's Linn. Birkhill Shales. Lapworth's Collection.

3 *d.* Specimen in relief, showing no septum, reverse aspect. Morben Quarry, Derwentas, Machynlleth. Llandovery. G. J. Williams' Collection.

4 *a—d.—Petalograptus palmeus*, var. *ovato-elongatus* (Kurck). (Page 277.)

4 *a.* Typical specimen, showing virgula. Long Linn, Dobb's Linn. Birkhill Shales (zone of *Monog. gregarius*). Elles' Collection.

4 *b.* Specimen somewhat compressed. Branch Linn, Dobb's Linn. Birkhill Shales (zone of *Cephalog. cometa*). Elles' Collection.

4 *c.* Longer specimen, in partial relief. Dobb's Linn. Birkhill Shales. Geological Survey of Scotland, Edinburgh.

4 *d.* Incomplete specimen. Ibid.

5 *a—e.—Petalograptus minor*, Elles. (Page 279.)

5 *a.* Typical specimen, in full relief, showing no septum, reverse aspect, figured, Elles, 'Quart. Journ. Geol. Soc.,' 1897, vol. liii, pl. xiv, fig. 19. Skelgill, Skelgill Beds. Sedgwick Museum.

5 *b.* Similar specimen from same locality as fig. 5 *a.*

5 *c.* Typical specimen, obverse aspect. Long Linn, Dobb's Linn. Birkhill Shales (zone of *Monog. gregarius*). Elles' Collection.

5 *d.* Larger specimen on same slab as fig. 5 *c.*

5 *e.* Characteristic small specimen, showing growth-lines. Dobb's Linn. Birkhill Shales (zone of *Monog. gregarius*). Sedgwick Museum.

6.—*Petalograptus cfr. ovatus* (Barrande). (Page 278.)

Compressed specimen. Skelgill. Browgill Beds. Sedgwick Museum.

7 *a—e.—Petalograptus altissimus*, Elles and Wood, nov. (Page 281.)

7 *a.* Typical specimen, in relief. Ettrick Bridge End, Selkirk. Upper Birkhill Shales. Geological Survey of Scotland, Edinburgh.

7 *b.* Less complete specimen. Dobb's Linn. Upper Birkhill Shales (zone of *Rastrites maximus*). Lapworth's Collection.

7 *c.* Narrow specimen, in high relief, 200 yards S. of Parbryn Sands, Cardiganshire. Llandovery-Tarannon. O. T. Jones' Collection.

7 *d.* Distal fragment in relief. Llanystwmdwy near Criccieth. Llandovery-Tarannon (zone of *Monog. turriculatus*). Fearnsides' Collection.

7 *e.* Flattened specimen. Woopland. Gala Beds. Lapworth's Collection.

8 *a—e.—Petalograptus folium* (Hisinger). (Page 282.)

8 *a.* Typical specimen (faulted), reverse aspect, figured, Elles, 'Quart. Journ. Geol. Soc.,' 1897, vol. liii, pl. xiii, fig. 1. Belcraig, near Moffat. Birkhill Shales (zone of *Monog. gregarius*). Elles' Collection.

8 *b.* Fragment of proximal end. Main Cliff, Dobb's Linn. Birkhill Shales (band with *P. folium*). Elles' Collection.

PLATE XXXII—*continued.*

8 *a — e.*—*Petalograptus folium* (Hisinger)—*continued.*

    8 *c.* Narrower specimen. Ibid.

    8 *d.* Fragment of narrower specimen. Branch Linn, Dobb's Linn. Birkhill Shales (band with *P. folium*). Elles' Collection.

    8 *e.* Fragment of proximal end. Ibid.

9 *a—d.*—*Cephalograptus tubulariformis* (Nicholson). (Page 287.)

    9 *a.* Typical specimen, reverse aspect, figured ? Nicholson, 'Geol. Mag.,' 1867, pl. vii, fig. 12, and Elles, as *Cephalog. petalum*, 'Quart. Journ. Geol. Soc.,' 1897, vol. liii, pl. xiii, fig. 8. Frenchland Burn, near Moffat. Birkhill Shales. British Museum (Natural History), S. Kensington.

    9 *b.* Proximal end, reverse aspect, figured, Elles, ibid., fig. 7. Duffkinnell Burn, near Wamphray. Birkhill Shales. British Museum (Natural History), S. Kensington.

    9 *c.* Long specimen. Belcraig Burn, near Moffat. Birkhill Shales (zone of *Cephalog. cometa*). Elles' Collection.

    9 *d.* Typical specimen incomplete, figured Elles, as *Cephalog. petalum*, 'Quart. Journ. Geol. Soc.,' 1897, vol. liii, pl. xiii, fig. 6. Duffkinnell Burn, near Wamphray. Birkhill Shales. British Museum (Natural History), S. Kensington.

10 *a—d.*—*Cephalograptus cometa* (Geinitz). (Page 285.)

    10 *a.* Typical specimen partly in relief. Pary's Mountain, Anglesea. Llandovery. G. J. Williams' Collection.

    10 *b.* Very long specimen. Dobb's Linn. Birkhill Shales (zone of *Cephalog. cometa*). Lapworth's Collection.

    10 *c.* Well-preserved, very typical specimen. Dobb's Linn. Birkhill Shales. Geological Survey of Scotland, Edinburgh.

    10 *d.* Distal fragment, showing virgula, figured Elles, 'Quart. Journ. Geol. Soc.,' 1897, vol. liii, pl. xiii, fig. 10. Duffkinnell Burn, near Wamphray. British Museum (Natural History), S. Kensington.

11 *a—d.*—*Cephalograptus (?) acuminatus* (Nicholson). (Page 289.)

    11 *a.* Typical specimen. Dobb's Linn. Birkhill Shales (zone of *Cephalog. (?) acuminatus*). Lapworth's Collection.

    11 *b.* Smaller specimen, straighter. Duffkinnell Burn, near Wamphray. Birkhill Shales. Lapworth's Collection.

    11 *c.* Small specimen. Dobb's Linn. Birkhill Shales. Lapworth's Collection.

    11 *d.* Longer straight specimen. Dobb's Linn. Birkhill Shales (zone of *Cephalog. (?) acuminatus*). Sedgwick Museum.

12 *a—d.*—*Cryptograptus tricornis* (Carruthers). (Page 296.)

    12 *a.* Long typical specimen, poorly preserved. The Cornice, Hartfell. Hartfell Shales. Lapworth's Collection.

    12 *b.* Ibid.

    12 *c.* Smaller specimen. The Cornice, Hartfell. Hartfell Shales (zone of *Climacog. Wilsoni*). Elles' Collection.

    12 *d.* Small specimen. Blaen-y-delyn Quarry, Fishguard. Llanvirn Beds. F. R. C. Reed's Collection.

13 *a—c.*—*Cryptograptus tricornis,* var. *Schäferi,* Lapworth. (Page 299.)

    13 *a.* Typical specimen. Llandrindod Wells. Llandeilo. Lapworth's Collection.

    13 *b.* Wider specimen. Pencerrig, near Builth. Llandeilo. Sedgwick Museum.

    13 *c.* Same locality etc., as fig. 13 *a.*

14 *a—e.*—*Cryptograptus (?) antennarius* (Hall). (Page 300.)

    14 *a.* Typical specimen with very long basal spines. Outerside, Keswick. Skiddaw Slates. British Museum (Natural History), S. Kensington.

    14 *b.* Smaller specimen, figured Elles, 'Quart. Journ. Geol. Soc.,' 1898, vol. liv, p. 520, fig. 31 *a.* Outerside. Skiddaw Slates. Sedgwick Museum.

    14 *c.* Specimen showing additional spines, figured ibid., fig. 31 *c.* Ibid.

    14 *d.* Specimen showing virgular tube. Ibid.

    14 *e.* Small specimen (young), figured Elles, ibid., fig. 31 *b.* Ibid.

15 *a, b.*—*Cryptograptus Hopkinsoni* (Nicholson). (Page 299.)

    15 *a.* Typical specimen, mentioned, Elles 'Quart. Journ. Geol. Soc.,' 1898, vol. liv, p. 521. Outerside, Keswick. Skiddaw Slates. Sedgwick Museum.

    15 *b.* Distal fragment. Ibid. British Museum (Natural History), S. Kensington.

16 *a—e.*—*Petalograptus (?) phylloides,* Elles and Wood, nov. (Page 284.)

    16 *a.* Typical small specimen. Belcraig Burn. Glenkiln Shales (zone of *Nemag. gracilis*). Elles' Collection.

    16 *b.* Ibid.

    16 *c.* Narrower specimen. Ibid. Wood's Collection.

    16 *d.* Specimen showing sicula. Ibid.

    16 *e.* Young specimen with sicula. Dobb's Linn. Glenkiln Shales (zone of *Dicellog. patulosus*). Elles' Collection.

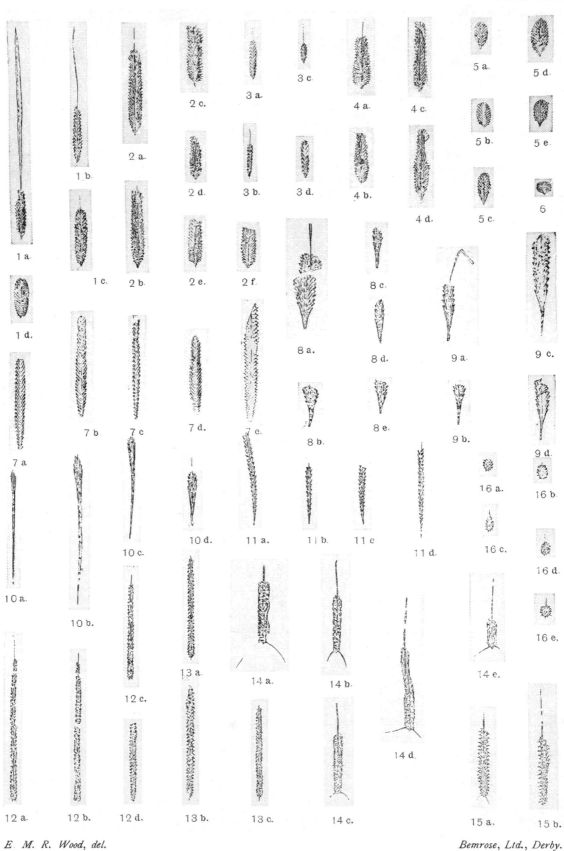

E. M. R. Wood, del.

Bemrose, Ltd., Derby.

PETALOGRAPTUS, CEPHALOGRAPTUS AND CRYPTOGRAPTUS.

PLATE XXXIII.

Genus **Glossograptus** and Sub-genus **Hallograptus**.

PLATE XXXIII—*continued.*

FIGS.

5 *a—e.*—*Glossograptus armatus*, Nicholson—*continued.*

    5 *c.* Scalariform view, showing septal spines, described Elles, ' Quart. Journ. Geol. Soc.,' 1898, vol. liv, p. 522. Thornship Beck. Skiddaw Slates (Ellergill Beds). Sedgwick Museum.

    5 *d.* Somewhat distorted specimen, showing long basal spines. Back Burn, ¾ mile W. of Nether Cog, Crawick Water. Glenkiln Shales. Geological Survey of Scotland, Edinburgh.

    5 *e.* Distal fragment, bi-profile view. Polmorlach Burn, Dumfries. Glenkiln Shales. Ibid.

6 *a—e.*—*Hallograptus mucronatus* (Hall). (Page 320.)

    6 *a.* Characteristic specimen, bi-profile view. Cairn Ryan. Glenkiln Shales. Sedgwick Museum.

    6 *b.* Distal fragment. Ibid.

    6 *c.* Typical specimen. Glenkiln Burn. Glenkiln Shales (zone of *Nemag. gracilis*). Elles' Collection.

    6 *d.* Smaller specimen. Cairn Ryan. Glenkiln Shales. Sedgwick Museum.

    6 *e.* Scalariform view, showing scopulæ. Glenkiln Burn. Glenkiln Shales. Lapworth's Collection.

7 *a—e.*—*Hallograptus mucronatus*, var. *inutilis* (Hall). (Page 322.)

    7 *a.* Incomplete specimen, bi-profile view. Half mile W. of Bencraff, Connemara. Arenig. Muff and Carruthers' Collection.

    7 *b.* Distal fragment, showing virgula. Ibid.

    7 *c.* Ibid.

    7 *d.* Narrower specimen. Ibid.

    7 *e.* Distal fragment. Ibid.

8 *a—e.*—*Hallograptus mucronatus*, var. *bimucronatus* (Nicholson). (Page 323.)

    8 *a.* Typical specimen, bi-profile view. Gairy near head of Garryhorn Burn, Carsphairn. Glenkiln Shales. Geological Survey of Scotland, Edinburgh.

    8 *b.* Long specimen, bi-profile view. Water of Deugh, a few yards below the moor. Glenkiln Shales. Geological Survey of Scotland, Edinburgh.

    8 *c.* Small specimen, bi-profile view. Glenkiln Burn. Glenkiln Shales. Lapworth's Collection.

    8 *d.* Specimen in scalariform view, showing scopulæ. Glenkiln Burn ? Glenkiln Shales. Lapworth's Collection.

    8 *e.*—Distal fragment, scalariform view, showing scopulæ. Polmorlach Burn, Dumfries. Glenkiln Shales. Geological Survey of Scotland, Edinburgh.

9 *a—d.*—*Hallograptus mucronatus*, var. *nobilis*, Elles and Wood, nov. (Page 324.)

    9 *a.* Well-preserved specimen, bi-profile view, showing septal strand and virgula. Burn W.N.W. of Low Glenling, seven miles W. by S. of Wigtown. Glenkiln Shales. Geological Survey of Scotland, Edinburgh.

    9 *b.* Specimen showing proximal end. Ibid.

    9 *c.* Long distal fragment, on same slab as 9 *b.* Ibid.

    9 *d.* Long specimen, sub-scalariform view with scopulæ. Ibid.

PLATE XXXIII.

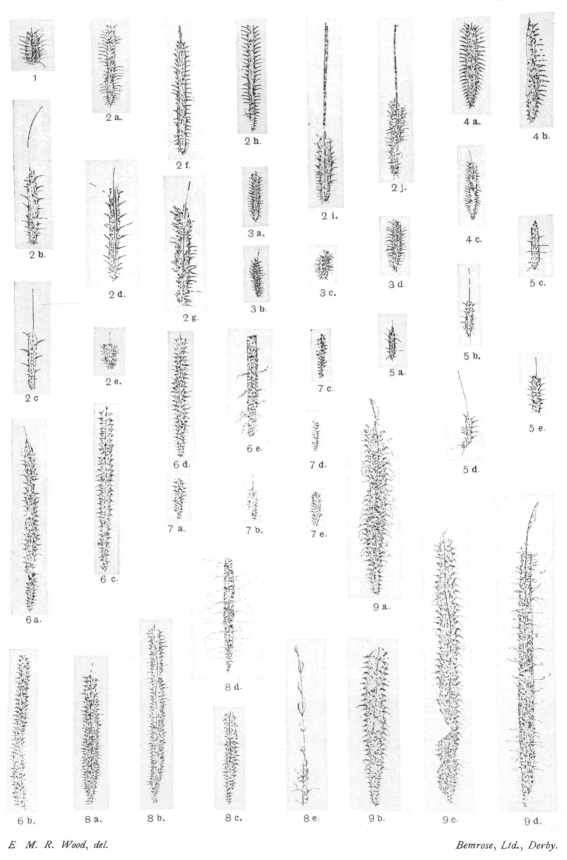

E. M. R. Wood, del.

Bemrose, Ltd., Derby.

GLOSSOGRAPTUS AND HALLOGRAPTUS.

PLATE XXXIV.

Sub-genera **Thysanograptus, Nymphograptus, Neurograptus, Gladiograptus, Plegmatograptus,** and **Gothograptus;** and Genus **Retiograptus.**

FIGS.

1 a—d.—*Thysanograptus Harknessi* (Nicholson). (Page 325.)
    1 a. Typical specimen with incomplete lacinia. Hartfell. Hartfell Shales. British Museum (Natural History). Specimen labelled by Nicholson. (? Type.)
    1 b. Broader form with more complete lacinia. Hartfell. Hartfell Shales (zone of *Climacog. Wilsoni*). Geological Survey of Scotland, Edinburgh.
    1 c. Sub-scalariform view, on same slab as fig. 1 b.
    1 d. Broad specimen, with incomplete lacinia. Hartfell. Hartfell Shales (zone of *Climacog. Wilsoni*). Geological Survey of Scotland, Edinburgh.

2 a—d.—*Thysanograptus Harknessi*, var. *costatus* (Lapworth). (Page 327.)
    2 a. Typical specimen. Cog Burn, a few yards above junction with Polroisk, Dumfriesshire. Glenkiln Shales. Geological Survey of Scotland, Edinburgh.
    2 b. Type specimen, with well-developed lacinia, figured Lapworth, "Grapt. Co. Down," 'Proc. Belfast Nat. Field Club,' pl. vi, fig. 26. Dobb's Linn. Hartfell Shales (zone of *Climacog. Wilsoni*). Lapworth's Collection.
    2 c. Specimen in very low relief. Oakwood, Pontesford, Shropshire. Llandeilo-Bala. Benson's Collection.
    2 d. Specimen with incomplete lacinia. Dobb's Linn. Hartfell Shales. Lapworth's Collection.

3 a—c.—*Thysanograptus retusus* (Lapworth). (Page 328.)
    3 a. Type specimen, figured Lapworth, 'Ann. Mag. Nat. Hist.' [5], vol. v, pl. v, figs. 24 a—d. Llandrindod Wells. Llandeilo. Lapworth's Collection.
    3 b. Specimen showing virgula, doubtfully referable to this species. Hartfell Spa? Glenkiln Shales? Lapworth's Collection.
    3 c. Broad specimen with virgula. Cwm Brith Bank, near Llandrindod Wells. Upper Arenig. Collection Miss C. Chamberlain.

4 a, b.—*Nymphograptus velatus*, Elles and Wood, nov. (Page 329.)
    4 a. Two specimens in association, with well-developed lacinia. Ettrick Bridge End, Selkirk. Hartfell Shales (zone of *Dicellog. anceps*). Geological Survey of Scotland, Edinburgh.
    4 b. Smaller fragment, with less perfect lacinia, profile view. Ibid.

5 a—c.—*Neurograptus fibratus* (Lapworth). (Page 331.)
    5 a. Type specimen, bi-profile view, ? figured Lapworth, 'Cat. West. Scott. Foss.,' pl. iii, fig. 62. Dobb's Linn. Hartfell Shales. Lapworth's Collection.
    5 b. Reverse of 5 a. Ibid.
    5 c. Specimen in scalariform view, showing scopulate processes. Ibid.

6 a—e.—*Neurograptus margaritatus* (Lapworth). (Page 332.)
    6 a. Typical specimen, with almost complete lacinia. Dobb's Linn. Hartfell Shales (zone of *Dicranog. Clingani*). Lapworth's Collection.
    6 b. Similar specimen with less perfect lacinia. Ibid.
    6 c. Specimen in scalariform view, on same slab as 6 a.
    6 d. Larger specimen, with incomplete lacinia. Hartfell Spa. Hartfell Shales. Wood's Collection.
    6 e. Distal fragment on same slab as 6 d.

7 a—d.—*Retiograptus Geinitzianus*, Hall. (Page 316.)
    7 a.—Specimen in low relief. Benan Burn, river Stinchar, Girvan. Glenkiln Shales. Geological Survey of Scotland, Edinburgh.
    7 b. Larger specimen, compressed. Birnock Water. Glenkiln Shales. Lapworth's Collection.
    7 c. Proximal fragment. Ibid.
    7 d. Part of ventral lattice. Ibid.

8 a—d.—*Gladiograptus Geinitzianus* (Barrande). (Page 336.)
    8 a. Well-preserved specimen in low relief, reverse aspect. Burn, Nether Stennis Water, 6½ miles N.N.W. of Langholm. Riccarton Beds. Geological Survey of Scotland, Edinburgh.
    8 b. Compressed specimen, showing reticula, obverse aspect. Pull Beck. Browgill Beds (zone of *Monog. crispus*). Sedgwick Museum (Coll. Marr).
    8 c. Smaller specimen from same locality as 8 a.
    8 d. Somewhat narrow specimen. Grieston Quarry, Innerleithen. Upper Gala Beds. Lapworth's Collection.

PLATE XXXIV—*continued.*

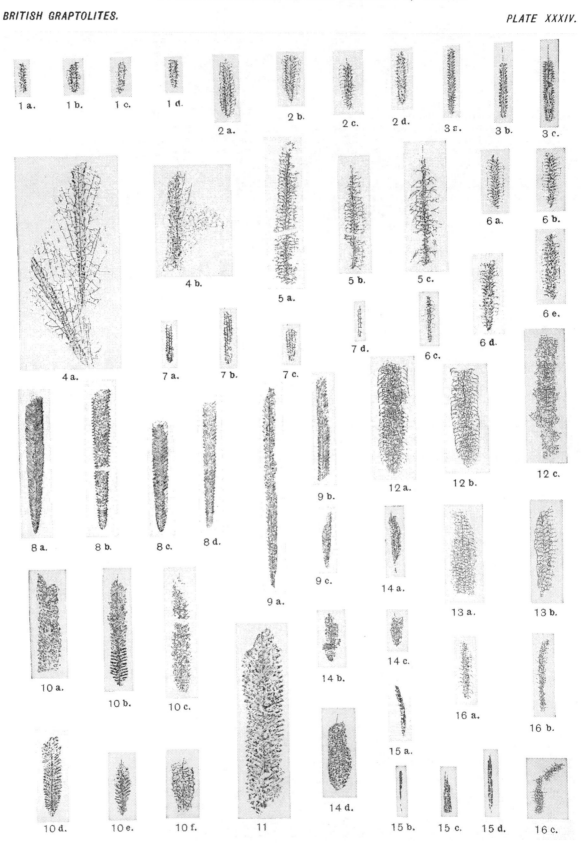

E. M. R. Wood, del.

Bemrose, Ltd., Derby.

THYSANOGRAPTUS, NYMPHOGRAPTUS, NEUROGRAPTUS, RETIOGRAPTUS, GLADIOGRAPTUS, PLEGMATOGRAPTUS AND GOTHOGRAPTUS.

PLATE XXXV.

Genera **Trigonograptus** and **Dimorphograptus**.

Figs.

1 *a—c.—Trigonograptus ensiformis* (Hall).   (Page 302.)

      1 *a*. Typical specimen with sicula? Mosedale Beck, Troutbeck. Skiddaw Slates. Sedgwick Museum.

      1 *b*. Distal fragment, narrower. Pont-y-Feni Quarry, 3 miles W. of St. Clears. Arenig. Geological Survey of England and Wales.

      1 *c*. Long, narrow specimen. Near Keswick. Skiddaw Slates. Sedgwick Museum.

2. *Trigonograptus ensiformis*, var. *lanceolatus* (Nicholson).   (Page 303.)

      Type specimen, figured Nicholson, 'Ann. Mag. Nat. Hist.,' 1869 [4], vol. iv, pl. xi, fig. 6. Ellergill, near Milburn. Skiddaw Slates (Ellergill Beds). British Museum (Nat. Hist.).

3 *a—d.—Dimorphograptus confertus* (Nicholson).   (Page 349.)

      3 *a*. Typical specimen, distorted by cleavage. Skelgill. Skelgill Beds (zone of *Dimorphog. confertus*). Sedgwick Museum.

      3 *b*. Well-preserved specimen. Main Cliff, Dobb's Linn. Birkhill Shales (zone of *Orthog. vesiculosus*). Elles' Collection.

      3 *c*. Specimen showing sicula and virgella. Urr Water, ½ mile S.W. Nether Glaisters, 7½ miles S.W. Dunscor. Birkhill Shales. Geological Survey of Scotland, Edinburgh.

      3 *d*. Well-preserved short specimen. Long Cliff, Dobb's Linn. Birkhill Shales. Lapworth's Collection.

4 *a—f.—Dimorphograptus confertus*, var. *Swanstoni* (Lapworth).   (Page 350.)

      4 *a*. One of type specimens, figured Lapworth, 'Geol. Mag.,' 1876, pl. xx, figs. 13 *a—c*. Coalpit Bay, Donaghadee. Birkhill Shales. Belfast Natural History Museum.

      4 *b*. Another typical specimen on same slab as 4 *a*.

      4 *c*. Ibid.

      4 *d*. On reverse side of same slab. Ibid.

      4 *e*. Longer specimen. Coalpit Bay, Donaghadee. Birkhill Shales. Lapworth's Collection.

      4 *f*. On same slab as 4 *e*.

5 *a—e.—Dimorphograptus decussatus*, Elles and Wood, nov.   (Page 352.)

      5 *a*. Typical specimen. Dobb's Linn. Birkhill Shales (zone of *Orthog. vesiculosus*). Elles' Collection.

      5 *b*. Somewhat smaller specimen, obverse view. Main Cliff, Dobb's Linn. Ibid.

      5 *c*. Reverse view. Same locality as 5 *a*. Ibid.

      5 *d*. Poorly preserved specimen. Same locality as 5 *b*. Ibid.

      5 *e*. Young specimen showing proximal end and virgella. On same slab as 5 *a*.

6. *Dimorphograptus decussatus*, var. *partiliter*, Elles and Wood, nov.   (Page 353.)

      Characteristic specimen. Main Cliff, Dobb's Linn. Birkhill Shales. Elles' Collection.

PLATE XXXV—*continued.*

Figs.

7 *a—d.—Dimorphograptus physophora* (Nicholson).   (Page 353.)

7 *a.* Long specimen with disc.   Dobb's Linn.   Birkhill Shales. Lapworth's Collection.

7 *b.* Similar specimen, sub-scalariform view.   Ibid.

7 *c.* Long and narrow specimen, showing sicula but no disc.   On same slab as 7 *a.*

7 *d.* Three young specimens in juxtaposition, showing sicula and beginnings of disc.   Dobb's Linn.   Birkhill Shales (zone of *Orthog. vesiculosus*).   Elles' Collection.

8 *a—d.—Dimorphograptus cfr. longissimus* (Kurck).   (Page 354.)

8 *a.* Typical specimen.   Keisley, E. Ridlaw.   Stockdale Shales (*Dimorphograptus* zone).   Marr's Collection.

8 *b.* Narrow specimen, obverse view.   Dobb's Linn.   Birkhill Shales (zone of *Orthog. vesiculosus*).   Elles' Collection.

8 *c.* Broad specimen.   Fruid Water, Tweedsmuir.   Birkhill Shales. Geological Survey of Scotland, Edinburgh.

8 *d.* Small specimen.   Quarter mile E. of Tarn Hows.   Stockdale Shales (zone of *Dimorphog. confertus*).   Sedgwick Museum.

9 *a—d.—Dimorphograptus erectus,* Elles and Wood, nov.   (Page 355.)

9 *a.* Typical specimen, with proximal vesicle.   Dobb's Linn.   Birkhill Shales (zone of *Orthog. vesiculosus*).   Elles' Collection.

9 *b.* Smaller specimen.   Ibid.

9 *c.* Incomplete specimen.   On same slab as 9 *b.*   Ibid.

9 *d.* Complete specimen.   Dobb's Linn.   Birkhill Shales (zone of *Orthog. vesiculosus*).   Elles' Collection.

10 *a—e.—Dimorphograptus extenuatus,* Elles and Wood, nov.   (Page 358.)

10 *a.* Typical specimen, somewhat distorted.   Coalpit Bay, Donaghadee.   Birkhill Shales.   Belfast Natural History Museum.

10 *b.* Ibid.

10 *c.* Ibid.

10 *d.* Smaller specimen.   Main Cliff, Dobb's Linn.   Birkhill Shales (zone of *Orthog. vesiculosus*).   Elles' Collection.

10 *e.* Young specimen.   Dobb's Linn.   Birkhill Shales.   Lapworth's Collection.

11 *a—c.—Dimorphograptus elongatus,* Lapworth.   (Page 357.)

11 *a.* Type specimen, figured Lapworth, 'Geol. Mag.,' 1876, pl. xx, figs. 12 *a, b.*   Dobb's Linn.   Birkhill Shales.   Lapworth's Collection.

11 *b.* Specimen on same slab as fig. 11 *a.*

11 *c.* Smaller incomplete specimen.   Dobb's Linn.   Lower Birkhill Shales.   Elles' Collection.

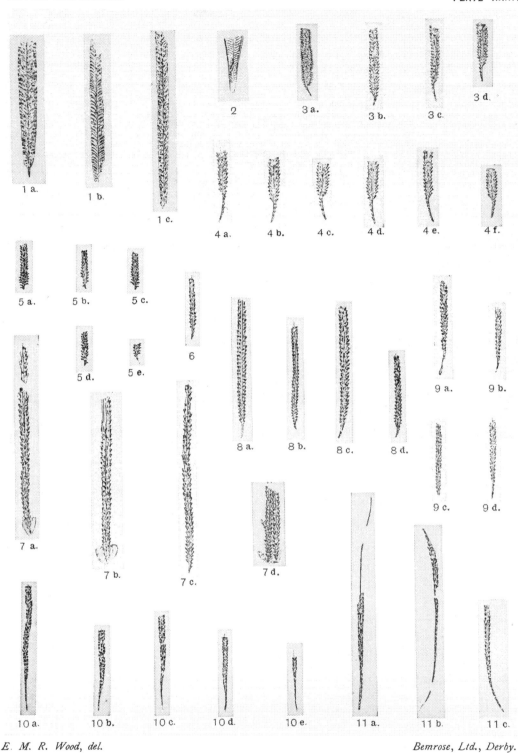

E. M. R. Wood, del.

Bemrose, Ltd., Derby.

TRIGONOGRAPTUS AND DIMORPHOGRAPTUS.

PLATE XXXVI.

Genus **Monograptus**, Geinitz.

Note.—The figures on Plates XXXVI—XLI (which should be studied with a hand lens) are approximately natural size, but there are slight variations.

FIGS.

1 *a—e.—Monograptus cyphus*, Lapworth.   (Page 362.)

> 1 *a*. Typical specimen, showing sicula.   Dobb's Linn, S. Scotland. Lower Birkhill Shales.   Lapworth's Collection.
>
> 1 *b*. Incomplete specimen, broadly recurved.   Waterfall, Long Burn, Dobb's Linn.   Lower Birkhill Shales.   Lapworth's Collection.
>
> 1 *c*. Incomplete specimen, preserved in relief.   Dobb's Linn.   Birkhill Shales.   Sedgwick Museum.
>
> 1 *d*. Specimen, with abrupt curvature of the proximal portion. Dobb's Linn.   Lower Birkhill Shales.   Lapworth's Collection.
>
> 1 *e*. Incomplete, but long specimen, (?) figured Lapworth, Geol. Mag., 1876, pl. xii, fig. 3 *a*.   Ibid.

2 *a—e.—Monograptus acinaces*, Törnquist.   (Page 364.)

> 2 *a*. Median fragment, in relief, figured Jones (as *M. rheidolensis*), Quart. Journ. Geol. Soc., vol. lxv, p. 534, fig. 19 *a*.   Rheidol Gorge, Pont Erwyd, Cardiganshire.   Rheidol Group.   Geological Survey of England and Wales, Jermyn Street (23708).
>
> 2 *b*. Reverse of above specimen, cast.
>
> 2 *c*. Fragment of proximal end.   Nant Fuches-wen, Pont Erwyd. Rheidol Group.   Jones' Collection.
>
> 2 *d*. Distal fragment in relief, figured Jones (as *M. rheidolensis*), Quart. Journ. Geol. Soc., vol. lxv, p. 534, fig. 19 *c*.   Rheidol Gorge. Geological Survey of England and Wales (23709).

3 *a—d.—Monograptus gregarius*, Lapworth.   (Page 365.)

> 3 *a*. Part of the type slab, showing the gregarious habit of this species.   Dobb's Linn.   Birkhill Shales.   Lapworth's Collection.
>
> 3 *b*. Complete specimen on same slab as above.
>
> 3 *c*. Ibid.
>
> 3 *d*. Small specimen, in relief.   Penwhapple Glen, Girvan.   Llandovery Beds.   Lapworth's Collection.

PLATE XXXVI—*continued*.

4 *a—d.*—*Monograptus bohemicus* (Barrande).   (Page 367.)

> 4 *a.* Typical specimen, figured Wood, Quart. Journ. Geol. Soc., vol. lvi, pl. xxv, fig. 27 A.   River Irfon, near Builth.   Lower Ludlow Shales.   Wood's Collection.
>
> 4 *b.* Specimen, showing sicula, figured Wood, loc. cit., fig. 27 B.   Aberedw Hill, near Builth.   Lower Ludlow Shales.   Wood's Collection.
>
> 4 *c.* Long specimen, with slightly curved distal portion.   Hospital Road, Builth.   Lower Ludlow Shales.   Sedgwick Museum.
>
> 4 *d.* Similar specimen.   River Irfon, near Builth.   Lower Ludlow Shales.   Wood's Collection.

5 *a—e.*—*Monograptus concinnus*, Lapworth.   (Page 368.)

> 5 *a.* Two specimens crossing each other, figured Lapworth, Geol. Mag., 1876, pl. xi, fig. 1 *c.*   Dobb's Linn.   Birkhill Shales. Lapworth's Collection.
>
> 5 *b.* Two similar specimens, one showing marked ventral curvature. Dobb's Linn.   Birkhill Shales.   Lapworth's Collection.
>
> 5 *c.* Median fragment, part in relief, part impression.   Waterfall, Dobb's Linn.   Birkhill Shales.   Lapworth's Collection.
>
> 5 *d.* More proximal fragment, mainly in relief.   Rheidol Gorge, 400 yards E.S.E. of Bryn-chwîth Farm, Pont Erwyd.   Geological Survey of England and Wales.
>
> 5 *e.* More distal fragment with marked curvature.   Dobb's Linn. Birkhill Shales.   Lapworth's Collection.
>
> 5 *f.* Very broad distal fragment of great length, partly in relief, provisionally referred to this species (the complete specimen is twice this length).   Rheidol Gorge, 420 yards E.S.E. of Brynchwîth Farm, Pont Erwyd.   Rheidol Group.   Geological Survey of England and Wales.

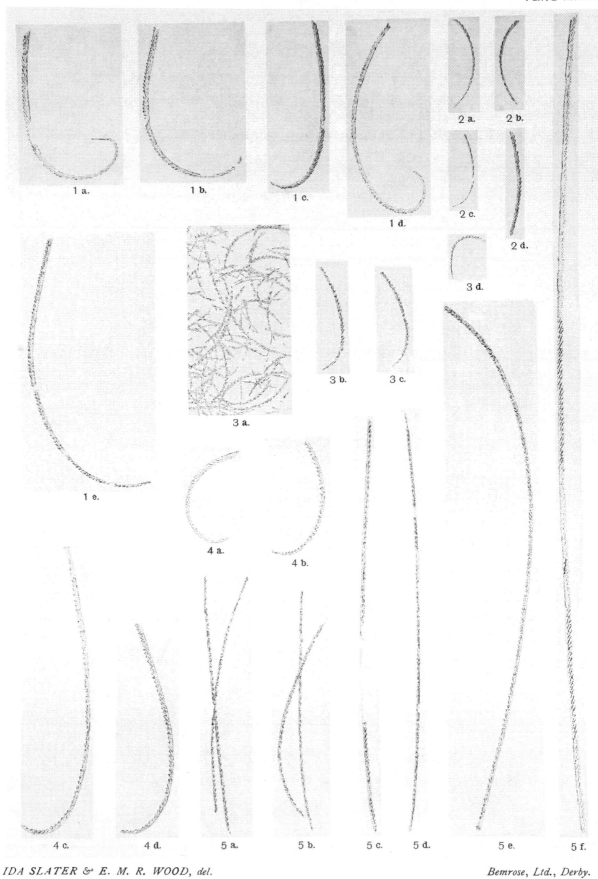

1 a.

1 b.

1 c.

1 d.

2 a.

2 b.

2 c.

2 d.

3 d.

3 a.

3 b.

3 c.

1 e.

4 a.

4 b.

4 c.

4 d.

5 a.

5 b.

5 c.

5 d.

5 e.

5 f.

IDA SLATER & E. M. R. WOOD, del.

Bemrose, Ltd., Derby.

MONOGRAPTUS.

PLATE XXXVII.

Monograptus—*continued.*

FIGS.

1 *a—e.—Monograptus Nilssoni* (Barrande).   (Page 369.)

    1 *a*. Typical specimen, figured Wood, Quart. Journ. Geol. Soc., vol. lvi, pl. xxv, fig. 28 A. Adferton, near Ludlow.   Lower Ludlow Shales.   Hopkinson's Collection.

    1 *b*. Specimen, showing double curvature.   Montgomery Road.   Lower Ludlow Shales. Sedgwick Museum.

    1 *c*. Broad distal fragment, figured Wood, loc. cit., fig. 28 B.   Elton-Evenhay Lane, near Ludlow.   Lower Ludlow Shales.   Wood's Collection.

    1 *d*. Long distal fragment.   Trefnant Brook, Long Mountain.   Lower Ludlow Shales. Wood's Collection.

    1 *e*. Fairly complete specimen, with abrupt proximal curvature.   Ibid.

2 *a—d.— Monograptus leptotheca*, Lapworth.   (Page 371.)

    2 *a*. Typical median fragment, internal cast in iron pyrites.   Dobb's Linn.   Birkhill Shales.   Lapworth's Collection.

    2 *b*. Proximal fragment, partly in relief.   Llanystwmdwy, near Criccieth.   Llandovery Beds.   Fearnsides Collection.

    2 *c*. Distal fragment, with slight dorsal curvature.   Rheidol Gorge, 440 yards S.S.E. of Bryn-chwîth Farm, Pont Erwyd.   Castell Group (base of zone of *M. convolutus*). Geological Survey of England and Wales.

    2 *d*. Long and broad distal fragment.   Pary's Mountain, Anglesea.   Llandovery Beds. Greenly's Collection.

3 *a—d.—Monograptus regularis*, Törnquist.   (Page 372.)

    3 *a*. Long distal fragment, impression.   Dobb's Linn.   Birkhill Shales.   Geological Survey of Scotland, Edinburgh.

    3 *b*. Distal fragment, mainly in relief.   E. side of old quarry, N.E. of Fagwr-fawr Farm, 2 miles E.N.E. of Pont Erwyd.   Castell Group (zone of *M. convolutus*).   Geological Survey of England and Wales.

    3 *c*. Fragment nearer proximal end, impression.   Pary's Mountain, Anglesea.   Llandovery Beds.   Greenly's Collection.

    3 *d*. Long fragment, cast.   Rheidol Gorge, 420 yards E.S.E. of Bryn-chwîth Farm, Pont Erwyd.   Rheidol Group.   Geological Survey of England and Wales.

4 *a—d.—Monograptus jaculum* (Lapworth).   (Page 373.)

    4 *a*. Proximal portion.   Dobb's Linn.   Upper Birkhill Shales.   Lapworth's Collection.

    4 *b*. Distal portion, impression.   Ibid.

    4 *c*. Long distal fragment, with slight dorsal curvature, impression.   Garple Linn, near Moffat.   Upper Birkhill Shales.   Lapworth's Collection.

    4 *d*. Specimen on same slab as fig. 4 *c*.

5 *a, b.—Monograptus variabilis* (Perner).   (Page 374.)

    5 *a*. Proximal portion, showing sicula.   Marsh, Conway.   Tarannon Shales.   Elles' Collection.

    5 *b*. Distal fragment.   Corner near Forge, Conway.   Tarannon Shales.   Elles' Collection.

6 *a—e.—Monograptus nudus* (Lapworth).   (Page 375.)

    6 *a*. Incomplete but typical specimen, figured Lapworth, Geol. Mag., 1876, pl. xii, fig. 1 *a*. Grieston Quarry, Innerleithen.   Tarannon Shales.   Lapworth's Collection.

    6 *b*. Proximal portion, showing sicula.   W. side of upper quarry, 550 yards E.S.E. of Fuches-gau Farm-house, Pont Erwyd.   Castell Group.   Geological Survey of England and Wales.

    6 *c*. Similar specimen, impression.   Nant Fuches-wen, Cardiganshire.   Castell Group. Jones' Collection.

    6 *d*. Distal fragment, partly in low relief.   Same locality as fig. 6 *b*.

    6 *e*. Proximal portion.   Gelli Stream, Newydd Fynyddog, near Llanbrynmair.   Tarannon Beds.   Wood's Collection.

PLATE XXXVII—*continued*.

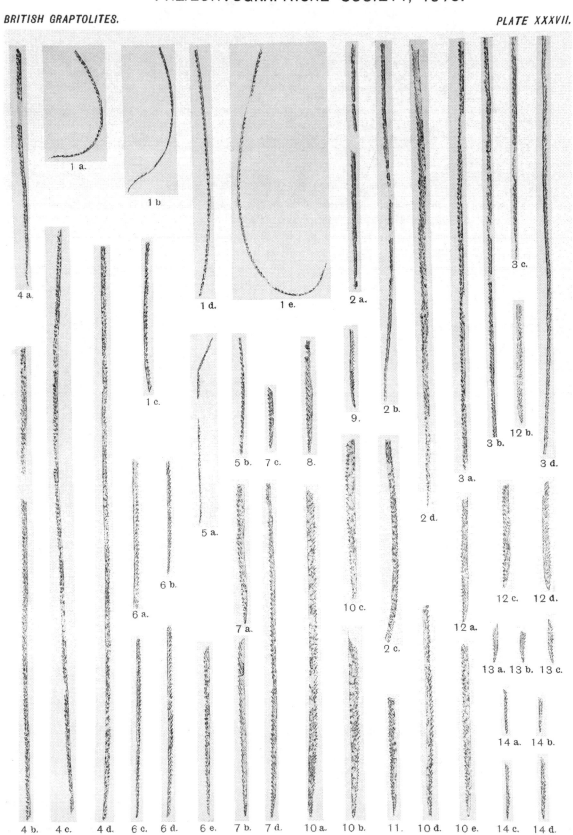

IDA SLATER & E. M. R. WOOD, del.

Bemrose, Ltd., Derby.

MONOGRAPTUS.

PLATE XXXVIII.

Monograptus—*continued.*

PLATE XXXVIII—*continued.*

FIGS.

6 *a—d.—Monograptus argenteus,* var. *cygneus* (Törnquist). (Page 389.)

> 6 *a.* Small specimen. Pary's Mountain, Anglesea. Llandovery Beds. Sedgwick Museum.
>
> 6 *b.* Specimen showing characteristic form, but badly preserved. Ibid.
>
> 6 *c.* Long distal fragment. Ibid.
>
> 6 *d.* Specimen distorted so as to resemble *M. limatulus.* Skelgill. Skelgill Beds. British Museum (Natural History), S. Kensington.

7 *a—d.—Monograptus limatulus,* Törnquist. (Page 390.)

> 7 *a.* Typical specimen. Grennan Point, N.W. of Port Logan, Wigtownshire. Birkhill Shales. Geological Survey of Scotland, Edinburgh.
>
> 7 *b.* Incomplete specimen. Skelgill. Skelgill Beds. Elles' Collection.
>
> 7 *c.* Long specimen. Dobb's Linn. Birkhill Shales (zone of *M. convolutus*). Elles' Collection.
>
> 7 *d.* Specimen much distorted, but showing long, slender proximal portion. Same locality as fig. 7 *b.*

8 *a—d.—Monograptus colonus* (Barrande). (Page 391.)

> 8 *a.* Typical specimen, mainly in relief, figured Wood, Quart. Journ. Geol. Soc., vol. lvi, pl. xxv, fig. 10 B. Helm Knot, Lake District. Coniston Flags. British Museum (Natural History), S. Kensington.
>
> 8 *b.* Rather narrow specimen, showing recurved proximal thecæ. Vicarage Road, Builth. Lower Ludlow Shales. Sedgwick Museum.
>
> 8 *c.* Broad specimen, in low relief, figured Wood, loc. cit., fig. 10 D. Adferton, near Ludlow. Lower Ludlow Shales. Hopkinson's Collection.
>
> 8 *d.* Characteristic specimen, figured Wood, loc. cit., fig. 10 *c.* River Irfon, Builth. Lower Ludlow Shales. Wood's Collection.

9 *a—c.—Monograptus colonus* (?), var. *ludensis* (Murchison). (Page 394.)

> 9 *a.* Complete specimen, figured Wood, loc. cit., fig. 11. Llanfair, Montgomeryshire. Ludlow Beds. Dr. Humphreys' Collection, Llanfair.
>
> 9 *b.* Fairly complete specimen. Llanfair, Montgomeryshire. Ludlow Beds. Wood's Collection.
>
> 9 *c.* Narrower specimen, showing sicula. Ibid.

10 *a—c.—Monograptus colonus,* var. *compactus,* Wood. (Page 393.)

> 10 *a.* Type specimen (somewhat broken), figured Wood, loc. cit., fig. 12. Elton-Evenhay Lane, near Ludlow. Lower Ludlow Shales. Wood's Collection.
>
> 10 *b.* Well-preserved specimen. Stormer Hall, near Leintwardine. Lower Ludlow Shales. Wood's Collection.
>
> 10 *c.* Small specimen. Elton-Evenhay Lane. Lower Ludlow Shales. Sedgwick Museum.

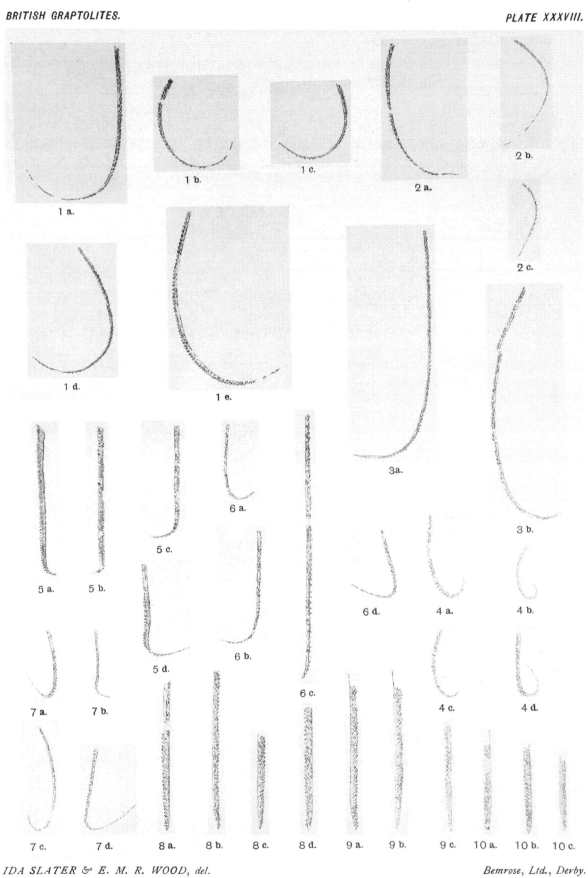

IDA SLATER & E. M. R. WOOD, del.

Bemrose, Ltd., Derby.

MONOGRAPTUS.

PLATE XXXIX.

Monograptus—*continued.*

PLATE XXXIX—*continued.*

FIGS.

6 *a—e.—Monograptus varians,* Wood (*continued*).

    6 *c.* Distal portion, showing virgula. Loc. cit., fig. 16 A. Stormer Hall, near Leintwardine. Lower Ludlow Shales. Wood's Collection.

    6 *d.* Complete specimen. Loc. cit., fig. 15. Elton Lane, near Ludlow. Lower Ludlow Shales. Wood's Collection.

    6 *e.* Complete specimen, cast, figured Lapworth (as *M. colonus*), Ann. and Mag. Nat. Hist. [5], vol. v, pl. iv, figs. 3 *b* and 3 *d,* and Wood, loc. cit., fig. 16 B. Mary Knoll, near Ludlow. Lower Ludlow Shales. Hopkinson's Collection.

7 *a—e.—Monograptus varians,* var. *pumilus,* Wood. (Page 396.)

    7 *a.* Type specimen, preserved so as to appear narrower than usual, figured Wood, loc. cit., fig. 17 B. Elton Lane, near Ludlow. Lower Ludlow Shales (zone of *M. scanicus*). Wood's Collection.

    7 *b.* Co-type, somewhat wider, figured Wood, loc. cit., fig. 17 A. Ibid.

    7 *c.* Small, but typical specimen. Elton Lane. Lower Ludlow Shales. Wood's Collection.

    7 *d.* Specimen on same slab as fig. 7 *c.*

    7 *e.* Complete specimen, on same slab as figs. 7 *c* and 7 *d.*

8 *a—f.—Monograptus leintwardinensis,* Hopkinson. (Page 401.)

    8 *a.* Type specimen, figured Wood, loc. cit., fig. 21 A. Church Hill Quarry, near Leintwardine. Lower Ludlow Shales (zone of *M. leintwardinensis*). Hopkinson's Collection.

    8 *b.* Co-type, figured Wood, loc. cit., fig. 21 B. Ibid.

    8 *c.* Long narrow specimen. Church Hill Quarry, near Leintwardine. Lower Ludlow Shales. Fearnsides' Collection.

    8 *d.* Broader specimen. On same slab as fig. 8 *c.*

    8 *e.* Short, broad specimen. On same slab as figs. 8 *c* and *d.*

    8 *f.* Characteristic specimen. On same slab as figs. 8 *c—e.*

9 *a—d.—Monograptus leintwardinensis,* var. *incipiens,* Wood. (Page 402.)

    9 *a.* Type specimen, figured Wood, loc. cit., fig. 22 A. Montgomery Road, Lower Ludlow Shales. Wood's Collection.

    9 *b.* Co-type, figured Wood, loc. cit., fig. 22 B. Long Mountain. Lower Ludlow Shales. Watts' Collection.

    9 *c.* Short broad specimen. Same locality as fig. 9 *a.*

    9 *d.* Very long specimen. On same slab as fig. 9 *b.*

10 *a—e.—Monograptus Sandersoni,* Lapworth. (Page 404.)

    10 *a.* Median fragment. Dobb's Linn. Birkhill Shales. Lapworth's Collection.

    10 *b.* Proximal fragment, with strong ventral curvature, figured Lapworth, Geol. Mag., 1876, pl. xi, fig. 2 *e.* Dobb's Linn. Birkhill Shales. Lapworth's Collection.

    10 *c.* Distal fragment with conspicuous ventral curvature. Coalpit Bay, Donaghadee. Llandovery Beds. Sedgwick Museum.

    10 *d.* Distal fragment, with only slight curvature, in low relief. East bank of road from Pont Erwyd to Devil's Bridge, 50 yards E.N.E. of Bryn-chwith Farm. Rheidol Group. Geological Survey of England and Wales.

    10 *e.* Long distal fragment. Dobb's Linn. Birkhill Shales. Lapworth's Collection.

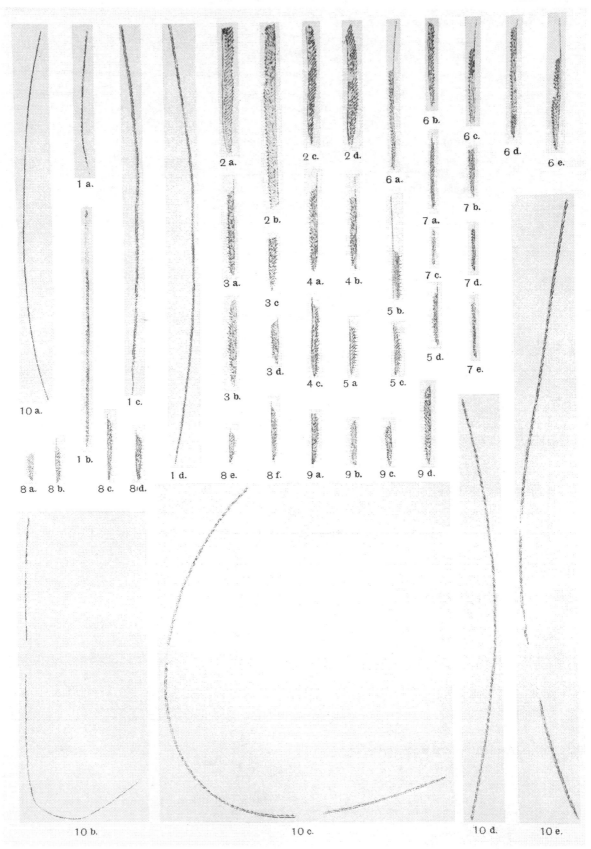

MONOGRAPTUS.

IDA SLATER & E. M. R. WOOD, del.

Bemrose, Ltd., Derby.

PLATE XL.

Monograptus—*continued.*

Figs.

1 *a—e.—Monograptus incommodus,* Törnquist. (Page 406.)

1 *a.* Long distal fragment. Skelgill. Skelgill Beds (zone of *Dimorphog. confertus*). Sedgwick Museum.

1 *b.* Proximal fragment, showing characteristic double curvature. Dobb's Linn. Birkhill Shales. Lapworth's Collection.

1 *c.* Distal fragment. Dobb's Linn. Birkhill Shales. Sedgwick Museum.

1 *d.* Small distal fragment, partly in relief. Rheidol Gorge. Rheidol Group. Geological Survey of England and Wales.

1 *e.* Median fragment. Dobb's Linn. Birkhill Shales. Lapworth's Collection.

2 *a—e.—Monograptus tenuis* (Portlock). (Page 407.)

2 *a.* Specimen on Portlock's type slab, showing general form of polypary. Limehill, Co. Tyrone. Upper Llandovery Beds. Geological Survey of England and Wales.

2 *b.* More distal fragment. On same slab as fig. 2 *a.*

2 *c.* Small proximal fragment, cart track, 10 yards west of entrance to lower quarry, 550 yards E.S.E. of Fuches-gau Farm-house, Pont Erwyd. Castell Group. Geological Survey of England and Wales.

2 *d.* Distal fragment. On same slab as fig. 2 *c.*

2 *e.* Proximal portion, labelled by Nicholson as *Monograptus discretus.* Skelgill. Skelgill Beds. British Museum (Natural History), S. Kensington.

3 *a—e.—Monograptus argutus,* Lapworth. (Page 408.)

3 *a.* Type specimen, showing general form, partly figured Lapworth, Geol. Mag., 1876, pl. x, fig. 13 *b.* Dobb's Linn. Birkhill Shales. Lapworth's Collection.

3 *b.* Proximal portion, showing curvature. S. side of old quarry, 270 yards N.E. of Gwen-ffrwd-uchaf Farm, Pont Erwyd. Base of zone of *M. convolutus.* Geological Survey of England and Wales.

3 *c.* Proximal portion, (?) figured Lapworth, loc. cit., fig. 13 *a.* Dobb's Linn. Birkhill Shales. Lapworth's Collection.

3 *d.* Proximal fragment. Rheidol Gorge, 440 yards S.S.E. of Brynchwîth Farm-house, Pont Erwyd. Base of zone of *M. convolutus.* Geological Survey of England and Wales.

3 *e.* Distal fragment, part in relief, part impression. On same slab as reverse side of fig. 3 *d.*

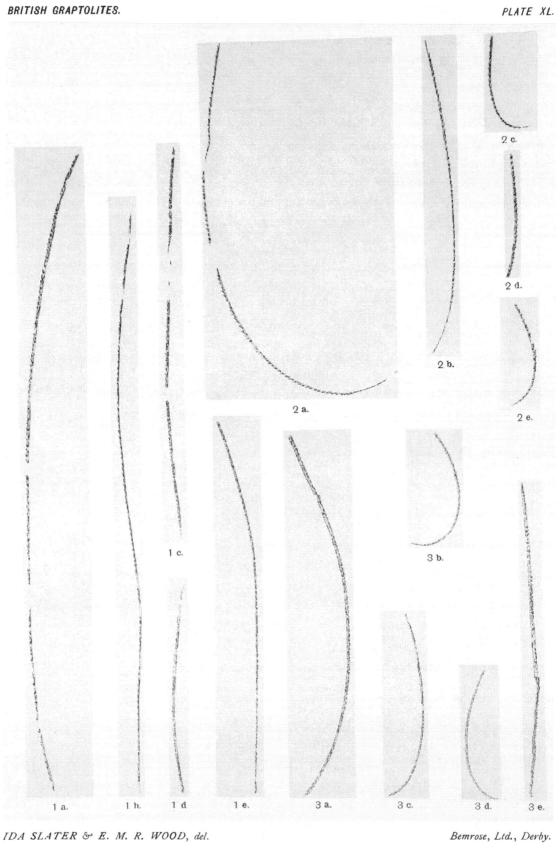

IDA SLATER & E. M. R. WOOD, del.

Bemrose, Ltd., Derby.

MONOGRAPTUS.

PLATE XLI.

Monograptus—*continued.*

PLATE XLI—*continued.*

Figs.

4 *a—d.—Monograptus vomerinus,* var. *crenulatus* (Törnquist). (Page 412)—
continued.

> 4 *c.* Proximal portion, in relief. Ibid.
>
> 4 *d.* Fragment showing proximal end. Williamshope (?), S. Scotland. Gala Beds. Lapworth's Collection.

5 *a—d.—Monograptus griestoniensis* (Nicol). (Page 413.)

> 5 *a.* Proximal fragment, specimen on Nicol's type slab. Grieston Quarry, S. Scotland. Gala Beds. Geological Survey of England and Wales (11,800).
>
> 5 *b.* Distal portion. Ibid. (11,801).
>
> 5 *c.* Proximal portion, showing considerable curvature. Grieston Quarry. Gala Beds. Sedgwick Museum.
>
> 5 *d.* Ibid.

6 *a, b.—Monograptus* cf. *griestoniensis.* (Page 414.)

> 6 *a.* Proximal portion. Tarannon River. Talerddig Grits. Wood's Collection.
>
> 6 *b.* Longer specimen, showing rigid form. Afon-cwm-calch, near Talerddig Railway Cutting. Talerddig Grits (upper part). Wood's Collection.

7 *a—e.—Monograptus crenularis,* Lapworth. (Page 414.)

> 7 *a.* Long median fragment. Dobb's Linn. Upper Birkhill Shales (zone of *Cephalog. cometa*). Lapworth's Collection.
>
> 7 *b.* Long distal fragment, impression. The Corrie, Dobb's Linn. Upper Birkhill Shales. Lapworth's Collection.
>
> 7 *c.* Distal fragment, part in relief, part impression. Ibid.
>
> 7 *d.* Proximal fragment. Ibid.
>
> 7 *e.* Ibid.

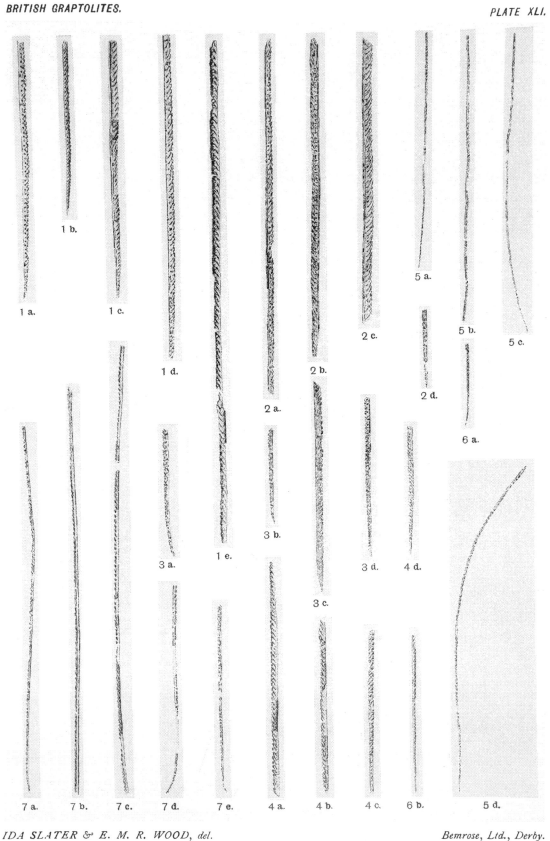

IDA SLATER & E. M. R. WOOD, del.

Bemrose, Ltd., Derby.

MONOGRAPTUS.

PLATE XLII.

Genus Monograptus.

1 *a*—*c*.—*Monograptus galaensis*, Lapworth. (Page 415.)

 1 *a*. Type specimen, showing sicula, figured Lapworth, Geol. Mag., 1876, pl. xii, fig. 5 *b*. Meigle Quarry, Selkirkshire. Gala Beds. Lapworth's Collection.

 1 *b*. Distal fragment, ? figured Lapworth, loc. cit. supra, fig. 5 *a*. Ibid.

 1 *c*. Incomplete proximal fragment, poorly preserved. Cliff, Ettrick Water, Ettrickbridge end, S. Scotland. *Rastrites maximus* band. Geological Survey of Scotland, Edinburgh.

2 *a*—*e*.—*Monograptus priodon* (Bronn). (Page 418.)

 2 *a*. Typical specimen, preserved mainly in relief. On same slab as one of figured specimens of *M. griestoniensis*. Grieston Quarry, S. Scotland. Gala Beds. Geological Survey of England and Wales. (11800).

 2 *b*. Long distal fragment. Grieston Quarry. Gala Beds. Lapworth's Collection.

 2 *c*. Smaller but well preserved specimen in low relief. Afon Tyn-y-rhos, Talerddig, near Llanbrynmair. Talerddig Beds. Wood's Collection.

 2 *d*. Small but wide specimen, showing proximal end preserved in full relief. Gwern-y-fed Fach, near Builth. Wenlock Shales. Sedgwick Museum.

 2 *e*. Wide distal fragment. Pencerrig, near Builth. Wenlock Shales. Sedgwick Museum (collected by Hopkinson).

3 *a*—*d*.—*Monograptus pandus* (Lapworth). (Page 421.)

 3 *a*. Proximal portion, probably referable to this species, (?) figured Lapworth, Graptolites of Co. Down, p. 129, pl. vi, fig. 3 *a*. Tieveshilly, Co. Down. Tarannon Beds. Lapworth's Collection.

 3 *b*. Typical distal fragment. Rawthey Bridge, Sedbergh. Browgill Beds. Sedgwick Museum.

 3 *c*. Similar wide fragment. Ibid.

 3 *d*. Somewhat narrower fragment. Ibid.

4 *a*—*d*.—*Monograptus Marri*, Perner. (Page 422.)

 4 *a*. Typical specimen, showing sicula. Tarannon River. Gelli Beds (zone of *M. crispus*). Wood's Collection.

 4 *b*. Small proximal fragment. Ibid.

 4 *c*. Long distal fragment, preserved partly in low relief. Tarannon River. Talerddig Beds (zone of *M. griestoniensis*). Wood's Collection.

 4 *d*. Somewhat flexed specimen. Same locality and horizon as fig. 4 *a*.

5 *a*.—*d*.—*Monograptus Flemingii* (Salter). (Page 425.)

 5 *a*. Typical specimen, figured Lapworth, Geol. Mag., 1876, pl. xx, fig. 8 *a*. Riccarton Junction, S. Scotland. Riccarton Beds. Lapworth's Collection.

 5 *b*. Similar specimen, preserved partly in low relief. Ibid.

 5 *c*. Proximal portion, figured Elles, Quart. Journ. Geol. Soc., vol. lvi, p. 403, fig. 14. River Irfon, near Builth. Wenlock Shales. Elles' Collection.

 5 *d*. Very long and somewhat flexed specimen. Ibid.

6 *a*—*d*.—*Monograptus Flemingii*, var. *primus*, Elles and Wood, nom. nov. (Page 426.)

 6 *a*. Type specimen, figured Elles, Quart. Journ. Geol. Soc., vol. lvi, p. 402, fig. 11. Dulas Brook, near Builth. Wenlock Beds (zone of *Cyrtog. rigidus*). Elles' Collection.

 6 *b*. Somewhat wider specimen, figured Elles, loc. cit. supra, fig. 11. Nant Prophwyd, near Builth (zone of *Cyrtog. rigidus*). Elles' Collection.

 6 *c*. Long and wide specimen. Nant Prophwyd, near Builth (zone of *Cyrtog. rigidus*). Elles' Collection.

 6 *d*. Shorter and narrower specimen. Ibid.

7 *a*—*d*.—*Monograptus Flemingii*, var. *compactus*, Elles and Wood, nom. nov. (Page 427.)

 7 *a*. Type specimen, figured Elles, Quart. Journ. Geol. Soc., vol. lvi, p. 403, fig. 13. River Irfon, near Builth. Wenlock Beds (zone of *Cyrtog. Lundgreni*). Elles' Collection.

 7 *b*. Slightly longer specimen with more reflexed proximal end, figured Elles, loc. cit. supra, fig. 13. Ackley Lane, Long Mountain. Wenlock Beds (zone of *Cyrtog. Lundgreni*). Elles' Collection.

 7 *c*. Small typical specimen. Llwynrhedith Quarry, Long Mountain, Wenlock Beds (zone of *Cyrtog. Lundgreni*). Elles' Collection.

 7 *d*. Larger specimen with more reflexed proximal end. Ibid.

8 *a*—*e*.—*Monograptus riccartonensis*, Lapworth. (Page 424.)

 8 *a*. Typical distal fragment, preserved as a cast. Penwhapple Glen, S. Scotland. Wenlock Beds. Lapworth's Collection.

 8 *b*. Typical proximal fragment, figured Lapworth, Ann. and Mag. Nat. Hist. [5], vol. 5, pl. iv, fig. 8 *c*. Elliotsfield, near Hawick. Riccarton Beds. Lapworth's Collection.

 8 *c*. Small proximal fragment, showing sicula, figured Elles, Quart. Journ. Geol. Soc., vol. lvi, pl. xxiv, fig. 5. Walcot Quarry, Chirbury. Wenlock Beds (zone of *Monog. riccartonensis*). Elles' Collection.

 8 *d*. Specimen showing typical "broken-backed" appearance. Llandrindod Wells. Wenlock Beds. Lapworth's Collection.

 8 *e*. Very long and stiff specimen. Balmae Shore, near Kirkcudbright. Wenlock Beds. Sedgwick Museum.

9 *a*, *b*.—*Monograptus cultellus*, Törnquist. (Page 423.)

 9 *a*. Typical specimen in relief. Road between Llanbrynmair and Talerddig, Montgomeryshire. Nant-ysgollen Shales. Wood's Collection.

 9 *b*. Larger fragment, doubtfully referable to this species. Tarannon River Gelli Beds. Wood's Collection.

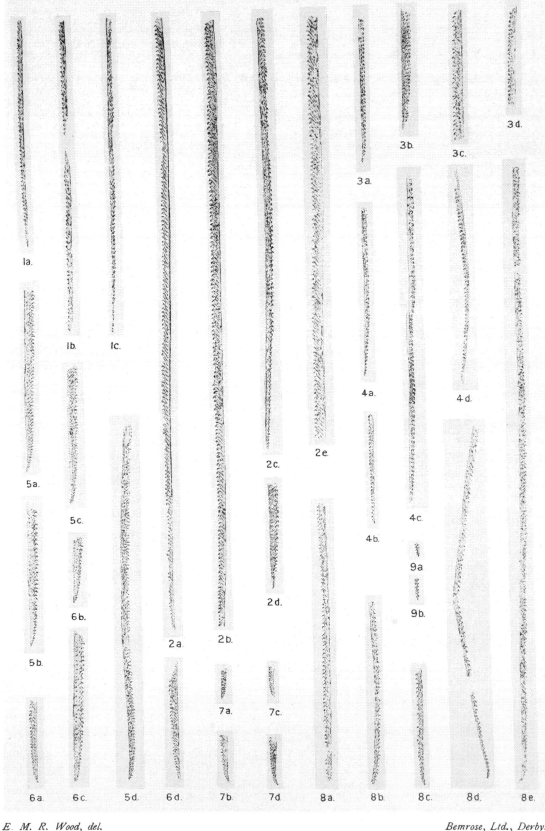

E. M. R. Wood, del.

Bemrose, Ltd., Derby.

MONOGRAPTUS.

PLATE XLIII.

Monograptus—*continued.*

FIGS.

1 *a—d.—Monograptus uncinatus,* var. *orbatus,* Wood. (Page 427.)

    1 *a.* Type specimen, figured Wood, Quart. Journ. Geol. Soc., vol. lvi, pl. xxv, fig. 23 A. Trefnant-Middletown Brook, Long Mountain. Lower Ludlow Shales. Wood's Collection.

    1 *b.* Typical specimen, figured Wood, loc. cit. supra, fig. 23 B. Ibid.

    1 *c.* Somewhat narrower specimen, partly figured Wood, loc. cit. supra, p. 476, fig. 20 *b,* Wren's Nest, Dudley. Wenlock Shales. Dr. Fraser's Collection.

    1 *d.* Broader specimen showing sicula. Trefnant-Middletown Brook, Long Mountain. Lower Ludlow Shales. Sedgwick Museum.

2 *a—c.—Monograptus uncinatus,* var. *micropoma* (Jaekel). (Page 428.)

    2 *a.* Typical specimen, figured Wood, Quart. Journ. Geol. Soc., vol. lvi, pl. xxv, fig. 24 A. Elton-Ludlow Road. Lower Ludlow Shales. Wood's Collection.

    2 *b.* Longer specimen, figured Wood, loc. cit. supra, fig. 24 B. Above Garbett's Hall, Long Mountain. Lower Ludlow Shales. Wood's Collection.

    2 *c.* Somewhat wider specimen, showing sicula. Stormer Hall, near Leintwardine. Lower Ludlow Shales. Sedgwick Museum.

3.—*Monograptus irfonensis,* Elles. (Page 429.)

    Type specimen, figured Elles, Quart. Journ Geol. Soc., vol. lvi, p. 407, fig. 19. River Irfon, near Builth. Wenlock Shales (zone of *Cyrtog. Lundgreni*). Elles' Collection.

4 *a—e.—Monograptus flexilis,* Elles. (Page 430.)

    4 *a* Type specimen, figured Elles, loc. cit. supra, p. 417, fig. 18. Moel Ferna, Dee Valley. Wenlock Shales (zone of *Cyrtog. Linnarssoni*). Elles' Collection.

    4 *b.* Imperfect specimen, on same slab as fig. 4 *a.*

    4 *c.* Incomplete but very long specimen, showing flexed form of polypary. Near Gill, Cautley, Yorkshire. Wenlock Beds (zone of *Cyrtog. symmetricus*). Miss Welch's Collection.

    4 *d.* Specimen with very long virgella. Middle Gill, Cautley Wenlock Beds. Welch's Collection.

    4 *e.* Similar specimen. Same locality as figs. 4 *a* and *b.*

5 *a—e.—Monograptus gemmatus* (Barrande). (Page 436.)

    5 *a.* Two specimens, (?) figured Lapworth (as *M. attenuatus*), Geol. Mag., 1876, pl. x, fig. 9 *b.* Dobb's Linn. Upper Birkhill Shales. Lapworth's Collection.

    5 *b.* Small curved fragment. Dobb's Linn. Upper Birkhill Shales (band of *Cephalog. cometa*). Wood's Collection.

    5 *c.* Broken fragments, preserved in relief. Parys Mountain, Anglesea. Llandovery Beds. G. J. Williams' Collection, Bangor.

    5 *d.* Two specimens crossing each other, figured Carruthers (as *M. capillaris*), Geol. Mag., 1868, pl. v, fig. 16. Belcraig Burn, near Moffat. Upper Birkhill Shales. British Museum (Natural History), South Kensington.

    5 *e.* Some flexed fragments. Duffkinnell Burn, Wamphray. Birkhill Shales. Geological Survey of Scotland.

6 *a—d.—Monograptus distans* (Portlock). (Page 433.)

    6 *a.* Typical specimen. Pomeroy, co. Tyrone. Llandovery Beds (zone of *Monog. Sedgwickii*). Geological Survey of England and Wales.

    6 *b.* Another specimen, on same slab as fig. 6 *a.*

    6 *c.* More distal fragment, on a slab which is the reverse side of that on which figs. 6 *a* and 6 *b* occur. Lapworth's Collection.

    6 *d.* Fragment with more closely set thecæ, but probably referable to this species. The Corrie, Dobb's Linn. Upper Birkhill Shales. Elles' Collection.

7 *a—c.—Monograptus acus,* Lapworth MS. (Page 431.)

    7 *a.* Large flexed specimen. Grieston-on-Tweed. Gala Beds. Sedgwick Museum.

    7 *b.* Distal fragment. Meigle Quarry, S. Scotland. Gala Beds. Lapworth's Collection.

    7 *c.* Proximal fragment. Cacra Bank, Tushielaw, Selkirkshire. Gala Beds. Lapworth's Collection.

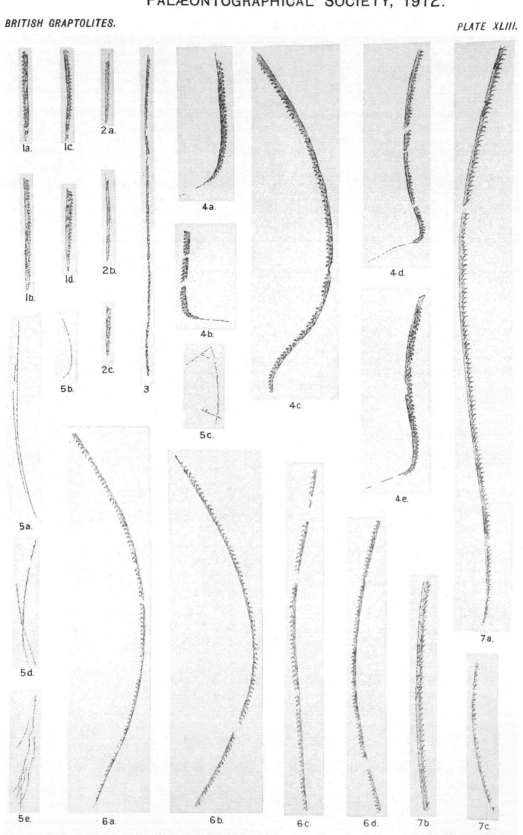

E. M. R. Wood, del.

Bemrose, Ltd., Derby.

MONOGRAPTUS.

PLATE XLIV.

Monograptus—*continued.*

Figs.

1 *a, b.—Monograptus Jaekeli,* Perner. (Page 435.)

    1 *a.* Distal fragment, with concave curvature. Builth Road, near Builth. Wenlock Shales (zone of *Cyrtog. Linnarssoni*). Elles' Collection.

    1 *b.* Distal fragment, with convex curvature. Ibid.

2 *a—d.—Monograptus scanicus,* Tullberg. (Page 433.)

    2 *a.* Specimen showing characteristic curvature, figured Wood, Quart. Journ. Geol. Soc., vol. lvi, pl. xxv, fig. 25 A. Aberedw Hill, near Builth. Lower Ludlow Shales. Wood's Collection.

    2 *b.* Distal fragment, figured Wood, loc. cit. supra, fig. 25 B. Ibid.

    2 *c.* Small proximal fragment. Elton-Ludlow Road, near Ludlow. Lower Ludlow Shales. Wood's Collection.

    2 *d.* Long distal fragment. Lletygynfach, Long Mountain. Lower Ludlow Shales. Lapworth's Collection.

3 *a—c.—Monograptus crinitus,* Wood. (Page 435.)

    3 *a.* Type specimen, figured Wood, loc. cit. supra, pl. xxv, fig. 26 A. Lower Winnington Lane, Long Mountain. Lower Ludlow Shales. Wood's Collection.

    3 *b.* Small fragment, figured Wood, loc. cit. supra, p. 481, fig. 23 c. On same slab as fig. 3 *a.*

    3 *c.* Portion of slab, showing general habit. Lower Winnington Lane. Long Mountain. Lower Ludlow Shales. Wood's Collection.

4 *a—e.—Monograptus turriculatus* (Barrande). (Page 438.)

    4 *a.* Typical specimen. Swindale Beck, Knock, Edenside. Browgill Beds. Sedgwick Museum.

    4 *b.* Specimen coiled in less regular spiral. Ibid.

    4 *c.* Small typical specimen. Ibid.

    4 *d.* Group of small specimens, flattened from above into a plane spiral. Ladhope, Galashiels. Gala Beds. Geological Survey of Scotland.

    4 *e.* Large, but imperfect, specimen. Bargany Pond Burn, $2\frac{1}{4}$ miles W. by S. of Dailly, Ayrshire. Gala Beds. Geological Survey of Scotland.

5 *a—d.—Monograptus discus,* Törnquist. (Page 439.)

    5 *a.* Large specimen, with open whorls. Portaferry, Co. Down. Tarannon Beds. Swanston's Collection, Belfast.

    5 *b.* Similar specimen. Pen-y-geulan, near Llanbrynmair Station. Gelli Beds. Wood's Collection.

    5 *c.* Group of small specimens, preserved as casts. Bryn-yr-odin, near Llangollen. L. J. Wills' Collection.

    5 *d.* Small, but typical, specimen. Meigle Quarry, Galashiels. Gala Beds. Lapworth's Collection.

6.—*Monograptus tortilis,* Linnarsson. (Page 440.)

    Distal fragment. Sedbergh, Yorkshire. Browgill Beds. Sedgwick Museum.

PLATE XLIV—*continued*.

FIGS.

7 *a, b.*—*Monograptus testis*, var. *inornatus*, Elles. (Page 446.)

    7 *a.* Type specimen, figured Elles, Quart. Journ. Geol. Soc., vol. lvi, p. 408, fig. 20 *a.* Garbett's Hall, Middletown Brook, Long Mountain. Wenlock Beds. Watts' Collection.

    7 *b.* Co-type, figured Elles, loc. cit. supra, fig. 20 *b.* Trewern Brook, near Middletown, Long Mountain. Elles' Collection.

8 *a—f.*—*Monograptus Halli* (Barrande). (Page 443.)

    8 *a.* Proximal end, presented in a sub-ventral view. Thirlstane Score, near Moffat. Birkhill Shales (band of *Rastrites maximus*). Lapworth's Collection.

    8 *b.* Typical distal fragment, showing apertural spines. Belcraig Burn, near Moffat. Band of *R. maximus*. Lapworth's Collection.

    8 *c.* Distal fragment in relief, apertural spines turned away and thus concealed from view. Thirlstane Score. Band of *R. maximus*. Lapworth's Collection.

    8 *d.* Fragment in relief, apertural margins partly embedded. Ibid.

    8 *e.* Wide distal fragment. Ettrickbridge End, S. Scotland. Band of *R. maximus*. Geological Survey of Scotland.

    8 *f.* Proximal fragment, preserved mainly as a low cast, probably referable to this species. Quarry S. side main road from Devil's Bridge to Rhayader, 200 yards S. of Bodcoll Farm. Castell Group. Geological Survey of England and Wales.

9 *a, b.*—*Monograptus M'Coyi*, Lapworth. (Page 446.)

    9 *a.* Type specimen, figured Lapworth, Graptolites of Co. Down, 1877, pl. vi, fig. 2. Tieveshilly, Co. Down. Belfast Natural History Museum.

    9 *b.* Wide fragment, possibly referable to this species. Belcraig Burn, near Moffat. Band of *R. maximus*. Lapworth's Collection.

10 *a—f.*—*Monograptus Sedgwickii* (Portlock). (Page 441.)

    10 *a.* Typical distal fragment, with conspicuous apertural spines. Pomeroy, co. Tyrone. Upper Llandovery Beds. British Museum (Natural History), S. Kensington.

    10 *b.* Proximal fragment, on same slab as fig. 10 *a.*

    10 *c.* Proximal fragment, showing sicula. Sundhope-on-Yarrow, S. Scotland. Upper Birkhill Shales. Lapworth's Collection.

    10 *d.* Broad specimen, with conspicuous apertural spines. Glenkiln Burn, S. Scotland. Upper Birkhill Shales. Lapworth's Collection.

    10 *e.* Specimen with slightly recurved proximal end. Duffkinnel Burn, Wamphray. Upper Birkhill Shales. British Museum (Natural History).

    10 *f.* Long fragment, with concave curvature, apertural spines concealed for the most part. On same slab as fig. 10 *e.*

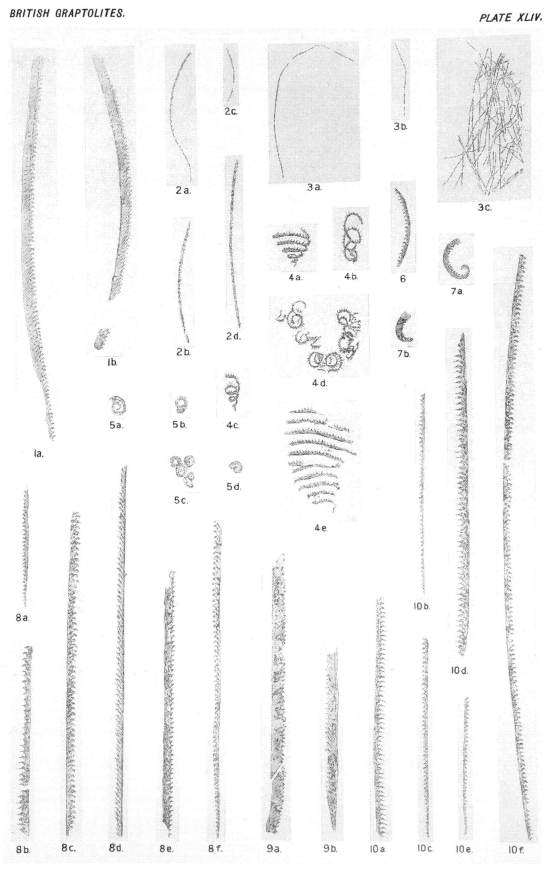

E. M. R. Wood, del.

Bemrose, Ltd., Derby.

MONOGRAPTUS.

PLATE XLV.

Monograptus—*continued.*

FIGS.

1 *a—f.—Monograptus lobiferus* (M'Coy). (Page 448.)

    1 *a.* Counter-impression of type specimen, figured M'Coy, Brit. Pal. Fossils, pl. i B, figs. 3 and 3 *a.* Moffat. Sedgwick Museum.

    1 *b.* Narrower and more proximal fragment, figured Lapworth, Geol. Mag., 1876, pl. xx, fig. 1 *a.* Craigierig, S. Scotland. Birkhill Shales. Lapworth's Collection.

    1 *c.* Long fragment, somewhat broken. Long Linn, Dobb's Linn. Upper Birkhill Shales. Elles' Collection.

    1 *d.* Wider distal fragment, irregularly flexed. Ibid.

    1 *e.* Proximal fragment, showing sicula, preserved in relief. Ten yards up stream from crest of anticline, N. of Fuches-gau Farmhouse. Castell Group. Geological Survey of England and Wales.

    1 *f.* Broad fragment, with unusually strong concave curvature. Duffkinnel, above Glanhall, S. Scotland. Birkhill Shales. Geological Survey of Scotland, Edinburgh.

2 *a—g.—Monograptus runcinatus,* Lapworth. (Page 450.)

    2 *a.* Proximal fragment, in relief, showing sicula. Glenkiln Burn, S. Scotland. Upper Birkhill Shales. Lapworth's Collection.

    2 *b.* Similar specimen, in low relief. River Twymyn, Llanbrynmair. Brynmair Beds. Wood's Collection.

    2 *c.* Longer proximal fragment. Glenkiln Burn. Upper Birkhill Shales. Lapworth's Collection.

    2 *d.* Similar specimen. Ibid.

    2 *e.* Long and curved distal fragment, preserved as a low cast, figured Lapworth, Geol. Mag., 1871, pl. xx, fig. 4 *c.* Ibid.

    2 *f.* Rather more rigid specimen. E. side Upper Quarry, 30 yards E. of Fuches-gau Farmhouse. Geological Survey of England and Wales.

    2 *g.* Two specimens crossing one another. W. side Upper Quarry, 550 yards E. of Fuches-gau Farmhouse. Geological Survey of England and Wales.

3 *a—g.—Monograptus runcinatus,* var. *pertinax,* Elles and Wood, nov. (Page 451.)

    3 *a.* Proximal fragment, showing sicula, preserved in relief. Gelli-dwyll Stream, below farmhouse, near Llanbrynmair. Brynmair Beds. Wood's Collection.

    3 *b.* Similar specimen. On same slab as fig. 3 *a.*

    3 *c.* Distal fragment, preserved mainly as a cast. Same locality.

    3 *d.* Broader fragment, with slight curvature, preserved in relief. Ibid.

    3 *e.* Long fragment imperfectly preserved. Ibid.

    3 *f.* Long fragment, probably referable to this variety. Buckholm Side, Galashiels. Gala Beds. Geological Survey of Scotland.

    3 *g.* Proximal fragment. Same locality as fig. 3 *a* and *c.*

4 *a—f.—Monograptus Becki* (Barrande). (Page 452.)

    4 *a.* Typical specimen, figured Lapworth, Geol. Mag., 1876, pl. xx, fig. 2 *a.* Meigle Quarry, Galashiels. Gala Beds. Lapworth's Collection.

    4 *b.* Similar specimen. On same slab as fig. 4 *a.*

    4 *c.* Long curved distal fragment. Same locality as above.

    4 *d.* Specimen with long tapering proximal portion. Portaferry, Co. Down. Tarannon Beds. Swanston's Collection, Belfast.

    4 *e.* Similar specimen. On same slab as fig. 4 *d.*

    4 *f.* Distal fragment. River Twymyn, near Llanbrynmair. Brynmair Beds. Wood's Collection.

5.—*Monograptus undulatus,* Elles and Wood, sp. nov. (Page 432.)

    Specimen, showing sicula, preserved in relief. E. side of quarry, N.E. of Fagwr-fawr Farmhouse, Pont Erwyd. Geological Survey of England and Wales.

6 *a—f.—Monograptus crispus,* Lapworth. (Page 456.)

    6 *a.* Type specimen, figured Lapworth, Geol. Mag., 1876, pl. xx, fig. 7 *a.* Meigle Quarry, Galashiels. Gala Beds. Lapworth's Collection.

    6 *b.* Proximal portion. On same slab as fig. 6 *a.*

    6 *c.* Proximal portion, showing sicula. Same locality as above.

    6 *d.* Complete specimen, showing characteristic form of polypary. Tieveshilly, co. Down. Gala Beds. Lapworth's Collection.

    6 *e.* Typical specimen. Portaferry, Ireland. Belfast Natural History Museum.

    6 *f.* Large fragment. Same locality as fig. 6 *d.*

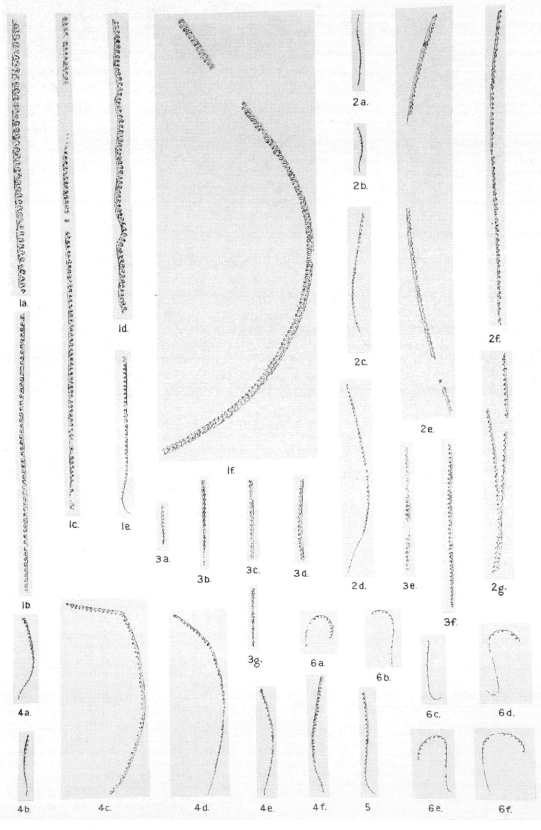

la.

ld.

lf.

2a.

2b.

2c.

2e.

2f.

2d.

3e.

2g.

3a.

3b.

3c.

3d.

3f.

lc.

le.

lb.

3g.

6a.

6b.

6c.

6d.

4a.

4b.

4c.

4d.

4e.

4f.

5

6e.

6f.

E. M. R. Wood, del.   *Bemrose, Ltd., Derby.*

MONOGRAPTUS.

PLATE XLVI.

Monograptus—*continued.*

PLATE XLVI—*continued.*

FIGS.

7 *a—d.—Monograptus dextrorsus*, Linnarsson.   (Page 460.)

   7 *a.* Long distal fragment. Afon Cwm Calch, near Talerddig, Montgomeryshire. Upper Talerddig Beds. Wood's Collection.

   7 *b.* Irregularly bent specimen. Ibid.

   7 *c.* Long curved specimen. Ibid.

   7 *d.* Specimen showing proximal end, probably referable to this species. Llanystwmdwy. Fearnsides' Collection.

8 *a, b.—Monograptus knockensis*, Elles and Wood, sp. nov. (Page 462.)

   8 *a.* Distal fragment. Knock, Edenside. Browgill Beds. Marr's Collection.

   8 *b.* Smaller fragment. Ibid.

9 *a, b.—Monograptus remotus*, Elles and Wood, sp. nov.   (Page 461.)

   9 *a.* Type specimen. Rigg Burn, River Esk, 6 miles N.N.W. of Langholm. Gala Beds. Geological Survey of Scotland.

   9 *b.* Co-type specimen. Ibid.

10 *a—d.—Monograptus millepeda* (M'Coy).   (Page 465.)

   10 *a.* Typical specimen, preserved in relief. Skelgill, Lake District Skelgill Beds. Sedgwick Museum.

   10 *b.* Specimen, preserved in partial relief. Skelgill. Skelgill Beds. British Museum (Natural History), S. Kensington.

   10 *c.* Well preserved, but incomplete specimen. Skelgill. Skelgill Beds (zone of *Monog. argenteus*). Marr's Collection.

   10 *d.* Large, but incomplete specimen, figured Lapworth (as *M. lobiferus*, var. *Nicoli*), Geol. Mag., 1876, pl. xx, fig. 1 *c.* Dobb's Linn. Birkhill Shales. Lapworth's Collection.

11 *a—f.—Monograptus Clingani* (Carruthers).   (Page 463.)

   11 *a.* Type specimen, figured Carruthers, Geol. Mag., 1868, pl. v, fig. 19 *a.* Duffkinnel (?). Birkhill Shales. British Museum (Natural History), S. Kensington.

   11 *b.* Co-type specimen, figured Carruthers, loc. cit. supra, fig. 19 *b.*

   11 *c.* Characteristic specimen, showing sicula. Moory Syke, S. Scotland. Birkhill Shales. Lapworth's Collection.

   11 *d.* Very long specimen, with broadly recurved proximal portion, figured Lapworth, Geol. Mag., 1876, pl. xx, fig. 3 *a.* Dobb's Linn. Birkhill Shales. Lapworth's Collection.

   11 *e.* Small specimen. Dobb's Linn. Sedgwick Museum.

   11 *f.* Specimen, preserved in partial relief, thecæ somewhat embedded, E. side Quarry, N.E. of Fagwr-fawr Farm, Pont Erwyd (zone of *Monog. convolutus*). Geological Survey of England and Wales.

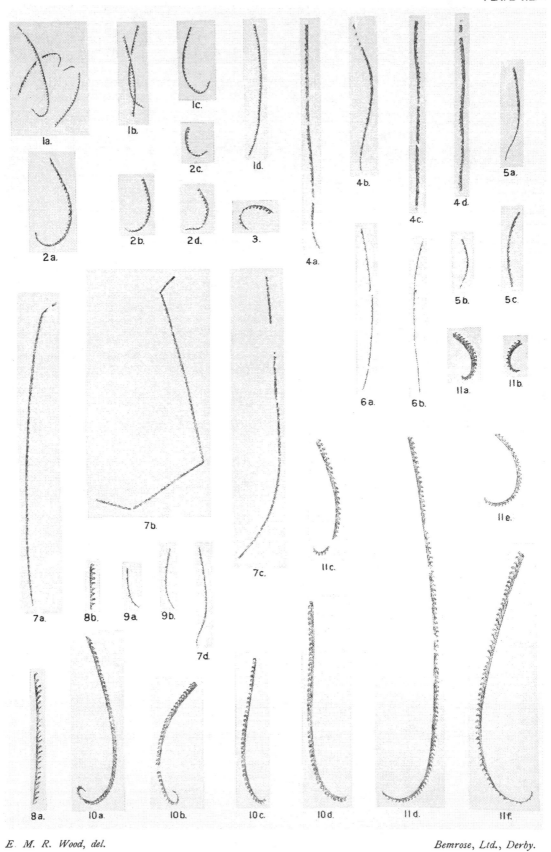

E. M. R. Wood, del.

Bemrose, Ltd., Derby.

MONOGRAPTUS.

PLATE XLVII.

Monograptus—*continued.*

Figs.

1 *a—d.—Monograptus convolutus* (Hisinger). (Page 467.)

    1 *a.* Very large specimen, showing *Rastrites*-like character of the proximal thecæ. Figured Carruthers, Geol. Mag., 1868, pl. v, fig. 1 *a.* Dobb's Linn. Birkhill Shales. Geological Survey of England and Wales.

    1 *b.* Smaller, but well preserved specimen. Pishnack Burn, Wee Queensberry. Birkhill Shales. Geological Survey of Scotland.

    1 *c.* Incomplete specimen, but showing well the *Rastrites*-like character of the proximal thecæ. Plewlands Burn, Raehills. Birkhill Shales. Geological Survey of Scotland.

    1 *d.* Less well preserved specimen. Duffkinnel Burn, Wamphray. Birkhill Shales. British Museum (Natural History), S. Kensington.

2 *a, b.—Monograptus delicatulus,* Elles and Wood, sp. nov. (Page 478.)

    2 *a.* Typical specimen. Sundhope-on-Yarrow. Birkhill Shales (zone of *Monog. gregarius*). Lapworth's Collection.

    2 *b.* Good specimen, showing character of proximal thecæ. The Corrie, Dobb's Linn. Birkhill Shales. Elles' Collection.

3 *a—e.—Monograptus decipiens,* Törnquist. (Page 469.)

    3 *a.* Two specimens crossing one another. Long Linn, Dobb's Linn. Birkhill Shales (zone of *Monog. gregarius*). Elles' Collection.

    3 *b.* Specimen broadly curved. E. side quarry, N.E. of Fagwr-fawr Farm, Pont Erwyd. Geological Survey of England and Wales.

    3 *c.* Specimen irregularly curved, preserved in full relief. Dobb's Linn. Birkhill Shales. Geological Survey of Scotland.

    3 *d.* Specimen with rather longer thecæ. Dobb's Linn. Birkhill Shales. Sedgwick Museum.

    3 *e.* Small specimen. Clanyard Bay, Port Logan. Birkhill Shales. Geological Survey of Scotland.

4 *a—f.—Monograptus triangulatus* (Harkness). (Page 471.)

    4 *a.* Specimen preserved in full relief. 430 yds. S.S.E. of Bryn-chwîth Farmhouse, Pont Erwyd. Zone of *Monog. triangulatus.* Geological Survey of England and Wales.

    4 *b.* Fairly complete specimen. Donaghadee. Birkhill Shales. Belfast Natural History Museum.

    4 *c.* Large fragment. Dobb's Linn. Birkhill Shales. Geological Survey of Scotland.

    4 *d.* Fairly complete specimen. Ibid.

    4 *e.* Specimen, preserved mainly as a cast. Rheidol Gorge, Pont Erwyd. Jones' Collection.

    4 *f.* Very large specimen, irregularly bent, preserved mainly as a cast, with thecæ partly embedded. Dobb's Linn. Birkhill Shales. Dairon's Collection, Kelvingrove Museum, Glasgow.

5 *a—d.—Monograptus triangulatus,* var. *major,* Elles and Wood, nov. (Page 472.)

    5 *a.* Typical specimen. Garple Linn, near Moffat. Birkhill Shales. British Museum (Natural History), South Kensington.

    5 *b.* Small proximal fragment. Dobb's Linn. Birkhill Shales. Elles' Collection.

    5 *c.* Small fragment, preserved as a cast. 430 yds. S.S.E. of Bryn-chwîth Farmhouse. Geological Survey of England and Wales.

    5 *d.* Specimen showing great length of thecæ. Dobb's Linn. Birkhill Shales. Elles' Collection.

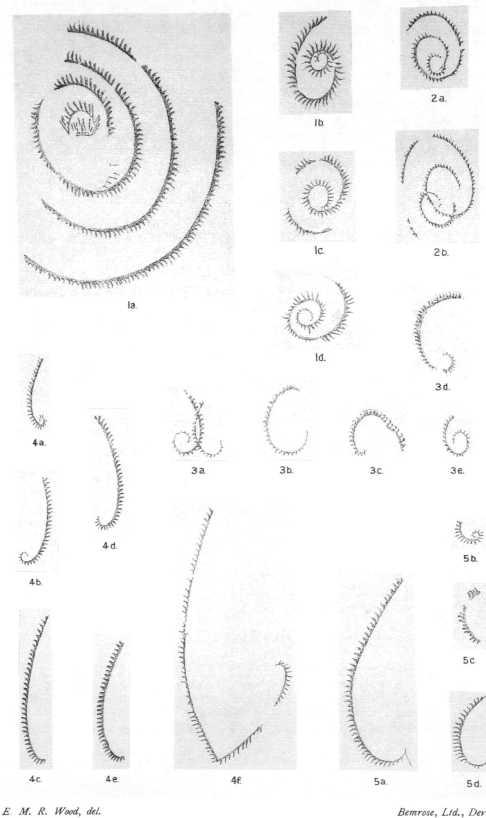

E. M. R. Wood, del.

Bemrose, Ltd., Derby.

MONOGRAPTUS.

PLATE XLVIII.

Monograptus—*continued.*

PLATE XLVIII—*continued.*

Figs.

5 *a—d.*—*Monograptus fimbriatus*, var. *similis*, Elles and Wood, var. nov. (Page 483.)
    5 *a.* Small specimen, showing short proximal end. Dobb's Linn. Birkhill Shales. Lapworth's Collection.
    5 *b.* Somewhat wider specimen. Ibid.
    5 *c.* Small specimen showing sicula. Dobb's Linn. Birkhill Shales. Sedgwick Museum.
    5 *d.* Long specimen, but incomplete proximally. On same slab as fig. 5 *b.*

6 *a—d.*—*Monograptus planus* (Barrande). (Page 484.)
    6 *a.* Specimen, preserved in partial relief. Plas-bach Stream, near Llanbrynmair. Junction beds of the Dolgau and Nant-ys-gollen Shales. Wood's Collection.
    6 *b.* Small specimen, showing proximal end. Dobb's Linn. Geological Survey of Scotland.
    6 *c.* Characteristic specimen. Hebblethwaite Gill, Lake District. Browgill Beds. Sedgwick Museum.
    6 *d.* Larger and broader specimen. Dobb's Linn. Geological Survey of Scotland.

7 *a—d.*—*Monograptus spiralis* (Geinitz). (Page 475.)
    7 *a.* Small specimen, showing proximal thecæ. Tarannon River, Montgomeryshire. Gelli Beds. Wood's Collection.
    7 *b.* Large specimen with characteristic irregular curvature. Portaferry, Co. Down. Gala Beds. Belfast Natural History Museum.
    7 *c.* Incomplete specimen. Thirlstane Score, near Moffat. Band of *Rastrites maximus.* Lapworth's Collection.
    7 *d.* Very regularly curved specimen, but probably referable to this species, true thecal form concealed for the most part. Nant Fuches-wen, Pont Erwyd. Jones' Collection.

8 *a—c.*—*Monograptus proteus* (Barrande). (Page 477.)
    8 *a.* Specimen, showing typical form of polypary. Bont Dolgadfan, near Llanbrynmair. Brynmair Beds. Wood's Collection.
    8 *b.* Similar specimen, but with longer proximal portion. Portaferry, Co. Down. Tarannon Beds. Swanston's Collection, Belfast.
    8 *c.* Group of distal fragments, showing form of thecæ. Gelli-dywyll Stream, near Llanbrynmair. Brynmair Beds. Wood's Collection.

9 *a, b.*—*Monograptus circularis*, Elles and Wood, sp. nov. (Page 479.)
    9 *a.* Typical specimen. Plewlands Burn, Raehills. Upper Birkhill Shales. Geological Survey of Scotland.
    9 *b.* Less well-preserved specimen. Ibid

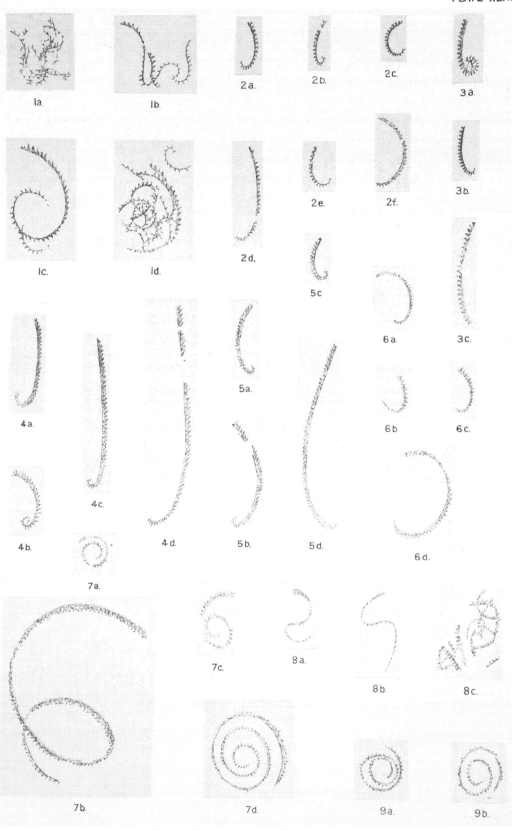

E. M. R. Wood, del.

Bemrose, Ltd., Derby.

MONOGRAPTUS.

PLATE XLIX.

Monograptus—*continued.*

Figs.

1 *a—e.—Monograptus communis* (Lapworth). (Page 480.)

  1 *a.* Type specimen, figured Lapworth, Geol. Mag., 1876, pl. xiii, fig. 4 *a.* Dobb's Linn. Birkhill Shales. Lapworth's Collection.

  1 *b.* Well-preserved specimen, in partial relief. Rheidol Gorge, Pont Erwyd. Jones' Collection.

  1 *c.* Typical specimen. Dobb's Linn. Birkhill Shales. Sedgwick Museum.

  1 *d.* Form with wide proximal curve, intermediate in character between *M. communis* and *M. millepeda.* Ibid.

  1 *e.* Curiously curved specimen, provisionally referred to this species, preserved in partial relief. Dobb's Linn (?) British Museum (Natural History), South Kensington.

2 *a—c.—Monograptus communis,* var. *rostratus,* Elles and Wood, var. nov. (Page 481.)

  2 *a.* Typical specimen, showing long proximal thecæ. Dobb's Linn. Birkhill Shales. Lapworth's Collection.

  2 *b.* Well preserved and typical specimen. Dobb's Linn. Birkhill Shales. Geological Survey of Scotland.

  2 *c.* Specimen with more abrupt proximal curvature. On same slab as fig. 2 *b.*

3 *a—c.—Monograptus intermedius* (Carruthers). (Page 485.)

  3 *a.* Portion of slab, showing mode of occurrence. Sundhope-on-Yarrow, S. Scotland. Birkhill Shales. Lapworth's Collection.

  3 *b.* Single proximal fragment. Ibid.

  3 *c.* Distal fragment. Ibid.

4 *a—c.—Monograptus involutus* (Lapworth). (Page 478.)

  4 *a.* Portion of slab, showing mode of occurrence. Sundhope-on-Yarrow. Birkhill Shales. Lapworth's Collection.

  4 *b.* Single specimen. Dobb's Linn (?). Birkhill Shales. Lapworth's Collection.

  4 *c.* Small fragment, showing form of polypary. W. side old quarry, N.E. of Fagwr-fawr, 2 miles E.N.E. of Pont Erwyd. Geological Survey of England and Wales.

5 *a—c.—Monograptus* cfr. *elongatus,* Tornquist (Page 486).

  5 *a.* Fragment. Llanystwmdwy, near Criccieth. Llandovery Beds. Fearnsides' Collection.

  5 *b.* Small fragment. Ten yds. up stream from crest of anticline, N. of Fuches-gau Farmhouse, Pont Erwyd. Geological Survey of England and Wales.

  5 *c.* Small proximal fragment. Rheidol Gorge, 440 yds. S.S.E. of Bryn-chwîth Farmhouse, Pont Erwyd. Geological Survey of England and Wales.

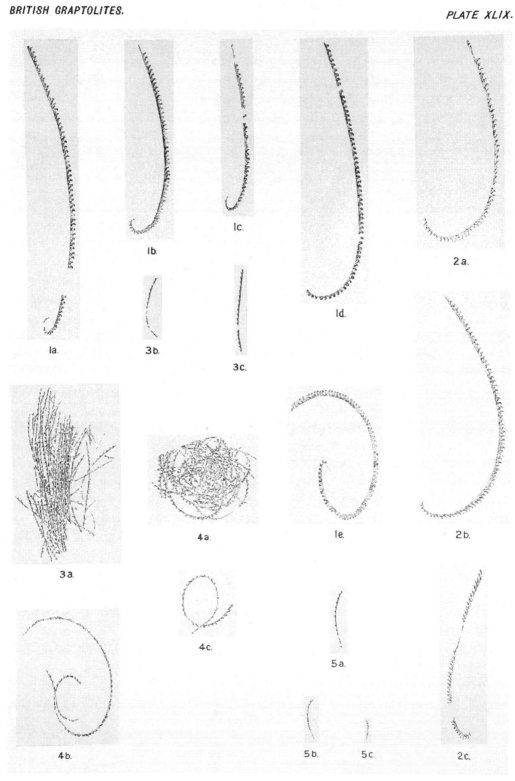

E. M. R. Wood, del.

Bemrose, Ltd., Derby.

MONOGRAPTUS.

PLATE L.

Genus **Monograptus** (**Rastrites**).

Figs.

a—e.—*Monograptus (Rastrites) peregrinus* (Barrande).   (Page 488.)

    1 a. Characteristic specimen.   Dobb's Linn.   Birkhill Shales (zone of *Monog. gregarius*).   Elles' Collection.

    1 b. Small fragment.   Dobb's Linn.   Birkhill Shales (zone of *Monog. convolutus*).   Elles' Collection,

    1 c. Fairly complete specimen.   Ibid.

    1 d. Portion of slab, showing the method of occurrence.   Coalpit Bay, Donaghadee.   Llandovery Beds.   Belfast Natural History Museum.

    1 e. Fragment with shorter thecæ.   Craigdarkes Hill, Dunscore.   Birkhill Shales.   Geological Survey of Scotland.

2 a—g.—*Monograptus (Rastrites) longispinus* (Perner).   (Page 489.)

    2 a. Large and typical specimen, showing sicula.   Grennan Point, Dumbreddan Bay, S. of Portpatrick.   Birkhill Shales.   Geological Survey of Scotland.

    2 b. Large, but less nearly complete specimen, and with more open curvature.   Coalpit Bay, Co. Down.   Birkhill Shales.   Belfast Natural History Museum.

    2 c. Smaller specimen, with shorter thecæ.   Dobb's Linn.   Birkhill Shales.   Lapworth's Collection.

    2 d. Specimen with more open curvature at the proximal end.   Garple Linn, near Moffat.   Birkhill Shales.   Geological Survey of Scotland.

    2 e. Smaller specimen.   Dobb's Linn.   Birkhill Shales.   Sedgwick Museum.

    2 f. Long distal fragment, somewhat distorted.   Mealy Gill, Coniston (zone of *Monog. fimbriatus*).   Skelgill Beds.   Sedgwick Museum.

    2 g. Small specimen, preserved in relief, and showing the reflexed apertural terminations.   430 yds. S.S.E. of Bryn-chwîth Farmhouse, Pont Erwyd.   Geological Survey of England and Wales, Jermyn Street.

3 a—d.—*Monograptus (Rastrites) setiger*, Elles and Wood, nov.   (Page 490.)

    3 a. Proximal fragment, showing part of sicula.   430 yds. S.S.E. of Bryn-chwîth Farmhouse, Pont Erwyd.   Geological Survey of England and Wales.

    3 b. Distal fragment.   Skelgill Beck, Lake District.   Skelgill Beds (zone of *Monog. fimbriatus*).   Sedgwick Museum.

    3 c. Distal fragment, showing the irregular curvature.   Ibid.

    3 d. Distal fragment.   Same locality, etc., as fig. 3 a.

4 a—f.—*Monograptus (Rastrites) hybridus* (Lapworth).   (Page 491.)

    4 a. Typical specimen.   Dobb's Linn.   Birkhill Shales.   Lapworth's Collection.

    4 b. Specimen showing the typical declination of the thecæ.   Skelgill Beck.   Skelgill Beds.   Sedgwick Museum.

    4 c. Fragment showing proximal end.   E. side of quarry, N.E. of Fagwr Fawr Farmhouse, 2 miles E.N.E. of Pont Erwyd.   Geological Survey of England and Wales.

    4 d. Two fragments in juxtaposition.   Dobb's Linn.   Birkhill Shales.   Geological Survey of Scotland.

    4 e. Distal fragment.   Belcraig Burn.   Birkhill Shales.   Sedgwick Museum.

    4 f. Distal fragment with irregular curvature.   Dobb's Linn.   Birkhill Shales.   Sedgwick Museum.

5 a—d.—*Monograptus (Rastrites) approximatus*, var. *Geinitzi* (Törnquist).   (Page 492.)

    5 a. Fragment with proximal end enrolled; thecæ showing " phleoid " development.   10 yds. up stream from crest of anticline, 30 yds. N. of Fuches-gau Farmhouse, Pont Erwyd.   Geological Survey of England and Wales.

    5 b. Enrolled proximal fragment.   S. side of old quarry, 270 yds. N.E. of Gwen Ffrwd Uchaf Farm, Pont Erwyd.   Geological Survey of England and Wales.

    5 c. Few thecæ showing " phleoid " development.   Same locality, etc., as fig. 5 a.

    5 d. Larger fragment.   Dobb's Linn.   Birkhill Shales.   Sedgwick Museum.

6 a—e.—*Monograptus (Rastrites) maximus* (Carruthers).   (Page 494.)

    6 a. Typical specimens in association.   Thirlstane Score, near Moffat.   Upper Birkhill Shales.   Lapworth's Collection.

    6 b. Fragment near proximal end.   Riskinhope Burn, S. Scotland.   Upper Birkhill Shales.   Lapworth's Collection.

    6 c. Typical fragment.   Same locality, etc., as fig. 6 a.

    6 d. Very large specimen, with the thecæ at rather irregular intervals.   Kirkhope Linn, Ettrickbridge.   Geological Survey of Scotland.

    6 e. Specimen with shorter thecæ, possibly a proximal fragment.   Belcraig Burn.   Upper Birkhill Shales (band of *Monog. (R.) maximus*).   Miss McPhee's Collection, Glasgow University.

7 a—d.—*Monograptus (Rastrites) fugax* (Barrande).   (Page 493.)

    7 a. Typical fragment.   Duffkinnel, S. Scotland.   Birkhill Shales.   Lapworth's Collection.

    7 b. Proximal fragment, showing sicula.   Dobb's Linn.   Birkhill Shales.   Lapworth's Collection.

    7 c. Small fragment.   Belcraig Burn.   Birkhill Shales.   Geological Survey of Scotland.

    7 d. Specimens with irregular curvature.   Duffkinnel.   Birkhill Shales.   Lapworth's Collection.

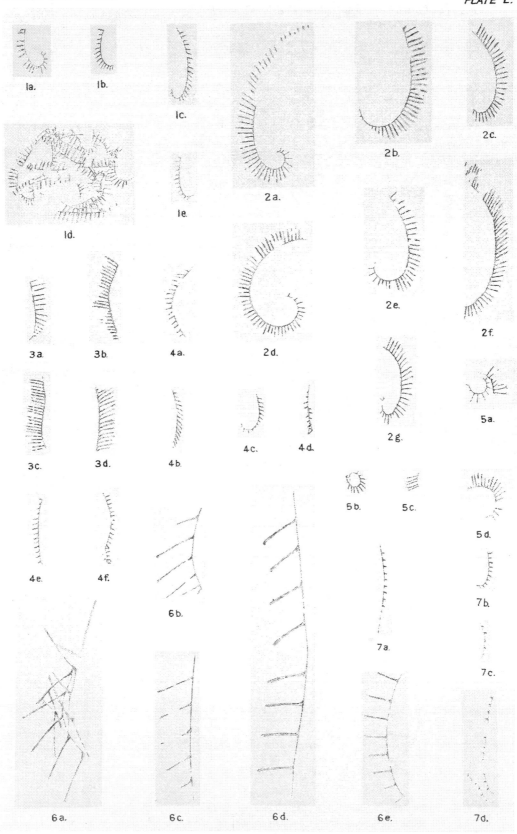

E. M. R. Wood, del.

Bemrose, Ltd., Derby.

MONOGRAPTUS (RASTRITES).

PLATE LI.

Genus **Monograptus** (**Rastrites**) and Genus **Cyrtograptus.**

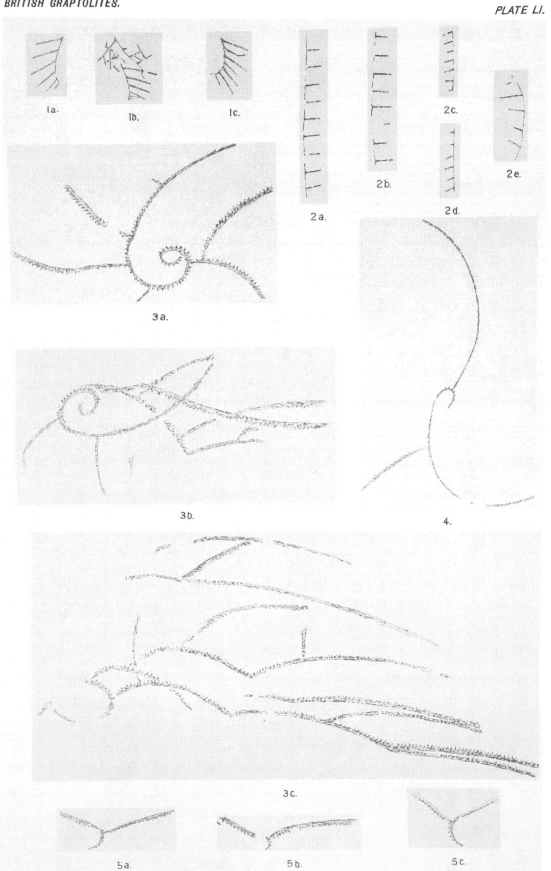

E. M. R. Wood, del.

Bemrose, Ltd., Derby.

MONOGRAPTUS (RASTRITES) AND CYRTOGRAPTUS.

PLATE LII.

Genus **Cyrtograptus.**

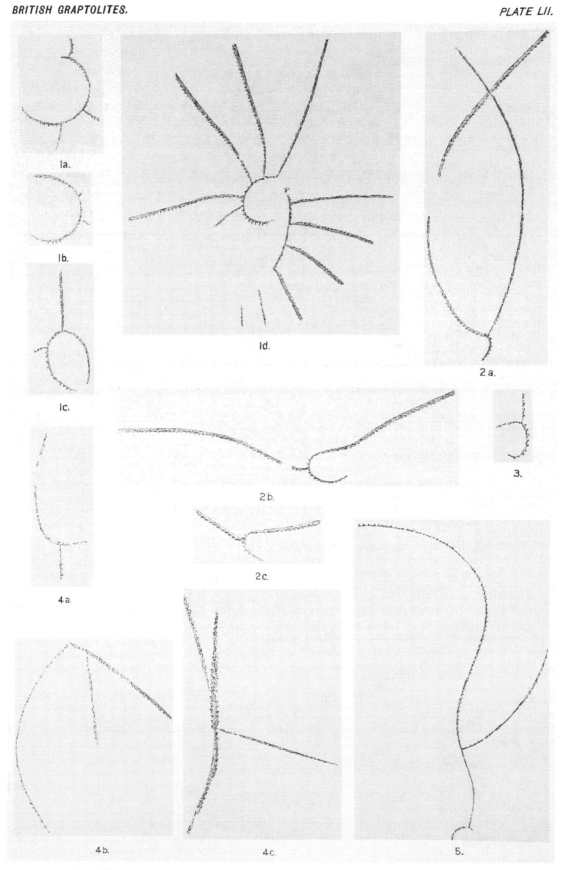

1a.

1b.

1c.

1d.

2a.

2b.

3.

2c.

4a.

4b.

4c.

5.

E. M. R. Wood, del.

Bemrose, Ltd., Derby.

CYRTOGRAPTUS.

Printed in the United States
By Bookmasters